History, Philosophy and Theory of the Life Sciences

Volume 22

More information about this series at http://www.springer.com/series/8916

Linda Deer Richardson

Benjamin Goldberg
Editor

Academic Theories of Generation in the Renaissance

The Contemporaries and Successors of Jean Fernel (1497-1558)

Springer

Linda Deer Richardson
YMCA George Williams College
London, UK

Editor
Benjamin Goldberg
Department of Humanities and Cultural
University of South Florida
Tampa, FL, USA

ISSN 2211-1948 ISSN 2211-1956 (electronic)
History, Philosophy and Theory of the Life Sciences
ISBN 978-3-319-69334-7 ISBN 978-3-319-69336-1 (eBook)
https://doi.org/10.1007/978-3-319-69336-1

Library of Congress Control Number: 2017960184

Printed on acid-free paper

This Springer imprint is published by Springer Nature
The registered company is Springer International Publishing AG
The registered company address is: Gewerbestrasse 11, 6330 Cham, Switzerland

Plato, Hippocrates, Aristotle and Galen in concert: from Symphorien Champier's Symphonia Platonis cum Aristotele: & Galeni cum Hippocrate, Paris, 1516 (Courtesy Bibliothèque Numérique de Lyon (NUMELYO))

For my family:
Warren, Ruth, Terry Elizabeth and Douglas

Abstract

The question which underlies this thesis is the relationship between medicine and natural philosophy in the sixteenth and early seventeenth centuries. It attempts to draw out and compare the arguments of the two groups of practitioners, and to place them in a proper intellectual and historical context. The questions I have chosen to examine are those concerned with academic theories of generation. These questions overlap the subject areas of medicine and natural philosophy; and they have not, to my knowledge, been dealt with in this way by modern scholars. Modern accounts of Renaissance generation theory suffer from 'peripheral vision effect'; their true focus is elsewhere, usually in a later, more recognisably modern period. Typically, they concentrate on the achievements of Renaissance anatomists, and emphasise the conflict between the two groups, doctors and natural philosophers, as followers of Galen and Aristotle respectively.

In contrast, this thesis looks at two groups of theoretical texts: commentaries on Aristotle's De generatione animalium, in Part II; treatises of theoretical medicine which dealt with generation in the context of the 'naturals', or physiology, in Part III. Both were based on the same set of classical authorities, reviewed in Part I. They were based, too, on the same methods, and the same conventions as to how knowledge was to be obtained. Moreover, they formed complementary parts of the same teaching tradition, that of the arts school. They benefitted alike from the new texts and editions of Plato, Hippocrates, Aristotle and Galen made available by humanist scholars and translators; and from similar new editions of Averroes and other major commentators.

Contemporaries were aware of potential conflicts among these authorities. Both groups made major efforts to resolve these conflicts through comparison and interpretation of classical texts, using earlier commentators, in particular Averroes and Peter of Abano, as models. Thus, in the argument of this thesis, Renaissance doctors and philosophers shared authorities, methods, questions and the commitment to concord which the Frontispiece represents.

Preface: On the Study of Medicine and Philosophy in the Renaissance

This book allows, for the first time, the brilliant dissertation written by Linda Allen Deer (now Richardson) to be widely available to a scholarly audience. Her thesis, which she defended in 1980 at the University of London-Warburg Institute, is entitled *Academic Theories of Generation in the Renaissance: The Contemporaries and Successors of Jean Fernel*. Deer Richardson published very little from this work and, soon after completing her dissertation, left academia. It is, at the time of this writing, 35 years old. Why publish it now? To answer this, let me provide some personal context and background.

In 2010, I was busy attempting to finish my dissertation, though it would take me two more years to defend it. My topic was the conception of natural philosophy and methodology of William Harvey, the physician who discovered that the blood moves around the body in a circuit. Key to my work were Harvey's investigations of generation, the (1651) *Exercitationes de generatione animalium*, a work of great importance and influence in its day, but which has not fared terribly well in the ensuing centuries – most modern historians of science and medicine have not paid it nearly the attention lavished on his (1628) *Exercitatio anatomica de motu cordis et sanguinis in animalibus*.[1] While the topic of generation has had a long record of discussion by historians,[2] much of this work has been burdened by anachronism and whiggishness, illustrated most clearly by the fact that these histories are of *embryology*, a topic and term that does not come into its modern connotation until the nineteenth century, though of course the term *embryo* is in use much earlier. (Indeed, Linda Deer Richardson herself, writing three and a half decades ago, complains of these very sins – some things change little.) So while there is extremely valuable research on the topic of generation by Walter Pagel, Vivian Nutton, A. J. Pyle, Charles Schmitt and others,[3] there is still a need for detailed and sensitive work analysing this area in detailed historical and philosophical context, on topics ranging from exploring the works of understudied figures such as Fernel and others to determining lines of influence and reaction between authors to figuring out the exact notions of causality at play in their theories of generation.

So, while writing my own thesis, I encountered a number of citations to her dissertation. Along with my advisor, James Lennox, we set out to discover what had happened to her and to *Academic Theories of Generation in the Renaissance*. Eventually we managed to get in touch with her and get ahold of the thesis itself – a copy of the original, typewritten document. My reaction was twofold: first, I was convinced I would never have made it as a scholar in the pre-computer era, owing to the necessity of having to write by hand all the Greek characters, and, worse still, the arduous process of footnoting in a time before word processors. And second, the historiographical situation described above being the case, I was overwhelmed and overjoyed by the existence of such a sensitive piece of scholarship addressing the very issues with which I was struggling (and more besides). Though it is over 500 pages long, I read it all in 2 days; so excited was I at its depth, its erudition, its bringing to the fore of obscure and understudied figures and its deep relevance to my own project. Deer Richardson's work, though it focuses on the extremely important figure of Jean Fernel, is, in fact, an in-depth study of the topic of generation that ranges over figures far beyond Fernel. Indeed, she goes far beyond studying just generation, and her research provides thorough and valuable discussions of many worthy topics, including methodological differences between *philosophi* and *medici*; the varying epistemologies of, among and between physicians and philosophers; theories of matter and the elements; methods of interpretation and integration of classical sources; modes of philosophical argument and organisation; traditions of commentary and compendia; Aristotle; Galen; Averroes; university training; and more. I can think of no higher compliment than to say that I wish I had written this work myself, and it gives me no small pleasure to help publish this book and make it more easily accessible to the broader academic community.

Despite the fact that it was completed somewhat long ago, *Academic Theories of Generation in the Renaissance* is a cutting-edge work. To this end, let me provide some very basic statistics that will demonstrate the current scholarly situation (noting that I am here making an effort to pay attention not just to Anglophone work but work in other European languages as well). Using EBSCOhost's database of histories of science, technology and medicine, currently the best resource for finding work relevant to those topics, there are *only two* results when you search for "Fernel" and "generation": Hiro Hirai's (2011) book *Medical Humanism and Natural Philosophy* and the work you are now reading.[4] If you search for "Fernel", there are just 45 entries *in totum* (some of which are duplicates and some of which are reviews of Forrester and Henry's translations of Fernel).[5] While forty-odd publications are not nothing, they are also not much: compared to work discussing Descartes at 2418 citations, or to Harvey at 1081, or even to Vesalius at 417. If one does the same search in the same database on some other figures discussed *at length* and *in detail* in Deer Richardson's thesis (again with the occasional duplicate), we find the situation is just the same, the numbers exceptionally low:

0 on Felix Accorambonius
0 on Johannes Argenterius
0 on Pompeio Caimo
0 on Gaspar Contarenus or Contarini
0 on Sebastian Paparella
3 on Jacobus Sylvius

8 on Jean Riolan
12 on Johannes Baptista Montanus
32 on Agostino Nifo
44 on Cesare Cremonini
55 on Zabarella
84 on Bernardino Telesio

Deer Richardson also mentions in varying detail, but does not focus upon, a number of other authors worth mentioning: Thomas Fienus (0 citations), Fortunius Licetus (0 citations), Johannes de Penna (0 citations), Ponce de Santa Cruz (physician to Philip of Spain) (2 citations), Franciscus de Zanellis de Bononia (0 citations), Jacob Schegk (8 citations), Niccolò Leoniceno (33 citations), Arcangelo Piccolomini (36 citations) and Julius Scaliger (36 citations). These numbers reveal that in the ensuing decades since Deer Richardson defended her thesis, very little work has been done on the topics and figures that she discusses and that, indeed, her work is some of the only available on a number of these individuals. We are doubly blessed that her work does not just discuss these individuals and their ideas but does so in loving detail, with deep historical sensitivity and even a bit of much appreciated dry humour. Deer Richardson's thesis contains a profoundly detailed study of, in particular, the relationship between the material constituents of generation (sperm, blood and so forth), the elemental theory of the philosophers, the physician's account of the temperaments and the theologian's understanding of the soul (and, indeed, the philosophers' and physicians' ideas as well). It is thus all the more important and valuable that her work is now being published and made available to a wider audience: it remains a cutting-edge work on a wide range of topics important in the study of anatomy, physiology, medicine and natural philosophy in the Renaissance.

Before I move on to outline and discuss the details of Deer Richardson's *Academic Theories of Generation*, I want to discuss how we, as historians, study these topics and what we might learn from this (now) decades-old work. In her conclusion, Deer Richardson notes two important themes, continuity and conciliation, and I believe these to be important lessons for us as well. The issue of the presence or absence of continuity between periods and ages is, of course, a large-scale historiographical problem, with good arguments on both sides of the debate and with the whole issue perhaps coming down to temperament: lumpers who stress commonalities and splitters who emphasise differences.[6] Deer Richardson comes down firmly on the side of the lumpers, and she provides numerous arguments, examples and instances of the ways in which the works and figures she analyses reflect a number of kinds of continuity. She emphasises, for example, continuity between the medieval and the Renaissance eras in terms of topics, formats, approaches, pedagogy and sources, and indeed, she argues for an overall continuity in the natural philosophies of the sixteenth and the first half of the seventeenth century, as well as for continuity between doctors and natural philosophers. The last is especially important, as it is something not noted very often in the work of historians of philosophy and medicine today.[7] The disciplines of the history of medicine and the history of philosophy are farther apart in their study of Renaissance and early modern work than is warranted by the historical actors and issues they study, to their mutual detriment. The modern disciplinary identities of historians today make a hash of the period they study, cutting and marking boundaries in a way that often does a disservice to their material. There is a great deal to be learned and explored by recognising the important continuity between natural philosophy and medicine, and I hope the present volume will serve as both a powerful argument and incitement to that end, spurring work that explores how physicians and philosophers interacted, disagreed, used sources and conceived of the God and the natural world.

The continuity between medicine and philosophy brings us to the second theme that Deer Richardson notes in her conclusion: conciliation. Deer Richardson has in mind the ways in which Renaissance thinkers almost universally attempted to make coherent the bodies of ancient wisdom that formed the foundation of their education and scholarly research, and they almost all tried to reconcile the various opinions and doctrines found in their classical sources. This is not, as Deer Richardson notes, to deny honest academic debate nor the existence of real controversies. Instead, what is emphasised throughout her work are the ways in which the differences between Renaissance writers very often comes down to subtle shades of opinion, to delicate differences in emphasis and to careful and considered selections of particular classical works and passages therein. The most important aspect of this conciliatory mode is found in Deer Richardson's argument that historians (such as Jacques Roger and Walter Pagel) have often overplayed the differences and hostility between those supposedly warring camps of Galenists and Aristotelians, an idea which has been subsequently substantiated in detail in the work of a number of later historians, especially Nancy Siraisi.[8] Part of the problem here is that, in order for us to notice this, we often cannot take the work of Renaissance thinkers at face value: for instance, despite his constant declaration of allegiance to Aristotle, and his attack on particular doctrines of Galen, William Harvey is deeply Galenic in his approach and method.[9]

Historiographically speaking, then, Deer Richardson's work is deeply relevant and worthwhile, especially if one aims at fostering a useful rapport between the histories of philosophy and medicine, allowing these two disciplines to interact and begin to contribute to each other's projects and conceptions of their subjects. This is not to say that *Academic Theories of Generation* is trendy: in many ways, this thesis is a work of very old-fashioned intellectual history, and it does not have much to say by way of social or cultural history, nor are any post-modernisms or profound theoretical lenses deployed. This is no criticism, as I believe there is room for intellectual history among its more popular brethren; indeed, there is even a need for it, especially given the dearth of research on so many of the figures and ideas that Deer Richardson discusses in detail. The trend in studies of generation written in the ensuing decades after Deer Richardson's work has usually focused on (a) small-scale analyses of specific theoretical aspects of generation theory and the metaphysics thereof (for instance, the work of Andreas Blank on biomedical ontology) and upon particular individuals and their theories (for instance, the work of Hiro Hirai for a meticulous study of the specifics of Jean Fernel and Jacob Schegk) or (b) on the social, cultural and gendered aspects of the study of reproduction (for instance,

Katherine Park's work on the idea and role of "mothers" and "motherhood" in the study and discussion of generation).[10] There are very few works like *Academic Theories of Generation*, focusing on the large-scale dynamics of generation theory, and, often, the focus is on the period after the Renaissance, the early modern or Enlightenment periods, or on the nineteenth century. Further, the lack of theoretical language in Deer Richardson's thesis makes it accessible to a wide range of scholars, across periods and even disciplines. And, finally, there is an interesting way in which a few aspects of Deer Richardson's ideas prefigure some recent academic trends, for instance, in its deployment of quantification in her discussion of the commentary and compendia tradition. I shall note some of the interesting details of this discussion below, but for now I merely want to emphasise that Deer Richardson's work is quite sophisticated historiographically and provides ample grist for our historical mills.

In the remainder of this introduction, I want to provide an overview and outline of this work and highlight just a few of its important and interesting contributions to the histories of philosophy and medicine. First is its organisation: the thesis is divided into three sections; the first deals with the classical texts used by and commented upon by Renaissance scholars, focusing especially upon Galen, Aristotle and Averroes (her discussion of this last is especially worthwhile). In particular, Deer Richardson shows how the issues that occupied Renaissance physicians and philosophers can be found in these ancient sources as a series of questions about generation that must be answered, thus providing, as she describes it, the "raw materials" for Renaissance theoretical treatments of the phenomena of reproduction. The second section focuses on natural philosophers, mainly expounding and explaining the natural world in general and generation in particular using primarily Aristotle's philosophy and a small number of Aristotelian texts, primarily his *Generation of Animals*. The third and final section (besides the conclusion) discusses the physicians, of which Fernel is the major figure and, indeed, model for other scholars contemporary and after him.

It is a major strength of Deer Richardson's work that she focuses so heavily on the sources of Renaissance philosophers and physicians, as their commentary and interpretation of classical sources formed a major part of their method of understanding nature. The first part of her work is extremely valuable for its direct and comprehensive discussion of the various sources and editions of the main classical texts used by Renaissance thinkers (including Plato, Hippocrates, Galen and Aristotle, as well as a very valuable discussion of pneuma and the pre-Socratics), even providing an overview of the timing at which these works were available. While her discussion of some of the secondary literature is somewhat dated, and more recent work has surpassed some of these secondary sources (for which see the bibliography below), the discussion itself is clear and quite helpful, as it both organises the discussion of how to think about the approach to generation in the Renaissance and informs the reader about the state of Renaissance literature. She offers a number of useful discussions of various ways of categorising generation theories (one seed vs. two seed, preformationist vs. epigeneticist, etc.) while, overall, rejecting these as being oversimplified and often anachronistic. In particular,

Deer Richardson provides a nuanced discussion of the various ins and outs of these theories and, focusing especially upon their contradictions, both internally within themselves and externally between the various systems. The focus is centred on the ancient debate as it was understood and translated in the Renaissance, paying attention in particular to various Latin phrases and terminology important in these debates and always keeping an eye on the original Greek terminology.

The second part focuses on Renaissance natural philosophers and their interpretations and uses of Aristotle's *Generation of Animals*, which was – by far – the single most important source and resource for all Renaissance theoretical discussions, a fact that I think is, perhaps, still not fully appreciated by modern scholars. In this section she focuses on a number of commentators on this work, noting similarities and differences in their style, language, interpretation and so on, discussing in detail the commentaries of Felix Accorambonius, Agostino Nifo and Cesare Cremonini and the "anti-commentary" of Bernardino Telesio. It is here that Deer Richardson notes the questions which formed the core of Renaissance philosophical commentaries (and, as she explains in the third part, most commentaries by physicians as well), which included issues such as the origin of semen, the mechanisms of heredity and the role of each parent in determining resemblance, as well as questions about what the material of which the foetus is constructed and questions about what patterns of growth and development it follows. Extremely interesting here, and unusual both at the time of its writing and even today, is Deer Richardson's simple, but quite enlightening, numerical analysis of commentaries on Aristotle. In some ways, Deer Richardson was doing the sort of work done in the digital humanities *avant la lettre*. This analysis gives us an overall sense of the lay of the land, a useful guide in attempting to evaluate and understand the place of particular ideas and works. More specifically, using 349 commentators from Charles Lohr's survey of Renaissance commentators, A through C, as a sample,[11] Deer Richardson notes some very interesting statistics: under half (150) wrote commentaries on at least one of the six natural philosophical works (*Physica*, *De anima*, *De generatione et corruptione*, *De caelo*, *Meteorologica* and the *Parva naturalia*), of which the most impressive are those of the Coimbra Jesuits. Of these commentaries most were on the *Physica* (117), followed by *De anima* (71), *De generatione et corruptione* (62), *De caelo* (45), *Meteorologica* (42) and finally, well in last place, the *Parva naturalia* (14, with 4 of these on one book only).

More relevant, however, are those commentators who would call themselves *philosophi et medici*, commentators who had taught, practised or at least studied medicine. Of these there are 29 in her sample, almost all of whom were trained in Italy or France. Many of these were highly influenced by what has been called the Paduan Averroist tradition, but which Deer Richardson complicates and analyses in detail, providing a very useful discussion of the way in which Averroes' ideas, and his approach to the topic, of generation deeply influenced a wide variety of Renaissance thinkers, sometimes indirectly through the work of European scholars who were more directly influenced. Even more relevant, and quite revealing, is the fact that there is but one manuscript commentary on the *De generatione animalium*, that of

Cremonini. Even though Lohr's census is incomplete, this fact is an extremely important one to note when trying to understand the overall tradition and importance of various Aristotelian works. Deer Richardson uses this brief numerical analysis to place the *De generatione* in its historical context and argues that it must be understood in terms of the Aristotelian corpus as a whole and, in particular, that it must be understood in terms of those commentaries that tried to make these works and the *De generatione* (and the other animal works) into a coherent, systematic whole, the focus clearly not on the *De generatione* itself.

The last section turns to works of theoretical medicine and what Deer Richardson calls the "compendia" tradition of Jean Fernel. The treatment here of a number of issues is absolutely fascinating, especially her long and detailed discussion of elements and temperaments and of spirits and innate heat. In the background here are issues of the soul and its materiality (or lack thereof), issues which became especially pressing after the Pope's *Apostolici Regiminis* after the Lateran Council in 1513, which affected all learned commentaries discussing the nature of the soul and the animation of the seed and foetus, especially those scholars whose Averroism was obvious. Here Deer Richardson emphasises especially the common structure of this tradition, which becomes even more interesting in light of the previous chapter on the natural philosopher's tradition. These physicians were also dealing with legacy of the ancients, but turned more often to Galen as a model and resource than to Aristotle, though she notes how the fundamental philosophical system remains deeply Aristotelian even when authors pledge their allegiance to Galen. This discussion and analysis thus allows Deer Richardson to do a rather fine-grained comparison and contrast of similarities and differences between the two traditions, arguing, as I noted above, that both groups ultimately had a great deal in common.

Everything is contextualised historically and linguistically, and the structure of what she calls the compendia tradition is analysed in terms of its Renaissance and early modern model, Jean Fernel, but always keeping in mind its origins in the medieval and Arabic traditions. Here she looks at the commentaries or treatises of four authors: Fernel (1542), Gaspar Contarenus or Contarini (1483–1542: published 1548), Jacobus Sylvius (1550) and Johannes Baptista Montanus (1498–1531: published 1554). She makes a number of interesting points here, including that the four elements of the Greeks formed a shared idea between doctors and philosophers, whereas the concept of temperaments was more exclusively the domain of the physicians. She further points out that the theories of both these minimal physical constituents, blends or whatever we might want to label them were, at the end of the day, entirely insufficient to understand the structure and function of the human body (noting that the focus for the physicians, unlike for Aristotle and later for Harvey, was almost exclusively upon humans and not animals in general). She also discusses the theory of mixtio, borrowed by doctors from philosophers and which has become an important topic in current studies of Renaissance thought. Deer Richardson describes the distinctions between elements and temperaments, focusing especially on heat, innate heat and their relation to the celestial heat of the heavens and their various and obscure categorisations and deployments in these specific philosophers and their ancient sources, as well as the convoluted terminology used by late medieval and Renaissance physicians to discuss the makeup, constitution and activities of the body, as well as how physicians can come to determine them in their patients through observation.

She comes to three important conclusions on the basis of her analysis: first, that all of her examples assume a four-element theory, ultimately founded upon the ideas of Aristotle – Paracelsus' chemical concepts and the various neo-Platonic ideas discussed by historians like Walter Pagel are not mentioned, even in passing, by any of her sources. Second, she argues that there are number of confusions to be found in her sources, in particular, confusions among three possible forms of the four-element theory. This confusion starts with Aristotle, who uses both qualitative change and change in density to explain transmutation and who also suggests that the higher elements like fire are closer to form and the denser like earth closer to matter. Importantly, Deer Richardson diagnoses this problem as one stemming from the desire to make coherent the inherited doctrines of all the ancients, who very often did not agree with Aristotle's own, confused, understanding of the basic constituents of things. Finally she notes the suggestion, made most strongly by Fernel, that the elements in their classical forms are not enough to explain the structure of the world and fare even worse in explaining living things and their activities. Thus another layer of explanation is called for temperaments, celestial heat, innate heat, spirits and soul, again a confusing mixture of terms and concepts, all of which call for going quite far beyond the elements, and again, Deer Richardson's discussion here is exemplary and is one of the most complete discussions of this topic available even today. I shall only note a few highlights: first she describes in detail how temperament is used by her sources and argues that it is, in fact, a theoretical means of comparing and then explaining individual and specific differences between animals and plants; in particular, in the context of generation, it helps explain the differentiation of the simple parts (blood, flesh, etc.). She notes further that the temperament system and Aristotle's conception of the active and passive are at odds. Perhaps the most interesting discussion is her analysis of Fernel's concept of spiritus and innate heat and the important roles these play in his physiology. Often she clarifies what is very confusing in Fernel (or in fact is contradictory) and, by pointing this out, does us a great service, for instance, helping us understand the complex relation between spiritus, heat and semen, as well as related matters. I note that, as always, Deer Richardson keeps a constant eye on the larger context of these discussions, and she compares Fernel's ideas on innate heat to those of Johannes Argenterius, Jean Riolan as well as Jacopo Zabarella, Sebastian Paparella and Pompeio Caimo and on the spirits to those of Julius Delphinus, Dominicus Bertacchius and Johannes Bronzerius. Her discussions here are extremely relevant to understanding fundamental issues about materiality, spirituality, God, the soul and the various and conflicting relationships between all of these ideas among Renaissance thinkers, grounded by her discussion of these figures and placing it in a larger historical context. Importantly, these are all central topics of early modern philosophy, but here we see how they play out not in the more familiar Cartesian context, but instead in a very sophisticated (if confusing) hybrid Galenic-Aristotelian-Hippocratic medical context. This is an important, and often ignored, part of the background needed to understand how we end up at Descartes, where, for instance, heat has lost its celestial connotations and no longer has any innate differences in living creatures making it different from, as Descartes notes, "damp hay".

My brief summary here truly only skims the surface of Deer Richardson's discussion – I have left out innumerable interesting and important aspects of her work that touches on God, on causality, on methodology, on reduction and on many other valuable and fascinating subjects, large and small, obscure and well-known. Her work is one that will benefit scholars of both the Renaissance and early modern periods and will enlighten both historians of medicine and philosophy. I can only hope that *Academic Theories of Generation* is studied half so well as she studied Fernel and those other Renaissance thinkers she writes about in such great detail.

Some Useful References on Jean Fernel and His Contemporaries

N.B.: This bibliography does not pretend to be comprehensive nor complete, and I have doubtlessly forgotten to include important and worthwhile works. Instead, I have gathered 60 articles and books that I have personally found valuable and which I believe will help supplement, complement and extend the work of Linda Deer Richardson. I have tried to pay special attention to non-English works, as these scholars are often missed or ignored by Anglophone scholars, at their peril and great loss. I have omitted any relevant works published prior to Richardson's thesis, as most of those references can be found cited there. I have also chosen not to include references dealing with Fernel's and his contemporaries' uses of classical sources, as these are too numerous to cite and would make the bibliography quite unwieldy, though there has recently been a great deal of quality work on this topic, especially concerning, for example, the influence of Averroes on Renaissance and early modern thought. Those sources found in Deer Richardson's thesis, however, remain an excellent place to start.

1. Aucante, Vincent. "La théorie de l'âme de Jean Fernel." *Corpus* 41 (2002): 9–42
2. Baader, Gerhard. "Jacques Dubois as a practitioner." *The Medical Renaissance of the Sixteenth Century*. Eds. Andrew Wear, Roger French, and Iain M. Lonie (Cambridge University Press, 1985): 146–154
3. Badaloni, Nicola. "Sulla costruzione e la conservazione della vita in Bernardino Telesio (1509–1588)." *Studi Storici: Rivista Trimestrale dell'Istituto Gramsci* 30 (1989): 25–42.
4. Bakker, Paul JJM, Sander W. De Boer, and Cees Leijenhorst (eds.). *Psychology and the Other Disciplines: A Case of Cross-disciplinary Interaction (1250–1750)*. (Brill, 2012).
5. Benoit, Robert. "Conceptions médicales à l'Université de Paris d'après les cours de Jean Riolan à la fin du XVIe siècle." *Histoire, Économie et Société* 14.1 (1995): 25–50.
6. Bianchi, Massimo Luigi. "Occulto e manifesto nella medicina del Rinascimento: Jean

a. Fernel e Pietro Severino." *Atti e memorie dell'accademia Toscana di scienze e lettere* 47
b. (1982): 183–248

7. Blank, Andreas. "Julius Caesar Scaliger on Plant Generation and the Question of Species Constancy." *Early Science and Medicine* 15 (2010): 266–286.
8. Blank, Andreas. "Daniel Sennert on Poisons, Epilepsy, and Subordinate Forms." *Perspectives on Science: Historical, Philosophical, Social* 19.2 (2011): 192–211.
9. Blank, Andreas. "Julius Caesar Scaliger on Plants, Species, and the Ordained Power of God." *Science in Context* 25.4 (2012): 503–523.
10. Bondì, Roberto. "Telesio e la tradizione magico-ermetica." *Rivista di filosofia* 88.3 (1997): 461–473.
11. Bono, James J., "Reform and the Languages of Renaissance Theoretical Medicine: Harvey versus Fernel." *Journal of the History of Biology* 23 (1990): 341–87.
12. Bono, James J. *The word of God and the languages of man: interpreting nature in early modern science and medicine*. (University of Wisconsin Press, 1995).
13. Bylebyl, Jerome. "The manifest and the hidden in the Renaissance clinic." *Medicine and the Five Senses*. Eds. William F. Bynum and Roy Porter. (Cambridge University press, 1993): 40–60.
14. Céard, Jean. "La physiologie de la mémoire, selon le médecin Jean Fernel." *Corpus: revue de philosophie* 41 (2002): 119–33.
15. Clericuzio, Antonio. "Spiritus vitalis: studio sulle teorie fijisiologiche da Fernel a Boyle." *Nouvelles de la République des lettres* 8.2 (1988): 33–84.
16. Cunningham, Andrew. "The Pen and the Sword: Recovering the Disciplinary Identity of Physiology and Anatomy before 1800 I: Old Physiology--the Pen." *Studies in History and Philosophy of Biological and Biomedical Sciences* Part C, 4 (2002): 631–665.
17. Cunningham, Andrew. "The pen and the sword: recovering the disciplinary identity of physiology and anatomy before 1800: II: Old anatomy—the sword." *Studies in History and Philosophy of Science Part C: Studies in History and Philosophy of Biological and Biomedical Sciences* 34, 1 (2003): 51–76.
18. Dessì, Cristina. "Marsilio Ficino, Jean Fernel e lo spiritus." in *Filosofijia, scienza, storia,* ed. Antonio Cadeddu (FrancoAngeli, 1995): 203–19.
19. Dessì, Cristina. "Le curiose ricerche di Jean Riolan sulle scuole di medicina di Parigi e di Montpellier: la difesa della tradizione e la critica della medicina chimica." *Annali della Facoltà di scienze della formazione dell'Università di Cagliari* 23 (2000): 117–146.
20. de Calan, Ronan. *Généalogie de la sensation. Physique, physiologie et psychologie en Europe, de Fernel à Locke* (Honoré Champion, 2012).
21. Di Liscia, Daniel A., Eckhard Kessler, and Charlotte Methuen. *Method and order in Renaissance philosophy of nature: the Aristotle commentary tradition*. (Ashgate, 1997).

22. Ferretto, Silvia. *Maestri per il metodo di trattar le cose : Bassiano Lando, Giovan Battista da Monte e la scienza della medicina nel XVI secolo.* (Cleup, 2012).
23. Forrester, John M. and John Henry. *The Physiologia of Jean Fernel (1567).* (Transactions-American Philosophical Society 93.1, 2003).
24. Forrester, John M. and John Henry. *Jean Fernel's On the Hidden Causes of Things: Forms, Souls, and Occult Diseases in Renaissance Medicine* (Brill, 2005).
25. Garozzo, Salvatore. "De morbi et eorum causis : dal trattato Universa medicina di Ioannis Fernelii." *Memorie e rendiconti* 4.1 (1991): 147–186.
26. Guerrini, Anita. "Experiments, Causation, and the Uses of Vivisection in the First Half of the Seventeenth Century." Journal of the History of Biology 46 (2013): 227–254.
27. Henry, John. "The fragmentation of Renaissance occultism and the decline of magic." *History of science* 46 (2008): 1–48.
28. Henry, John. "'Mathematics Made No Contribution to the Public Weal': Why Jean Fernel (1497–1558) Became a Physician." *Centaurus* 53.3 (2011): 193–220.
29. Henry, John. "Jean Fernel on Celestial Influences and the Reform of Medical Theory." In *Celestial Novelties on the Eve of the Scientific Revolution, 1540—1630*, eds. Patrick J. Boner and Dario Tessicini (Olschki, 2013): 133–158.
30. Hirai, Hiro. *Le concept de semence dans les théories de la matière à la Renaissance: de Marsile Ficin à Pierre Gassendi.* (Brepols, 2005).
31. Hirai, Hiro. "Alter Galenus: Jean Fernel et son interpretation platonico-christienne de Galien." *Early Science and Medicine* 10 (2005): 1–35.
32. Hirai, Hiro "Ficin, Fernel et Fracastor autour du concept de semence: aspects platoniciens de seminaria." In *Girolamo Fracastoro fra medicina, fijilosofijia e scienze della natura*, ed. Alessandro Pastore and Enrico Peruzzi (Olschki, 2006): 245–60.
33. Hirai, Hiro. "The Invisible Hand of God in Seeds: Jacob Schegk's Theory of Plastic Faculty." *Early Science and Medicine* 12.4 (2007): 377–404.
34. Hirai, Hiro. "Prisca Theologia and Neoplatonic Reading of Hippocrates in Fernel, Cardano and Gemma." in *Cornelius Gemma: Cosmology, Medicine and Natural Philosophy in Renaissance Louvain*, ed. Hiro Hirai (Rome: Serra, 2008): 91–104.
35. Hirai, Hiro. "Lecture neoplatonicienne d'Hippocrate chez Fernel, Cardan et Gemma." in *Pratique et pensée médicales à la Renaissance*, ed. Jacqueline Vons (Paris: De Boccard, 2009): 241–56.
36. Hirai, Hiro. *Medical humanism and natural philosophy: Renaissance debates on matter, life and the soul* (Brill, 2011).
37. Houtzager, H. L. "Vesalius contra Sylvius." *Scientiarum historia* 18.2 (1992): 123–126.
38. Kennedy, Leonard A. "Cesare Cremonini and the immortality of the human soul." *Vivarium* 18 (1980): 143–158.

39. Laird, W.R. "Venetischer Aristotelismus im Ende der aristotelischen Welt: Aspekte der Welt und des Denkens des Cesare Cremonini (1550–1631)." Renaissance Quarterly 52 (1996): 230–32.
40. Mahoney, Edward P. "Pico, Plato, and Albert the Great: The Testimony and Evaluation of Agostino Nifo." *Medieval Philosophy & Theology* 2 (1992): 165–192.
41. Michael, Emily. "The nature and influence of late Renaissance Paduan psychology. *History of Universities* 12 (1993): 65–94.
42. Mikkeli, Heikki. "Jacopo Zabarella (1533–1589): The Structure and Method of Scientific Knowledge." In: *Philosophers of the Renaissance*. Ed. Paul Richard Blum (Catholic University of America Press, 2010): 181–191.
43. Mulsow, Martin. *Frühneuzeitliche Selbsterhaltung: Telesio und die Naturphilosophie der Renaissance*. (Walter de Gruyter, 1998).
44. Mulsow, Martin. "Arcana naturae: Verborgene Ursachen und universelle Methode von Fernel bis Gemma und Bodin." In: *Der Naturbegrifff in der Frühen Neuzeit: Semantische Perspektiven zwischen 1500 und 1700*. Ed. Thomas Leinkauf (Niemeyer, 2005): 31–68.
45. Nutton, Vivian. "Montanus, Vesalius and the haemorrhoidal veins." *Clio Medica* 18 (1983): 33–36.
46. Ongaro, Giuseppe. "La controversia tra Pompeo Caimo e Cesare Cremonini sul calore innato." *Historia* 1 (2000): 87–110.
47. Ongaro, Giuseppe, Maurizio Rippa Bonati, and Gaetano Thiene, eds. *Harvey e Padova: atti del Convegno celebrativo del quarto centenario della laurea di William Harvey, Padova, 21–22 novembre 2002*, 39. (Antilia, 2006).
48. Palmieri, Paolo. "Science and Authority in Giacomo Zabarella." *History of Science* 45 (2007): 404–427.
49. Perfetti, Stefano. "'Metamorfosi di una traduzione. Agostino Nifo revisore dei De animalibus gaziani." *Medioevo* 22 (1996): 259–301.
50. Perfetti, Stefano. "Three Different Ways of Interpreting Aristotle's De Partibus Animalium: Pietro Pomponazzi, Niccolò Leonico Tomeo and Agostino Nifo," In: *Aristotle's Animals in the Middle Ages and Renaissance*, Eds. C. Steel, P. Beullens, and G. Guldentops. (Leuven University Press, 1999).
51. Perfetti, Stefano. *Aristotle's Zoology and Its Renaissance Commentators, 1521–1601*. (Leuven University Press, 2000).
52. Pozzo, Riccardo. "Res considerata and modus considerandi rem: Averroes, Aquinas, Jacopo Zabarella and Cornelius Martini on reduplication." *Medioevo* 24 (1998): 151–175.
53. Raimondi, Francesco Paolo. "La filosofia naturale di G. Zabarella e la scienza moderna : connessioni e divergenze." *Physis* 31.2 (1994): 371–391.
54. Regier, Jonathan. "Kepler's Theory of Force and His Medical Sources." *Early science and medicine* 19.1 (2014): 1–27.
55. Rommevaux, Sabine. "A treatise on proportion in the tradition of Thomas Bradwardine: The De proportionibus libri duo (1528) of Jean Fernel." *Historia Mathematica* 40.2 (2013): 164–182.

56. Schmitt, Charles. "Aristotle Amongst the Physicians," In: *The Medical Renaissance of the Sixteenth Century*, Eds. Andrew Wear, Roger French and I.M. Lonie (Cambridge University Press, 1985): 1–15.
57. Schuhmann, Karl and Theo Verbeek. "Le concept de matière chez Bernardino Telesio." *Discorsi A* 9.2 (1989): 262–281
58. Schweikardt, Christoph. *Theoretische Grundlagen galenistischer Therapie im Werk des Giessener Arztes und Professors Gregor Horst (1578–1636): ein Vergleich zu Jean Fernel (1497–1558), dem Leibarzt des französischen Königs Heinrich II.* (PhD dissertation, 1997).
59. Secret, François. "De Mésué à Hieronymus Rubeus, en passant par Giovanni Mainardi et Jacques Dubois." *Chrysopoeia* 5 (1992–1996): 453–466.
60. Zanier, Giancarlo. "Platonic Trends in Renaissance Medicine." *Journal of the History of Ideas, 48* (1987): 509–519.

University of South Florida Benjamin Goldberg
Tampa, FL, USA

Notes

1. For instance, Roger French's *William Harvey's Natural Philosophy* contains no chapter dedicated to this work, nor does generation appear in the index; French, Roger (1994), *William Harvey's Natural Philosophy*, Cambridge University Press. Recent scholars have started paying more attention to this work, however. For instance, see Lennox, J. G. (2006), "The Comparative Study of Animal Development: William Harvey's Aristotelianism", in *The Problem of Animal Generation in Early Modern Philosophy*, ed. J. E. H. Smith, Cambridge University Press; Ekholm, K. J. (2008), "Harvey's and Highmore's Accounts of Chick Generation. *Early Science and Medicine*, *13*(6), 568–614; and my own (2013), "A Dark Business, Full of Shadows: Analogy and Theology in William Harvey", *Studies in History and Philosophy of Biological and Biomedical Sciences* 44: 419–432. At the end of this preface, I provide an extended (though by no means comprehensive!) bibliography.
2. Going back at least to Joseph Needham's (1959), *A History of Embryology*, Cambridge University Press
3. Pagel, Walter (1967), *William Harvey's Biological Ideas*; S. Karger (1976), *New Light on William Harvey*; Nutton, Vivian (1990), "The Anatomy of the Soul in Early Renaissance Medicine", in *The Human Embryo*, ed. G. R. Dunstan, University of Exeter Press; Pyle, A. J. (1987), "Animal Generation and the Mechanical Philosophy: Some Light on the Role of Biology in the Scientific Revolution", *History and Philosophy of the Life Sciences* 9: 225–254; Schmitt, Charles 1989, "William Harvey and Renaissance Aristotelianism: A Consideration of the Preface to De generatione animalium", in *Reappraisals of Renaissance Thought*, ed. Charles Webster, Variorum

4. Hirai's work here is exemplary in how he has pushed forward on many of the topics explored in this thesis, and his work is well worth deep study. See especially his (2011) *Medical Humanism and Natural Philosophy: Renaissance Debates on Matter, Life and the Soul*, Brill. Further, I make no claim here that EBSCOhost exhausts all that might have been written on these figures and topics, or that I have exhausted every relevant search permutation, but it does, at least, cover the work that is easily accessible.
5. Forrester, John M., and John Henry (2003), *The Physiologia of Jean Fernel (1567)*, Transactions-American Philosophical Society 93.1, and their (2005) *Jean Fernel's On the Hidden Causes of Things: Forms, Souls, and Occult Diseases in Renaissance Medicine*, Brill
6. The origin of these terms is found, I believe, in Charles Darwin's discussion of taxonomy and species in his letter to J. D. Hooker in July of 1857.
7. This is something I have been deeply interested and concerned with in my own work, and I have recently edited a book with Evan Ragland and Peter Distelzweig on just this issue; please see our volume of essays: (2015) *Early Modern Medicine and Natural Philosophy*, Springer.
8. See, for instance, her (1987) *Avicenna in Renaissance Italy: The Canon and Medical Teaching in Italian Universities After 1500*, Princeton University Press. I think, in general, this lesson is one that has been well learned by many (though by no means all) specialists in Renaissance philosophy and intellectual culture, but has not, perhaps, been taken up more widely.
9. For which see my (forthcoming) "William Harvey on Anatomy and Experience", *Perspectives on Science*, 24:3
10. See Hirai *op. cit.* and his (2007) "The Invisible Hand of God in Seeds: Jacob Schegk's Theory of Plastic Faculty", *Early Science and Medicine* 12.4: 377–404; Blank, Andreas (2010), *Biomedical Ontology and the Metaphysics of Composite Substances: 1540-1670*, Philosophia Verlag; and Park, Katharine (2006), *Secrets of Women: Gender, Generation, and the Origins of Human Dissection*, Zone Books.
11. Lohr, Charles H. (1974), "Renaissance Latin Aristotle Commentaries: Authors A-B", *Studies in the Renaissance*, 21, 228–289, and (1975) "Renaissance Latin Aristotle Commentaries: Authors C", *Renaissance Quarterly*, 28, 689–741

Acknowledgements

These come in two categories. The first is thanks for those who helped me prepare my thesis in the late 1970s. I thank:

- Professor D.P. Walker, my supervisor, for patient and illuminating guidance;
- Professor M.A. Screech, Drs. Walter Pagel, Charles Schmitt, Vivian Nutton and James Bono for valuable discussions and bibliographical aid;
- Staff and scholars of the Warburg Institute and the Wellcome Institute for the History of Medicine for comradeship, patient help with enquiries, and the encouragement I needed to see the task through.

The second group thanks those who were instrumental in the publication of my thesis in 2017:

- Dr James G. Lennox, Professor,Department of History and Philosophy of Science, University of Pittsburgh.
- Jim is the main reason this thesis was published. He persuaded me that it was a worthwhile contribution to knowledge, with praise I blush to think of; and then persuaded a publisher, Springer, of the same thing.
- Dr Benjamin Goldberg, Department of Humanities and Cultural Studies, University of South Florida, Tampa.

As Benny writes in his preface, he is the other reason my thesis was published, because it was his interest that led him, with Jim, to track it down and read it. I am extremely grateful to him for contributing his preface.

Many people from Springer have been involved in publishing this work, but I especially want to thank:

- Dr Charles T. Wolfe, Editor, Springer series in History, Philosophy and Theory of the Life Sciences, who has been an enthusiastic and amazingly patient editor; and
- Ms Christi Lue, formerly Senior Editorial Assistant, Springer Science and Business Media B.V., a fixer of infinite patience.

Thank you all.

Contents

Editors Photos

Linda Deer Richardson

Benjamin Goldberg

Abbreviations and Editorial Conventions

Primary sources listed in the bibliography, including translations and modern editions, are referred to by author, short title, place and date at the first reference and thereafter by author and short title with date if necessary to make the reference clear. Full references are given in the bibliography.

Journal articles and other secondary sources are given in full in the first reference, and short title references for works cited repeatedly are given in brackets after the first reference. The major classical sources and their Renaissance editions are dealt with as follows:

Aristotle: Cited by Bekker number. References to works are given in the notes in the standard abbreviations of the Oxford edition, and this is the English translation used unless otherwise stated. Latin citations are taken from the translation of Gaza, Venice, 1476 (and later editions).

Galen: Cited by the Kühn edition, (1964–5) as Kühn with volume and page. Unless attributed, English translations are my own. For the major generation works, Kühn's Latin versions are taken from the translation of Cornarius, and thus parallel those of the Aldine edition of 1541 which I have used as a standard reference.

Hippocrates: Cited by the edition with French translation of Littré, (1839–1861) as Littré with volume and page. English translations of The Seed and On the Nature of the Child are by I.M. Lonie, from Hippocratic Writings, edited by G.E.R. Lloyd, Penguin (1978). Other English translations are from the same edition unless given as my own. Latin citations are from the Opera Omnia edited by Cornarius, Basle, 1546, unless otherwise noted.

Plato: Cited by the ed. princeps. references. The English translation of the Timaeus is by Desmond Lee (Penguin, 1974). The Latin translations consulted were those of Chalcidius (1520) and Ficino (1484–5).

A survey of major Renaissance editions of generation texts by these four authors is given in Appendix I.

Chapter 1
Introduction

Abstract The introduction gives the shape and contents of the book as a whole and includes a literature review (of works published by 1980). This review suggests that none of the writers is interested in the Renaissance for its own sake. Most of them concentrate on one group of writers, the anatomists, or alternatively the authors of obstetric treatises. No attempt is made to set their descriptions into any other theoretical context than that of modern embryology.

The book distinguishes four main types of sixteenth century writing on generation:

1. Practical treatises and anthologies dealing with the diagnosis of pregnancy and the care of women in childbirth. These are the most common.
2. Anatomical texts which include a description of the parts serving generation and of the development of the foetus.
3. Textbooks of theoretical medicine which include generation, and commentaries or monographs by medical writers.
4. Commentaries on the *Generation of Animals* of Aristotle and related works, usually by natural philosophers.

Virtually all other historical accounts base themeselves on the first two categories, though without distinguishing them from the last two. This book focuses on the last two groups.

This study began from the attempt to understand one particular Renaissance treatise on generation, Jean Fernel's "De hominis procreatione atque de semine",[1] and to place it in a proper intellectual and historical context. This led me back to the classical authorities on which the work was based, in their Renaissance translations and editions. Coming forward again, I wanted to be able to compare Fernel's work with those of other contemporary medical writers, and with another group of Renaissance writers on generation, the natural philosophers, with whom he seems to have been familiar.

The work thus falls into three sections. The first deals with the major classical texts on generation, which provided the raw materials for treatment of theoretical questions in the late sixteenth and early seventeenth centuries. The second part

L. Deer Richardson, *Academic Theories of Generation in the Renaissance*,
History, Philosophy and Theory of the Life Sciences 22,
https://doi.org/10.1007/978-3-319-69336-1_1

considers the natural philosophical contribution, primarily a relatively small number of commentaries on Aristotle's Generation of Animals, which was one of the group of Aristotelian texts explaining the workings of the natural world. This section also includes monographs which consider particular questions arising from the commentary tradition. The third part looks at the work of medical writers, of whom Fernel is taken as the major example, who were themselves trained in this tradition of natural philosophy. In particular, it examines the treatment of generation in textbooks of theoretical medicine written on the model of Jean Fernel's Physiologia.

The supposition on which this approach is based is that, despite the advances in anatomy and botany which have received the major share of attention from historians,[2] the core of the medical curriculum before 1650 was based on the study and commentary of classical texts, in particular the relevant works of Plato, Aristotle, Galen, Hippocrates and their commentators. These formed the basis for the teaching programme in theoretical medicine or Theoria, which together with Praxis made up the syllabus. Moreover, theoretical medicine had important links with another branch of the academic curriculum, the arts course. Medical students were almost invariably required to be arts graduates; which meant that they had, or should have, received a thorough grounding in classical languages and the principles of Aristotelian natural philosophy.

Such training placed the doctor in a special relationship with his teaching colleagues in the arts faculty, and was bound to affect his approach to the basic questions of medicine and what the sixteenth century had already begun to call "physiology". One of the most important of these topics in theoretical medicine was the problem of generation, which included a group of "classic questions" debated by doctors and natural philosophers alike. The topic of generation draws on so many of the central controversies of this debate that it serves as a useful touchstone. And generation was itself considered by contemporaries as a central problem for both theoretical medicine and natural philosophy. Despite this, no general study of academic theories of generation exists for the century between Fernel and Harvey.

Moreover, the approach which I propose to take in this study is substantially different from that of other historians who have written on the subject. The Renaissance has been ignored, to a greater or lesser extent, by the authors of the standard general histories of generation theory, or as they prefer to call them, histories of embryology. This may be in part because, as with so many aspects of medical history, generation theory has until relatively recently been the preserve of embryologists. Embryology textbooks of the 1920's and 1930's, such as those of Marshall (1922) and Huxley and De Beer (1934), might have a few pious words of historical introduction.[3] These were amplified in general histories written by embryologists (or at least anatomists) which reached a peak in the 1930's.

One of the earliest of these historical accounts does, however, consider the period before 1650 in some detail. This is Bruno Bloch's Die geschichtlichen Grundlagen der Embryologie bis auf Harvey (1904). In his introduction,[4] Bloch points out that history of embryology, especially for this earlier period, is a virtually untouched field, except when considered as a branch of the history of anatomy or physiology or in general histories of medicine such as those of Pagel and Neuburger. Outstanding embryologists, such as Wolff, von Baer and Pander, who have successfully argued for the independence of their speciality, have treated its historical development only sketchily and without attempting to draw together general principles. Bloch himself presents embryology as a sub-division of anatomy, in three chronological sections: antiquity from Hippocrates to Galen; Galen to 1500; the sixteenth and early seventeenth centuries, up to, but not including, Harvey's De Generatione Animalium.

In the final chapter, which covers much the same chronological period as this study, Bloch argues that the sixteenth century was characterised by a revolution in anatomy and in its sister-science, embryology. In his account of the "great reforming revolution" in sixteenth century embryology, Bloch looks almost entirely at the work of anatomists, especially those, like Aldrovandi, Coiter and Fabricius, who examined hens' eggs or mammalian foetuses. The outstanding characteristic, and chief virtue, of Bloch's Renaissance embryologist is as an observer. Apart from a passing reference to humanism and the suggestion that the writings of Hippocrates acted as a "lever" in the embryological revolution, Bloch devotes very little space to classical theories or their development in the Renaissance. His judgment of Capivaccius is typical of his generally unsympathetic attitude to medical theory:

> No better was the treatment of the same questions by the Paduan professor Hieronymo Capivaccio. He confessed himself to be with the Galenists on all points, but tried to mediate between Aristotle and the Greek physician in some controversies – certainly without much success.[5]

Such attempts at mediation are an important theme of what follows. But Bloch does make one observation which at least suggests that anatomical discoveries need to be interpreted in their theoretical context. The Italian anatomist Aldrovandi not only followed Hippocrates in his daily examination of fertile eggs, but he did so in order, like Capivaccius, to try to determine one of the outstanding controversies between the Galenist and Aristotelian positions: that of the order of formation of parts. Aldrovandi decided, with Aristotle, in favour of the heart – a choice which causes Bloch to make his only admission that anatomical "experiment" itself is not entirely value-free.[6]

One of the earliest accounts in English is E.S. Russell's Form and Function (1916).[7] Russell's book is concerned less with embryology per se than with the study of morphology and comparative anatomy. Accordingly, after an introductory chapter on Aristotle, he devotes two pages to the next 2000 years. His next real historical event is the invention of the microscope, though he mentions in passing two groups of writers in the Renaissance, "the great Italian anatomists" (Vesalius, Fallopius and Fabricius) and "the first systematists", Rondelet, Aldrovandi and Gesner.[8]

The other three "standard English surveys of the history of embryology",[9] which appeared in the 1930's, are like Russell in neglecting the Renaissance and in emphasising those aspects of sixteenth and seventeenth century "embryology" which seem to anticipate modern ideas: in particular, the investigations of certain anatomists. F.J. Cole's Early Theories of Sexual Generation (1930) is not quite "early" enough: it begins in the second half of the seventeenth century, with the discovery of the spermatazoon and, later, the mammalian ovum.[10] A.W. Meyer, a contemporary American embryologist, is similar. The Rise of Embryology (1939), his main work in the field, has a chapter on "Early Historic Ideas of Generation", which ends with Galen. "After Galen" he writes, "came a long period of silence."[11] Later chapters are organised thematically, and deal for the most part with debates, such as epigenesis versus preformation, of the eighteenth and nineteenth centuries. Meyer's Preface suggests that his work was intended as an inspiration to students of embryology, and offers the lesson that "although chance contributed a few facts, observation and experiment were the real keys to progress. Metaphysics revealed no new facts whatsoever and often led to long-continued error".[12]

The most important, and indeed still respected, of these 1930's textbooks is undoubtedly Joseph Needham's History of Embryology (1934, second edition 1959).[13] Needham's stature as a biologist and historian of science is beyond dispute, and there have been no surveys of comparable breadth to supersede his History in its revised edition. Needham's work, like Meyer's, clearly filled a need, felt by embryologists,[14] for a general account of the progress of embryological knowledge, which would be an encouragement to students. It begins with "primitive" ideas on generation and devotes one chapter each to Antiquity, the period from Galen to 1600, the seventeenth, and the eighteenth centuries. Both in the detail he gives for the earlier periods and in his approach Needham is less obviously biased towards the modern than Meyer. But if we exclude the section on Leonardo (whose anatomical work was not known to his contemporaries) there are only twenty pages covering the period from the end of the Middle Ages to the middle years of the seventeenth century. Once again, the writers whom Needham praises, as signalling "the end of the twilight period" are all anatomists.[15] And again it is their use of "experiment" – either dissection or the observation of hens' eggs – which Needham finds praiseworthy. What he seeks in these Renaissance writers, and rarely finds, is "genuine works of research". On Fernel, for instance, he remarks that he "followed Aristotle and Galen in nearly all particulars, and made no real contributions to embryology".[16]

Needham also has a section on Renaissance obstetrics, and apart from the anatomists, almost the only other group of writers he mentions are the authors of what I have called "women's books".[17] He thus includes almost none of the theoretical discussions of generation with which I shall be concerned, and, as we have seen with Fernel, is generally unsympathetic to the theoretical approach. Moreover, he compares writings of various types and levels of sophistication with no apparent awareness of the anomalies this creates. For instance, he can claim that the illustrations of Rueff's De conceptu, a popular German treatise on obstetrics, give us insight into "the general ideas … at this period about mammalian embryology" held by a very wide range of authors. To take Rueff's naive and literal renderings of the classical descriptions of foetal development as an accurate guide to the thinking of academically trained doctors and natural philosophers seems a curious inversion of priorities: rather as though one were to look to a modern textbook of obstetric nursing for a guide to the latest research in genetics. Moreover, Needham's list of writers on generation is itself very mixed.[18]

Among more recent (1942, 1966) writers, Howard Adelmann comes closest to perpetuating the attitude of the 1930's general histories, in two massive and important studies of the work of Fabricius and Malpighi. Adelmann, like Needham and Meyer, is an embryologist, and his attitude is clear from the historical introductions to the two books. The Renaissance section of "A Brief Sketch of the History of Embryology before Fabricius" concentrates entirely on anatomists.[19] Volume II, Part 2 of Malpighi, "The Main Currents of Embryological Thought before Malpighi" looks at the same group of anatomists, beginning with Volcher Coiter: those whom Adelmann calls "followers of the Hippocratic injunction" to observe foetal development systematically, by daily examination of hens' eggs. Adelmann dismisses almost all the writers on generation between De natura pueri and the seventeenth century for failure, not simply to observe, but to observe in this particular manner:

> The sum of observations on the embryos of various animals contained in the literature as late as Galen is certainly not contemptible; yet these observations were almost entirely fortuitous and at best lack the continuity requisite for comprehension. On such incomplete and accidental observation early theory was based. … The author of the Hippocratic treatise De natura pueri recognized, to be sure, the importance of systematic, comparative observation when he urged the daily inspection of the hen's egg throughout incubation as a clue to the understanding of human development;… yet no one took the advice of the Hippocratic author for over two milleniums.[20]

Similarly, in his chapter on the sixteenth and seventeenth centuries, he devotes a paragraph to the theoretical questions which were discussed by Renaissance medical writers: questions such as the origin of the semen, the role of the two sexes, and the order of development of parts. But the attempts to answer these questions with reference to classical authority are of no interest to Adelmann, and he concludes scathingly:

> We should gain little from continuing this analysis of the unprofitable discussion of worn-out issues. The important fact is that none of these men followed the Hippocratic injuction to investigate systematically at close intervals the development of the chick, and none had investigated methodically the whole progress of development in any other animal.[21]

Adelmann, then, divides his subjects into reactionaries and "those who were able to throw off … the shackles of authority and precedent; apparently rebelling against all a priori reasoning and all authority, they presented a naked set of observations unencumbered by philosophical and authoritarian baggage".[22]

Adelmann's definition of profitable activity is a very narrow one, and as Wightman reminds us, his picture of the unencumbered anatomical observer is to say the least unrealistic.[23] His approach has been criticised even by reviewers who go some way towards sharing his point of view.[24] Applied to Malpighi himself, Adelmann's method produces valuable results, but he fails altogether to come to terms with the fact that all historians of science are dealing with ideas and theories which have been superseded; to call them "worn-out issues" does nothing to help us understand them. The theoretical questions which Adelmann rejects as "unprofitable discussion" are the central subjects of this book.

Other recent studies in this field, which Frederick Churchill has called "The History of Embryology as Intellectual History", can be dealt with more briefly.

Elizabeth Gasking's Investigations into Generation is a series of essays, beginning with Harvey's publication of De Generatione Animalium in 1651. She claims that this work inaugurated a new period in generation theories, by proving from dissection that the pregnant uterus did not contain a mass of seed mingled with menstrual blood, as virtually all earlier writers had assumed.[25]

Ian Maclean's Woman Triumphant: Feminism in French Literature 1610–1652 uses literary sources to examine some aspects of the social position of women in seventeenth century France. Apart from a brief mention of the medical definition of women in his introductory chapter, he does not touch on the questions considered in this study; chronologically, too, his focus is rather later.[26]

Pierre Darmon's Le mythe de la procréation à l'âge baroque is not about generation theory itself, but about the "myths" surrounding generation: those concerning sterility, parthenogenesis, emboîtement, in the late seventeenth and eighteenth centuries. His treatment consists in grouping quotes on a given topic without much regard for chronology or for the weight or tone of the work cited, and his approach tends to emphasise "sensational" aspects of the subject.[27] The result is disappointingly lightweight, and, in the two pages he devotes to systems of generation between Hippocrates and Descartes, he is vague, sneering and ignorant:

> En dehors de quelques observations, l'embryologie ne suscite que des opinions fausses, des recherches stériles ou des nouveautés fantaisistes.

Descartes' Traité de l'homme et De la formation du foetus is itself a late example of this "immobilisme ... en plein XVIIe siècle".[28]

There are, however, two major recent works on generation theory by French writers, which need to be considered. Both Jacques Roger and François Jacob view the history of biology, in particular the history of ideas about generation and heredity, not so much as a triumphant progress from error to science as a series of discrete stages. Jacob's first stage ends with the sixteenth century, while Roger looks back into the sixteenth century from the period 1600–1670 in the introductory section of his book. But both writers' views of generation theory in the Renaissance is coloured by the fact that their main interest lies elsewhere: for Jacob in the late eighteenth and nineteenth century and up to the present day; for Roger in the second half of the eighteenth century, in particular the work of Diderot and Buffon.

The theme of François Jacob's La logique du vivant is "the transformations which have progressively modified the way of considering the nature of living things" and in particular the way in which thinking about questions of heredity and reproduction have changed.[29] Jacob rejects the "genealogical" approach to history of science, which extrapolates backwards, seeking the thread which connects theories of today with their "ancestors" in the past, in order to uncover the "royal road" to biological truth. Instead, he sees biological thought as undergoing a series of "transformations": a succession of different logical and observational ways of perceiving, within which the science of each stage developed. The most important of these "transformations" was the one occasioned by acceptance of the theory of evolution, with all that it implied[30]; and the motor for these changes of perceptual frame, which have something in common with the "scientific revolutions" of Thomas

Kuhn, and, according to one analysis, with the ideas of Michel Foucault,[31] was most often the access to a new level of observation – although new tools, such as the microscope, were less important than new ways of looking at the organism and of formulating questions about it. These levels were first, at the end of the seventeenth century, that of "visible surfaces", i.e. gross anatomical structure; towards the end of the eighteenth century, the level of chromosome and gene; and finally the fourth, inner layer of the nucleic acid molecule.[32]

Strictly speaking, the Renaissance forms a fifth stage, before level one. For according to Jacob there is a fundamental difference between the point of view of Antiquity and the Renaissance and that of the seventeenth century, the "classic age". The idea of generation held by sixteenth century writers was of a creation which required the direct intervention of divine powers, mediated by the soul and by innate heat. But because of their immediate dependence on divine will, no system of laws of nature, which could distinguish between the contingent and the necessary, was possible.[33] Similarly, there was no true idea of species; only a hierarchy of beings in which mineral shaded imperceptibly into plant and plant into animal. Each generation was a "new creation" which, like artistic creation, was independent of all other events and had no necessary links with the past. The evidence for this, according to Jacob, lies in the admission of the possibility of spontaneous generation – which allows two equally possible ways of achieving the same result – and in the acceptance of monsters as possible "natural" variations in the course of events.[34] By contrast, the seventeenth century saw the establishment of laws and of the concept of species:

> La génération d'un étre ne peut plus alors constituer un événement isolé, unique, indépendant. Elle devient l'expression d'une loi qui témoigne de la régularité de l'univers.[35]

Jacob's is a stimulating book, and it has the merit of taking seriously the ideas, as well as the observations, of Renaissance writers on generation. The Renaissance, however, is not his main concern: it serves merely as the pre-scientific age before the first, classical, age of biology proper. I hope it will be shown in what follows that the concept of a radical transformation between the sixteenth and early seventeenth centuries is untenable, and that his interpretation of the earlier period is somewhat fanciful – if there is a break, I would want to place it around 1650, with the theories of Harvey and Descartes, the discovery of spermatazoa and the development of new mechanical models of generation by Gassendi and Borelli.

Jacques Roger's study of biological ideas in eighteenth century France[36] is similarly divided into contrasting periods. The purpose of the long prefatory section, "The End of the Renaissance (1600–1670)" is to act as a setting, and a foil, for the biological developments of the later eighteenth century, which are Roger's real interest. It is true that Roger has examined a larger number of treatises on generation theory from the late sixteenth and seventeenth centuries than any of the other writers considered here, and from a bibliographical standpoint his study is valuable. But I cannot agree with Churchill that Part One of Les sciences de la vie "also forms in its own right an excellent 150-page portrait of the medical and biological concerns of the seventeenth century".[37]

For although it is more detailed, it is little more sympathetic in its tone, despite Roger's claim to look at his subjects from the inside: "C'est leur langage que nous parlons". This is clear from the question with which he begins the book: did Molière lie? Was Thomas Diafoirus a true portrait of the seventeenth century doctor? According to Roger, the answer is that Molière did not lie, although he may have exaggerated a little. Seventeenth century medicine showed "surprising scientific immobility" and "indifference to research". Doctors used outdated teaching methods, were sunk in complacency and absorbed in questions of corporate privilege. A "new scientific spirit", when it came, had to be imported from the outside, from the new philosophy and physics.[38] Roger's account thus fits into a familiar "Whiggish" pattern: why was there not modern medical theory in the seventeenth century? or rather, why did French medicine appear to lag behind contempory physics and philosophy? Answer: because there were obstacles to scientific progress, which need first to be assumed, then described.[39]

As well as external explanations based on the social position and corporate identity of the French physician in the seventeenth century, Roger offers internal ones, based on the nature of the theories themselves. Here he is rather more interesting: Renaissance attempts to explain the phenomena of generation were hampered, he suggests, by a concept of "nature" which was both rigid and vague, and confused by the bewildering number of possible levels of meaning at which explanation could be offered.[40]

In the second chapter, "Le combat contre les ombres", he deals with particular theoretical problems as they were dealt with by typical doctors and philosophers. The main battle he describes is not against shadows, however, but between the doctors and philosophers themselves, whom he divides into two warring camps, followers of Aristotle and Hippocrates-Galen. Earlier, he had argued that seventeenth century physicians faced a clear choice between traditional authorities and Paracelsus, and here he argues again for an either-or position. It is the question of female semen "which divides the biologists of the seventeenth century into two irreconcilable camps"; and the two doctrines form up opposite one another, "without possibility of conciliation".[41]

In what follows, I hope to be able to argue that this is an unhelpful and oversimplified model, and that important attempts at conciliation were made by medical and philosophical writers, in an attempt to bring together divergent authorities into one harmonious and reasonable account. And Roger himself suggests, at the end of Part One, that this "warring camps" approach obscures certain important similarities. In the first place, there were attempts to combine elements from the different explanations into a synthesis: but the result was a kind of patchwork, a "harlequin costume", in Roger's view. Some writers diverged from the party line in various ways; moreover, the medical tradition itself rested on an Aristotelian one, and "Galenists" like Fernel and Riolan were heavily influenced by Aristotle. Also, according to Roger, Aristotelians and Galenists "shared a general state of mind in the face of one of the essential problems of biology". Both refused to accept a simple spiritual or material explanation of generation, and both were painfully aware of the complexity of vital phenomena. Considered together, their approach was not so much "false" as "sterile", and it is with some relief that he turns to the opinions of other, more progressive writers who, in the second half of the seventeenth century, led "the search for clear ideas".[42]

Based on this review of earlier work in the field, I should like to suggest that previous attempts to study generation theory in the Renaissance suffer from a group of typical shortcomings. In part, these seem to be caused by a sort of peripheral vision effect. None of these writers is interested in the Renaissance for its own sake: the centre point of their work is elsewhere, in the eighteenth and nineteenth century or in the achievements of modern embryology, for which they are attempting to construct a "genealogy", to use Jacob's term. Furthermore, most of them concentrate on one group of sixteenth and seventeenth century writers, the anatomists (or alternatively the authors of obstetric treatises). These – as with Bloch, Needham and Adelmann – are taken as the norm, and more theoretical accounts are either ignored or condemned as a falling off from the standard of "research" and "experiment" set by this small group. So one group of writers is made to stand for the whole, and no attempt is made to set even their descriptions into any other theoretical context than that of modern embryology.

The other feature which I would see as a failing is what one might call the "warring camps" approach, found at its most developed in Roger. This consists of dividing all dramatis personae into two groups – for Roger Aristotelians and Galenists, for Adelmann reactionaries and revolutionaries – and writing the intellectual history of the period as a conflict between them. It is true that there are controversies, for instance on the role of female seed or the order of formation of the parts. But, as I argue below, it is also possible to see the position of Aristotelians and Galenists, natural philosophers and doctors, as in some sense complementary; and to suggest that one theme of the debate on certain issues was the attempt to produce some sort of synthesis which would bring together conflicting points of view and relate them in some rational way. One such attempt, I would argue, was Jean Fernel's Physiologia.

Any study of generation theory in the Renaissance must of necessity be selective: there are an enormous number of treatises which relate to the subject, although many of them are popular works with very little real theoretical content. It seems sensible at this point to set out what type of sources I plan to use in this study: and equally what type I do not plan to include.

When I attempted to compare the range of Renaissance writings dealing with generation, I came to the conclusion that they did not all belong to the same tradition, as most modern writers on the subject seem to assume. This means, I think, that the form of the work needs to be noted as carefully as its content; otherwise, one is in danger of lumping together a wide variety of types operating on very different levels and asking very different questions. The effect of this is rather as though a modern survey of 'generation theory' tried to include everything from the research of Crick and Watson to Old Moore's Almanack and the advice columns of women's magazines.

Form, I think, needs to be considered in two ways: first, by noting an author's terminology and the "classic questions" around which he builds his argument –

which in turn gives a clue to its source; and second, by distinguishing different types of works, written for different purposes and based, once again, on different sources and authorities. I think it is possible to distinguish four main types:

1. Practical treatises and anthologies, in Latin or the vernacular, dealing with the diagnosis of pregnancy and the care of women in childbirth: the so-called "women's books". Renaissance examples include the midwifery textbooks of Rueff and Roesslin, the anthologies of Spachius and De Castro, the pseudo-Albertus Secreta Mulierum, the Expositio of James of Forli, and editions of Soranus and Trotula. In fact, the majority of Renaissance treatises which I have examined fall into this category, and a full list would be a long one.[43]
2. Anatomical texts, which include a description of the parts serving generation and of the development of the foetus.[44]
3. Textbooks of theoretical medicine which include the topic of generation among their subjects, and theoretical commentaries or monographs by medical writers.
4. Commentaries on the Generation of Animals of Aristotle and related works, usually by natural philosophers and intended as part of the teaching materials for the arts degree.

Virtually all historical accounts base themeselves on the first two categories, though without distinguishing them from the last two; my interest, conversely, is almost exclusively with the last two groups, which are theoretical in form as well as in content. The two theoretical groups differ from one another in their form and to some extent in their audience, but they share, I think, a set of questions, and a list of authorities, which are distinct from those of the two 'practical' groups.

Scanning the range of popular works commonly accepted as Renaissance treatises on generation, I found many to be hopelessly confused, repeating authorities at second and third hand and often printing conflicting positions side by side with no apparent awareness of the contradiction between them. I began by assuming that this mass of potboilers, designed to instruct midwives and surgeons on the management of pregnancy, were watered-down and confused popularisations of the debates in contemporary theoretical medicine and natural philosophy. But I have now come to think that they are better considered as a separate tradition, based, in general, on different texts or portions of the same text, addressing themselves to different, practical problems and in general answering a different range of questions. The questions which appear in the "women's books" – on the impediments to conception, the progress of pregnancy and the signs of approaching parturition, on monsters and uterine moles – can be traced back to sources such as Hippocrates' De natura pueri; Aristotle's History of Animals, Books VII and X, and the works which derive from it, such as Albertus Magnus' De Animalibus and Book III, Fen 20 and 21 of the Canon of Avicenna; and the pseudo-Aristotelian Problems.[45]

These questions are distinct from those in Renaissance texts which deal with the theory of generation, and the theoretical writings take their form and their detailed content from a different set of authorities. The questions they ask concern the nature of the principles of generation, the contributions of male and female parent, whether or not the seed and the embryo possess soul, the order of development and the "instruments" involved. Hippocrates' De natura humana and De genitura; the works of Galen, especially De semine, De foetuum formatione, de usu partium and more general theoretical or "physiological" texts[46] are especially important to the medical writers. Aristotle's De generatione animalium and related works by Galen, Greek commentators such as Philoponus and Alexander of Aphrodisias, and Averroes are major sources for natural philosophical discussions of the generation question in the Renaissance.

It is true that the "theoretical" questions appeared also in some of the "practical" texts in the Renaissance.[47] But it seemed to me that they could be studied in their purest form by looking at two particular groups of writings: commentaries on De generatione animalium made by Renaissance writers and available between the mid-sixteenth and the mid-seventeenth centuries; and treatises of theoretical medicine of similar date which contain a section on generation. These accordingly form the subject matter of this book. With the exception of the subsidiary question of inheritance, few of these theoretical problems have received the attention of historians of the Renaissance.[48] And as far as I know the attempt to discriminate between different kinds of sixteenth and seventeenth century texts, all dealing in some sense with the problem of generation, has not been made in this way before. Instead, as we have seen, historians seem often to have assumed that all Renaissance writers on generation were asking the same kinds of questions, not only as each other, but as modern biologists.

What are the elements which make up the theoretical tradition, and how do they come together? The evidence used, the ingredients as it were, are almost always taken from the opinions of one or another of the great classical philosophical and medical authorities. How these ingredients are put together depends on the purpose for which the work was written and the conventions or rules of method which govern its argument. The tradition of commentary is a very old one: Aristotle himself, in Generation of Animals, provides a critical commentary on his own predecessors, and the same is true of the even earlier Hippocratic work On the nature of man. The main lines of the Renaissance debate are already clear in the Galenic commentaries on Aristotle's theory of generation, especially in De semine, and are further developed by later commentaries, in particular Averroes, in the form which was to be characteristic in the Renaissance.

The group of textbooks of theoretical medicine which follow the same format as Fernel's Physiologia derive from a model which starts to take shape in the First Fen of Book I of the Canon of Avicenna. But Avicenna does not include generation among the "physiological" questions which he discusses in Book I, and his treatment of generation in Book III of the Canon resembles not the theoretical discussions of the Renaissance, but the dogma of the "women's books". Averroes' Paraphrase on the Generation of Animals comes much closer to a comprehensive discussion of the questions set by Galen and debated in the Renaissance. And his Colliget provides another model for textbooks of theoretical medicine. In some sense it is Peter of Abano in his book called Conciliator who appears to put the two pieces together: the topics or chapter headings of a work such as Physiologia and the "classic questions" of academic Renaissance generation theory.

The book therefore has three major sections, one for the classical sources, one for the medical writers and one for the natural philosophers. In the section on medical treatises, I approach the problem of generation as it appeared in discussions of a natural-philosophical question – the physical constitution of the natural world – which was at the same time basic to contemporary theoretical medicine and to theories of generation in particular. Chapters on the role of spiritus and heat in generation, on elements and temperaments and on the function of the soul provide examples of specific questions, basic to medical theories of generation, which were discussed in terms of a broader, natural philosophical tradition. In the second section of the thesis, I look at the same questions from a different angle: viewed through the writings of men who were writing primarily as natural philosophers, but who were looking at a problem, the theoretical explanation of generation, on which medical and philosophical interest converged.

The relationship between medicine and natural philosophy in the sixteenth and early seventeenth centuries is not simply a historian's question, but one which concerned contemporaries as well. In particular, the question of whether medicine was a scientia – a body of knowledge based on general philosophical principles and having a method – rather than a craft or art, based primarily on experience, produced debate among Renaissance doctors and natural philosophers. It is question three in the list of nearly 200 debate topics to be found in a work known as Conciliator, whose full title is "The book called conciliator of philosophical differences, especially in medicine". Conciliator was written around 1310 but was printed early and often in the Renaissance, and it provided a compendium of the pros and cons in various philosophical debates, especially as they applied to medicine.[49] Some of its questions, such as whether fire is hot, whether cold diseases are more dangerous than hot ones, whether more patients die at night or in the daytime, are hard to work up an enthusiasm for unless one has been trained up in a Renaissance arts course. But the 200 differentia and their continuing popularity do illustrate one aspect of the close relationship between philosophy and medicine. The question whether medicine is a science, or more accurately whether medicine has a philosophy, is of more fundamental importance. And it is one way of approaching the question of the relations of Renaissance philosophy and Renaissance medicine.

What then is the relation between medicine and philosophy in the Renaissance: between what a doctor did and thought, and the activities of his counterpart in the school of philosophy, or arts faculty? Despite the debates fossilized in a work such as Peter of Abano's Conciliator; despite the practical or craft side of medicine which is always difficult to relate to a philosophical framework, the relationship was clearly a very close one. The Renaissance doctor, whatever his doubts or prejudices, was as clearly a member of the "philosophical family" which Dante saw grouped around Aristotle, "the master of them who know" in his visit to Limbo in 1300, as the modern doctor is not. The title page of Symphorien Champier's Symphony of Plato with Aristotle and Galen with Hippocrates (1516) shows the relationship between philosophy and medicine as that of the players in a quartet of viols, each playing a different instrument but all sounding in tune.[50] This was an ideal: among their Renaissance followers; inevitably there were discords. In what follows, I should like to concentrate on two of these players, Aristotle and Galen: one as a logician and natural philosopher, the other as a medical philosopher, and both as seen through Renaissance eyes.

If for philosophy we read "Aristotelian philosophy" – and I would argue that for purposes of this question other brands are hardly in the running – we can see the links with medicine especially clearly. Contemporary philosophy, i.e. Renaissance Aristotelianism, provided the framework of medical education. According to John Randall, it was deeply implicated in the development of medical and scientific method.[51] It provides, then, the framework for medical theorizing and research. And it is, additionally, the foundation on which the systems of the Greek and Arabic authorities of Renaissance medicine themselves were built. In particular, Galenism as it was taught in Renaissance universities and published in medical treatises was based on the tradition of the logical and natural philosophical writings of Aristotle, even when it differed from them in detail.

In what follows, I should like to look briefly at the way in which medical education brought up the young doctor within this "philosophic family", and at the logical and methodological structure within which he did his work. I should then like to look at the area of contact between the subject matter of medicine and philosophy.

In speaking of the basis of medical education in philosophy, I am, of course, and deliberately, confining myself to a small and privileged group of the medical profession. Medicine was an academic discipline: that is, if one excludes the guild-trained surgeons and apothecaries, the barbers, midwives and untrained practitioners, the quacks, and those, such as the Paracelsians, who deliberately rejected the authorities of academic medicine. Doctors within this charmed circle received their training in the universities, and their medical training was almost universally a sequel to an undergraduate arts course firmly based in Aristotelian natural philosophy and logic. The arts course was seen as a preparation for all three of the professions, not medicine alone: prospective doctors, lawyers and theologians all received the same basic discipline before going on to their own specialities. Such a pattern of education was not new to the Renaissance: in many cases, both the university and its curriculum requirements dated from the Middle Ages. But some scholars have suggested that in the Renaissance the old debates took on a new twist, and a new urgency. R.R. Bolger, for example, in his book on the Classical Heritage, argues that

> It was the urgent zeal of the great professional categories to find solutions for their theological, legal and medical problems that made men put such energy into classical studies.[52]

Unfortunately, we know less about the content of university education in the Renaissance than we would like, and the evidence we do have is patchy and unevenly distributed. The details which we have all seem to point in much the same direction, however. The Bologna statutes of 1345 already required the doctoral candidate to be "competent in arts" (in artibus … sufficientem).[53] In 1517, the doctors of Montpellier agreed on an examination for candidates for admission to the study of medicine, in order, it appears, to keep out the ignorant types they had been getting in the past. "And they first laid down, that no one would be accepted as a matriculant, unless he were of sufficient attainment in logic and philosophy, as to seem capable of medicine."[54] An arts education, then, was not simply considered to be desirable for the prospective doctor: it was a necessity, and later on I hope to be able to show why it was so important, and why the doctors of Montpellier should have felt, as we certainly do not, that only a philosopher would be capable of medicine.

The same qualifications are present in the Paris statutes of 1598. The young scholar intending to study medicine had to know Greek, Latin and rhetoric, and to have had 2 years' of philosophy, which included logic, ethics, physics (i.e. natural philosophy) and metaphysics. In particular, he had to have studied Aristotle. He could then present himself for examination as a master of arts, before going on to take the degree of bachelor and doctor of medicine.[55]

These statutes come at the end of the sixteenth century, but they reflect a feeling of what was appropriate to the study of medicine with which professors in fourteenth century Bologna would have agreed. Yet these are reform statutes, prepared as part of Henry IV's plan to restore the standards of the university and hence its ancient glory. They replaced the statutes of 1350, and themselves remained in force, or at least in print, until the middle of the eighteenth century.

What are we to conclude from the unexpected longevity of the arts curriculum as a qualification for the study of medicine? It is possible to suggest, of course, that the requirement of logic and philosophy served as a prop for medical privilege, intended to keep the unlettered out and the student body small. It is also possible to suggest that university medical courses in the sixteenth century were the home of hidebound conservatism, with a profound disregard for what was appropriate to the practical instruction of doctors. Kristeller, however, writing on sixteenth century Aristotelianism, suggests otherwise: conservatism, he says, was

> not necessarily due to professional inertia as to the wealth and solidity of subject-matter contained in the Aristotelian writings, to which its critics for some time could not oppose anything comparable.[56]

John Randall is even more outspoken in his defence of the old philosophy:

> ... though the language is diverse, the whole great literature on method that fills the scientific writing of the seventeenth century is at bottom a series of footnotes to the Organon of Aristotle. Indeed, the more fully the record of late medieval and Renaissance thought is studied, the clearer it becomes that the most daring departures from Aristotelian science were carried on within the Aristotelian framework, and by means of a critical reflection on the Aristotelian texts ... The "father" of modern science, in fact, turns out to be none other than the Master of them that know.[57]

Randall's position might be considered an extreme one, and his attempt to trace the scientific method of the mechanical philosophers of the seventeenth century to the Paduan Averroists of the fourteenth, fifteenth and sixteenth has not won universal support.[58] He is writing, in any case, as he makes clear, in reaction to the tendency of earlier historians to look at the humanists and the neo-Platonists: both largely outside the universities: as the progenitors of modern science. Whatever one thinks of his argument as a whole, I think his observation that even departures from Aristotelian science were carried on within an Aristotelian framework, and by means of commentary on Aristotelian texts, is an illuminating one.

This same paradox, and the same tenacious survival, is true, I think, of Galen's reception in the Renaissance. And so it is to the medical side of the equation that I should now like to turn.

Owsei Temkin, in his classic study of Galenism,[59] points out that Galen's high place as an authority contributed to the relationship between medicine and natural philosophy in important ways – a relationship which he traces back to the medical school at Alexandria. In the first place, it is possible to speak of Galen as an Aristotelian; the scientific basis of his medicine comes largely from Aristotle. This includes, according to Temkin, his notion of research, his belief in the purposeful working of Nature, and his doctrine of the elements, tissues and organic parts. But at the same time, "Galenism" as it appeared in the Renaissance was itself a medical philosophy, opposed in certain details to Aristotelian doctrine and with its own authority, especially in the clinical sphere. The result, as Temkin describes it, was a problem of split authority, resulting in various compromises by Renaissance philosopher physicians. Major areas of controversy were biological, however, rather than philosophical: they included such questions as the seat and divisions of the soul; the relative importance of heart, brain and liver; and the relative function of the male and female seed in generation. Although Temkin sees the relationship between these two groups as a relatively complex one, he still describes it largely in terms of conflicts between those who were loyal to different authorities: Aristotelians, Galenists and the group of "naturalist" philosophers who followed Telesio. The main example he gives is the debate between those Aristotelians who argued for a single unitary soul and spirit centred in the heart, and those who followed Galen in arguing for three vital centres, heart, brain and liver.[60] In somewhat similar fashion, as we have seen, Roger argues for a conflict between sixteenth century Aristotelians and Galenists on the question of female semen.

I should like to suggest that neither identity nor conflict provides an entirely adequate explanation, and that the two Renaissance disciplines, medicine and natural philosophy, not so much split as overlapped: in their educational and professional roles, in the authorities to which they appealed, and finally in their subject area and the questions they chose to ask.

But before we turn to look at the area in which these two viewpoints overlapped, and sometimes conflicted, it may be worth looking briefly at the fortunes of Galenism in the Renaissance. To some extent it parallels that of the Aristotelian tradition which we have just considered. Renaissance Galenism was in many respects a continuation of a medieval tradition. Most of the important teaching works were known to medieval commentators, and had appeared in medieval translations, not only from Arabic, but in some cases directly from the Greek originals. In some cases, these medieval translations from Greek survived as the first printed editions, with only a little polishing of the style. But the humanist editors of the Renaissance, as Richard Durling points out in his study of editions of Galen, did contribute something new. They discovered and published certain important treatises which had been unknown before, and introduced a new and more critical spirit into their efforts to recover the Greek text and produce a more faithful translation from it. They also claimed a great deal for their new translations, and were less than accurate in condemning their medieval and Arabic predecessors as barbarians who had done nothing more than corrupt the purity of Greek medicine. It was perhaps natural, and certainly fashionable, for humanists to contrast the dark night of the Middle Ages, in which, according to Martin Acakia, medicine was "buried and overwhelemed in a great gloom", with the great light of Renaissance humanism.[61] Unfortunately, posterity has tended to accept the humanists' own high valuation of their achievements, and to undervalue those of their predecessors. The publication of the complete works of Galen by the famous Aldine press of Venice in 1525 was a great landmark and recognized as such; but it hardly justified the publisher's boast that "we have raised Galen up from the dead, completely restored"[62] and as it were disinfected after his passage through the hands of medieval and Arabic scholars.

So much for the one great myth of Renaissance medicine, the impression that no one in the West was able to read Galen before the fall of Constantinople. The other myth, which I should like to treat very briefly, is that with the publication of Vesalius' De Fabrica (On the Fabric of the Human Body) in 1543, Galenic medicine became a lame, if not a dead, duck. Temkin, in his analysis of the decline of Galenism in the West, suggests that the publication of anatomical works such as Vesalius' may have contributed to this decline: but that they did so less because of their hard-hitting critical text than because of the pictures, which brought out "the dilemma in representing traditional physiological concepts in visual form".[63] It is also worth pointing out that Vesalius' text can best be regarded as a commentary on the major Galenic work of physiology, On the Use of the Parts. It is a critical commentary, of course, but many of Vesalius' assumptions about the relation of structure and function are Galenic, and he paraphrases Galen repeatedly, often without acknowledgement. Even the most outspoken of Galen's Renaissance critics – if we exclude once more the Paracelsians – shared many of the same presuppositions as their hard-line Galenist colleagues, and in any case took Galenism seriously as a philosophy. Conducted between university-trained members of the arts and medical faculties, these battles were family quarrels as it were, donnish debates rather than struggles to the death.

Galen himself survived these battles remarkably well: far from becoming professorial mincemeat, he continued as a major authority in university medical courses, although the number of editions published dropped around 1560, after forty years or so of heavy production.[64] In 1586, the University of Bologna changed its set books as the result of a curriculum reform. Henceforth, only one text was to be permitted in the teaching of surgery: Galen.[65]

The time has come to look at the area of contact, whether harmonious or otherwise, between medicine and philosophy in the Renaissance. Clearly, certain aspects of philosophy were outside the professional concern of the doctor, just as the clinical side of medicine lay outside the interest or competence of the philosopher. But both Aristotle and Galen themselves had suggested that the realm shared by medicine and philosophy was both large and important, and that co-operation was both natural and necessary. Aristotle explained this interdependence in the introduction to his short work on sensation:

> But it behoves the Physical Philosopher to obtain also a clear view of the first principles of health and disease, inasmuch as neither health nor disease can exist in lifeless things. Indeed we may say of most physical inquirers, and of those physicians who study their art philosophically, that while the former complete their works with a disquisition on medicine, the latter usually base their medical theories on principles derived from the Physics.[66]

What Aristotle seems to be saying here, if I understand him correctly, is that the subject matter of medicine and natural philosophy is not the same, but complementary. One begins with the actions of animate beings, the other with inanimate: but each can learn from the other, and fundamentally the same philosophical principles apply to both.

Galen's viewpoint is rather different: the doctor and the natural philosopher may be looking at exactly the same phenomenon: what differs is their approach, and the questions they ask. One of his theoretical treatises, which was translated by Erasmus, is called "that the best doctor is also a philosopher".[67] Galen clearly saw himself as evidence for the truth of this statement. But he meant by this primarily a natural philosopher, or 'physical philosopher' as Aristotle called him: someone who understood the principles governing the actions of inanimate bodies in the natural world. For though he was well aware of the importance of questions such as the nature and substance of the soul to the medical debates of his own times, he preferred to side-step such questions as 'contentious' and not capable of certain answers:as well as largely irrelevant to the physician's task of healing. As he explained in the work On the Use of Parts, there were certain approaches which 'belonged to a work on natural philosophy' rather than to a work of medicine. The doctor's role in studying the human body was to relate structure to function in descriptive terms. The natural philosopher, on the other hand, might demonstrate that an organ 'had to' be constructed in a certain way for physical reasons; he might also explain how it happened that an organ performed a certain action, whether voluntary or involuntary. The example Galen gives here is the tensing of the penis:

> For that it does so happen when the hollow nerve is filled with pneuma is appropriate to the business now in hand; [i.e. medicine] how it happens belongs to a work on natural philosophy.[68]

Galen's natural philosopher, then, can claim to study the motions which govern the actions of living bodies as well as non-living ones: actions which the doctor will look at from a different perspective.

For Cesare Cremonini, professor of philosophy at Padua from 1591 to his death, and heir to a great tradition of Aristotelian commentary in natural philosophy and logic, the doctor must give way to the philosopher as his superior in the knowledge of the actions governing living things, including man. And man, as Cremonini defined him in his inaugural lecture of 1591, is the microcosm and epitome of the natural world; by understanding the composition of the human body, we can best come to understand the lesser bodies of the natural world.[69]

This definition of man as the prime concern of natural philosophy set a collision course with the area of study traditionally mapped out by medical writers. If natural philosophers and doctors claimed the same territory – the study of man – then areas of conflict, and of confusion, in their work were almost inevitable. Furthermore, since doctors were trained up as philosophers, it was natural for them to overstep the bounds of their own discipline and look at their subject through the eyes of the natural philosopher. Renaissance commentators on medical subjects, I would suggest, can be divided into two broad categories: those who accept Galen's separation of the concerns of medicine from those of philosophy, and concentrate on the former, usually for teaching purposes; and those who overstep the limits in an effort to find out how as well as that something exists. The latter are in many ways more interesting: but they are also more liable to get into the sort of trouble which Galen himself was concerned to avoid. They are also, it would appear, in the minority.

For although in theory the writings of both Aristotle and Galen contain topics which are of interest both to doctors and to philosophers, in practice, the subject area in which medicine and philosophy overlapped was not the major concern of either. Aristotle's logical, metaphysical, and physical treatises, which were included in the arts curriculum, inspired numerous commentaries. His biological works, which on the whole were not on the curriculum, did not. Much the same applies to Galen. The Renaissance medical course was divided into two areas, theoria and practica. But if we can judge from the evidence of the most popular editions, and the works chosen as set books for lecturing, practica received the lion's share of the attention. The nine most popular Galenic texts according to Durling include only three works of theory, all of which were in use as textbooks.[70] Only two of these methodological works, the Art of medicine and the Method of curing, (Methodus medendi) appear among the eight Galenic treatises which the professors of Montpellier chose for their lectures in 1537.[71] Neither is a philosophical work in the sense of On the opinions of Hippocrates and Plato, for example, or even Use of the parts. Since both these latter works are cited frequently in medical treatises, it suggests that philosophical medicine belonged on the whole to what one might call the realm of postgraduate medical studies, and to the world of the professor rather than the practitioner.

In the sixteenth and early seventeenth centuries there were thus two groups of practitioners, broadly speaking, who could be said to be concerned with the study of man: medical writers and natural philosophers. If we asked a sixteenth century doctor or natural philosopher what his method of gaining knowledge of man, scientia, was, he might well reply that it rested on three foundations: reason, authority and experientia. The opinions of classical authors provided a foundation and a framework for knowledge, but this had to be supplemented by logical reasoning to draw out the consequences of their statements and, sometimes to determine between them; and by experience: not "experimental method", but something more nearly akin to common observations or sensations (which might be repeated by any observer). The balance among these three, as Kristeller points out,[72] varied according to the individual.

If we think of method in the instrumental sense of "how one goes about it", a historian might want to distinguish two possible ways in which Renaissance writers went about the study of man. One, which applies particularly to medicine, was through observation: either clinical observation which might be written up as case notes, or the study of anatomy. The other, which applies to both philosophy and medicine, was through the editing, translating and commentary on texts. It is with this second method of study, and particularly the aspect of commentary on and conciliation of texts, that this study is concerned.

How did certain questions concerning generation – those on the origin of the seed, the function of the testes and of the heart, and the role of the two sexes – attain the status of "classic questions" for discussion and commentary by Renaissance doctors and natural philosophers alike?

In addition to the overlap of educational background, subject matter and authority which I have already suggested, I would like to bring out two further factors which I think may have contributed to this development. One is the increased availability of the texts and new translations of the works on generation by classical authorities, in particular Galen, Aristotle and Hippocrates, in the Renaissance.

The first thing which the humanist doctor, or natural philosopher, required was a reliable text of the classical work, both in Greek and in a modern translation purged of the "vulgar errors" of medieval writers. These were available for all the major authorities on generation by the mid 1520's.[73] The Gaza translation of Aristotle's works On Animals appeared in Venice in 1476, and the first edition of the Greek Opera came in an Aldine edition of 1497. The commentary on Generation of Animals attributed to John Philoponus (now considered to be by Michael of Ephesus) appeared in a Latin translation in 1526, together with a Greek text of the work itself.

Hippocrates' Opera appeared in Rome in translation in 1525, although individual works had appeared much earlier. A Greek first edition came out in the following year. Cornarius published new translations of the major works on generation, On geniture and On the nature of the child, in Basle in 1525.

Galen's Opera were published in Latin in 1490 and in Greek in 1525, both in Venice; in the 1530's, the new translations of On the seed (by Guinterius) and of On the formation of the foetus by Guinterius, Cornarius and Felicianus were published separately in Paris and Basle.

The other factor which I should like to emphasise, and which parallels the increased availability of classical texts, is the increase in the number of new texts and translations of the works of Averroes in the century between 1472 and 1575.

"Averroism" in the Renaissance has usually been associated with a particular logical method, sometimes identified with the methods of modern science, as in the "Randall thesis" and its later developments; or alternatively with unorthodox religious opinions, in particular the questioning of the immortality of the soul.[74] Without prejudice to these arguments, I should like to consider another, and perhaps more innocent, aspect of "Averroism". This is his role as an important commentator on the natural philosophical works of Aristotle, including Generation of Animals; and as the author of medical textbooks and treatises, of which Colliget is the best known. In both his natural philosophical and medical writings he discusses the problem of generation as it was treated by Aristotle and by Galen. I believe that in these discussions or commentaries he identifies the points of conflict which arise between Aristotle's and Galen's accounts, attempts to reconcile them using a mixture of reason, authority and observation and in so doing does much to determine the terms of the debate as it was argued out in the Renaissance.

Averroes' importance as a commentator on the writings of Aristotle has been recognised by several major Renaissance scholars, among them Kristeller, Cranz and Schmitt.[75] They are agreed in saying that the earlier attempts to define Averroes as a proto-freethinker or the father of modern science use the term "Averroist" imprecisely and in too narrow and partisan a sense. Cranz points out that Averroes seems to have been considered by his Renaissance editors "not so much as the exponent of a specific philosophical position but rather as part of the general learning of the times".[76] Kristeller and Schmitt make similar points, comparing Averroes with two other great "medieval" commentators, Thomas Aquinas and Albertus Magnus, and suggesting that all three were considered primarily as tools for the better understanding of nature and of philosophy – or more accurately for the better understanding of Aristotle as the ultimate authority in these questions.[77]

Speaking particularly of the Aristotelian tradition, Dr. Charles Schmitt has argued that by the 1540's new editions and textual criticism gave the foundation for an increased number of commentaries, prepared for the use of students. A similar argument might be made for the medical tradition, where, following the publication of a work such as Jean Fernel's Medicina (1542, 1554) the number of medical textbooks, commentaries and treatises seems to have increased similarly in the second half of the sixteenth century. These commentaries and textbooks form the material of the major parts of the argument, in Parts II and III. But I begin by considering the classical authorities on which they were based and the theory, or rather theories, of generation which they put forward.

Notes

1. Jo. Fernelii Ambiani, Medicina, Paris, 1554. Physiologia, Book vii.
2. An anatomical bias is evident in many of the works discussed below. Recent publications and theses which consider the anatomical and botanical component of medical education are those by R.G. Lewis, "Guillaume Rondelet and his circle", Oxford, 1967 (viewed as unpublished D.Phil. thesis submitted to Oxford University); Andrew Cunningham, "Aspects of the history of medical education in Britain in the seventeenth and early eighteenth centuries", London, 1974 (unpublished Ph.D. thesis) and "The Kinds of Anatomy", Medical History, 1975, vol.19(1); L.R. Lind, Studies in Pre-Vesalian Anatomy: Biography, Translations, Documents, Philadelphia, 1975; see also Andrew W. Wear, "Contingency and logic in Renaissance anatomy and physiology", London, 1975 (unpublished Ph.D. thesis) and "The spleen in Renaissance anatomy", Medical History, 1977, vol.21(1), pp.43–60.
3. F.H.A. Marshall, Physiology of Reproduction, London, 1922; J. Huxley and G.R. Beer, Elements of Experimental Embryology, Cambridge, 1934.

4. Bruno Bloch, Die Geschichtlichen Grundlagen der Embryologie bis auf Harvey, Halle, 1904 (reprinted from Abh.der Kaiserl. Leop.-Carol.Deutschen Akademie der Naturforscher, Nova Acta, Band LXXXII, Nr.3, pp.217–334) "Einleitung", pp.217–218.
5. Ibid., pp.290–334, esp. p.303: "Nicht Besseres lässt sich der Behandlung des gleichen Gegenstandes durch den Paduaner Professor Hieronymo Capivaccio sagen. Er bekennt sich in allen Stücken zur galensichen Schule, sucht aber in einigen Kontroversen zwischen Aristoteles und dem griechischen Arzte zu vermitteln – freilich ohne viel Glück."
6. Ibid., pp.305–7.
7. Edward Stuart Russell, Form and Function: A Contribution to the History of Animal Morphology, London, 1916.
8. Ibid., Chapter 2, pp.17–19.
9. Frederick B. Churchill, "The History of Embryology as Intellectual History", Journal of the History of Biology, iii, (1970)155–168; p.156. This is an essay review which compares the work of Roger, Adelmann and Gasking, all mentioned below, together with two collections of essays which deal entirely with a later period.
10. F.J. Cole, Early Theories of Sexual Generation, Oxford, 1930. Cole was Professor of Zoology at the University of Reading and also produced a History of Comparative Anatomy (1944).
11. Arthur William Meyer, The Rise of Embryology, Stanford, California and London, [1939], pp.17–27; 27. An earlier work is his Essays on the History of Embryology, reprinted with a few changes and omissions and with references from California and Western Medicine, volumes 35–37, 1931–2. This deals largely with the period post-Harvey but is similar in its approach.
12. Ibid., Preface, p.vii.
13. Joseph Needham, A History of Embryology, Cambridge, 1959 (first edition 1934).
14. Meyer, op.cit. (1939), complains about the lack of historical work in embryology in terms very similar to those of Bloch: Preface, p.vii.
15. Needham, op.cit., p.100. He mentions Vesalius ("who however had no opportunities for dissecting human embryos, and took hardly any interest in foetal development"), Aldrovandi, Fallopio, Aranzio, and Coiter; also Laurentius and Fabricius, who are seen as important chiefly as teachers of Harvey.
16. Ibid., p.109.
17. See below for my definition, and for the argument that these can be considered as a distinct group of writings, asking different questions and appealing to different authorities than theoretical generation treatises.
18. Needham, op.cit., p.111. "I think they [the woodcuts from Rueff, a series of seven which show a cloudlike coagulum in the shape of an egg, gradually interlaced by vessels and, in the last frame, with a fully formed infant inside the "egg"] show very clearly what the general ideas were at this period about mammalian embryology, and this affords us a precious insight into what was in the minds of writers such as Riolanus the elder, Mercurialis, Saxonia, Rondeletius,

Venusti, Holler and Vallesius. There are many points which their expositions of foetal growth and development leave vague, and without Rueff it would be difficult or impossible to imagine in what manner they imagined it to go on." [Riolan's work on generation and Mercurialis's come within the area of textbooks of theoretical medicine, Holler's in a textbook on the cure of internal disease. Rondelet's work is a history of fishes and Venusti's a vernacular compilation, rather more sophisticated than some but written for a lay audience. And Vallesius's is a treatise on controversies between Aristotle and Galen.]

19. The Embryological Treatises of Hieronymus Fabricius of Aquapendente ... The Formation of the Egg and of the Chick/De Formatione Ovi et Pulli/ ... The formed Foetus/De Formato Foetu/. A Facsimile Edition, with an Introduction, a Translation, and a Commentary by Howard B. Adelmann. ..., Ithaca, 1942, Chapter 2, pp.53–70, "The Fifteenth and Sixteenth Centuries".
20. Howard B. Adelmann, Marcello Malpighi and the Evolution of Embryology, Ithaca and London, 1966, volume II, pp.747–8.
21. Ibid., p.755.
22. Ibid., p.756.
23. W.P.D. Wightman, The Emergence of Scientific Medicine, Edinburgh, 1971, p.30: "It is sometimes said that the 'facts' of anatomy must be established before any appeal can be made to physiology – that is, function. It is highly improbable that such an ideal could be realised even by an ideally unprejudiced observer. For what are the 'facts' of anatomy? Thick tubes, thin tubes, sheets enclosing body fluid easily mistaken for tubes, the whole liable to be obscured by blood. Without some degree of interpretation there are no meaningful facts".
24. For instance, Churchill, op.cit., pp.165–174, especially 171–2, where he quotes the passage cited above (Note 22) and agrees that "such a criticism of sixteenth and seventeenth century embryology is unquestionably valid" but goes on to argue that "to summarily reject the subject is to leave a fragmented picture" and that "with careful scrutiny we certainly can gain from the analysis of 'worn-out issues'".
25. Elizabeth Gasking, Investigations into Generation, 1651–1828, London, 1967. Churchill's comments are fairly critical: op.cit., pp.174–177.
26. Ian Maclean, Woman Triumphant: Feminism in French Literature 1610–1652, Oxford, 1977, Chapter 1, pp.6–12. The few medical works cited are almost all "women's books", e.g. Israel Spachius's Gynaecea (1597). I have not viewed *The Renaissance notion of woman* (Cambridge University Press, 1980) but assume that this uses a similar approach and similar sources.
27. Pierre Darmon, Le mythe de la procréation à l'age baroque, Paris, 1977. Darmon's approach to the question of generation is suggested by his other writings: a thesis on "Les Procédures de nullité du mariage pour impuissance" and other writings on adultery and the crime of seduction and on misogyny, mentioned in the dust jacket potted biography.
28. Ibid., p.51.
29. François Jacob, La logique du vivant: Une histoire de l' hérédité, Paris, 1970, p.18.

30. Ibid., pp.18–21.
31. Frederic L. Holmes, "A Review of Francois Jacob, La Logique du vivant. …" in William Coleman and Camille Limoges (editors), Studies in the History of Biology, Baltimore and London, 1977, p.209–218, especially pp.216–217.
32. Jacob, op.cit., pp.22, 24–25.
33. Ibid., p.27: "il n'existe pas encore de lois de la nature, pas plus pour régir la génération des animaux que le mouvement des astres".
34. Ibid., pp.27–37.
35. Ibid., p.37.
36. Jacques Roger, Les Sciences de la Vie dans la Pensée du XVIIIe Siècle: la Génération des Animaux de Descartes à l'Encyclopédie, Paris, 1963.
37. Churchill, op.cit., p.157.
38. Roger, op.cit., Chapter 1, pp.7,8, 13–19; 45.
39. A similar approach to French physicians of the seventeenth century appears to lie behind some work by L.W.B. Brockliss; see his "Medical teaching at the University of Paris, 1600–1720", Annals of Science, 35(3), May 1978, pp.221–251.
40. Roger, op.cit., pp.40–41.
41. Ibid., Chapter 2, especially pp.50, 58, 63.
42. Ibid., pp.91–95.
43. Jakob or Jacobus Rueff, Ein schön lustig Trostbüchle, Zürich, 1544. This was an official textbook for the midwives of Zurich. It appeared in Latin as De conceptu et generatione hominis … libri sex, Zurich, 1554 (and later editions); and in English as The Expert Midwife, London, 1637. Eucharius Rösslin (or Roesslin), Der Swangeren Frawen und Hebammen Rosengarten…., Frankfurt, 1513 (and later editions). This also was published in Latin as De partu hominis…, Frankfurt, 1532, in French, and in English as The Byrth of Mankynde, London, 1540; later editions as The Woman's Boke (which gives me the name for this group of texts). Israel Spach, Gynaeciorum …, Strasbourg, 1597 and Rodriguez de Castro, De universa mulierum medicina, Hamburg, 1603 are two of the later Renaissance anthologies; earlier examples are Caspar Bauhin, Gynaeciorum, Basle, 1586 and Caspar Wolf (Wolphius), Gynaeciorum; Basle, 1566. The bibliographies of Needham and Roger in particular give other examples. See Needham also for a selection of editions of De Secretis Mulierum, and Lynn Thorndike, A History of Magic and Experimental Science (Hereafter cited as Thorndike, History, with volume and page) vol.II, New York, 1923, pp.739–745 for details of the many fifteenth and sixteenth century editions of this popular and very traditional compendium. I have used Magni Alberti secreta mulierum et virorum:, [s.l.,s.n.] 1542. James of Forli's commentary on Avicenna's Canon, Book III, Fen 21 is both early and was evidently popular; the British Museum has three versions of 1479, 1489 and 1502. Rosslin's Rosengarten is based on the late classical text of Soranus (d.138), translated in the sixth century A.D. by Muscio ("Moschion" in Renaissance editions). A modern English translation is Soranus's Gynaecology, translated and with an introduction by Owsei Temkin, Baltimore, 1956. The translation is based on the

text reconstructed from later compilations by J. Ilberg in the Corpus Medicorum Graecorum, vol.IV. Trotula appears in a number of the Renaissance anthologies, including that by Bauhin.

44. Lind, Needham and Adelmann devote particular attention to anatomists; Lind, op.cit., p.19, lists "the Chief Pre-Vesalian Anatomical Works" and the major anatomists of the fifteenth and sixteenth centuries. The most important of those who touch on questions of generation are Alessandro Achillini, Annotationes Anatomicae, Bologna, 1520 (ibid., pp.49–51); Guilio Cesare Aranzio, De humano foetu libellus, Bologna, 1564 (and later editions); Alessandro Benedetti, Anatomice, Venice, 1502 (ibid., pp.96–101); Berengario da Carpi, Commentaria ... super Anatomia Mundini ..., Bologna, 1521; Volcher Coiter, ... Tabulae et Exercitationes, Nuremburg, 1573 (cf. Needham, op.cit., pp.251–252); Niccolò Massa, Liber Introductorius Anatomiae, Venice, 1536 (Lind, op.cit., pp.198–209); Gerolamo Mercuriale, Jacob Sylvius and Costanzo Varolio, Anatomiae..., Frankfurt, 1591. Gabriele Zerbi (or de Gerbi), Liber anathomiae corporis humani..., Venice, 1502 (cf. Lind, op.cit., p.147). Other major anatomical writers who concerned themselves with questions of generation in the sixteenth and early seventeenth centuries, mentioned by Needham and Adelmann, are Realdus Columbus, De Re Anatomica, Venice, 1559; Bartolommeo Eustachio, Opuscula anatomica, Venice, 1563; Gabriele Fallopio, Observationes Anatomicae, Venice, 1561; Fabricius ab Aquapendente (cf. Adelmann, op.cit.); Andreas Laurentius, Historia anatomica..., Frankfurt, 1599; Arcangelo Piccolomini, Anatomicae praelectiones, Rome, 1586; Jean Riolan the Younger, especially the Foetus humani historia, Paris, 1618; Andreas Vesalius, especially De Fabrica, Basle, 1543; and later writers such as Aldrovandi, the Bartholins and Malpighi as well as Harvey himself.

45. See Part I, below, and Appendix I for Aristotle and Hippocrates editions. Albertus Magnus, De Animalibus, Venice, 1519; Avicenna, Liber Canonis, facsimile of Venice, 1507; Hildesheim, 1964; Pseudo Aristotle, Problemata, from Opera Omnia, Venice, 1550, volume VII.

46. For example the texts on the natural faculties, elements, temperaments and the art of medicine: see below, Parts I and III.

47. For example in Jacobus Foroliviensis, Expositio ... supra [aureum] capitulum de Generatione embrionis, Venice, 1502 [earlier editions in 1479, 1489].

48. The classical period has been better served: see below, Part I, for references to the work of Balss, Lesky, Peck, Preus and Temkin among others. For a sensitive appreciation of certain of these theoretical questions in the sixteenth and seventeenth century, see Walter Pagel, whose enormous output (to 1972) is listed by Marianne Winder, "A Bibliography of the Writings of Walter Pagel", Science, Medicine and Society in the Renaissance: Essays to Honor Walter Pagel, edited by Allen G. Debus, London, 1972, vol.II, pp.289–326. Of particular relevance are "The Reaction to Aristotle in seventeenth century biological thought. Campanella, Van Helmont, Glanvill, Charleton, Harvey, Glisson and Descartes" in Science, Medicine and History. Essays on the evolution of scientific thought and medical history written in honour of Charles Singer, Oxford, 1953, I,

pp.489–509; Paracelsus: An Introduction to Philosophical Medicine in the era of the Renaissance, Basle and New York, 1958; the major writings on Harvey and Van Helmont; and his reviews of Sir Charles Sherrington, "The Endeavour of Jean Fernel", Bull.Hist.Med., 1946, 20:587–9. And see D.P. Walker, Spiritual and Demonic Magic from Ficino to Campanella London, 1958; "The Astral Body in Renaissance Medicine", Journal of the Warburg and Courtauld Institutes, 1958, 21:119–133; "Francis Bacon and Spiritus" in Science, Medicine and Society in the Renaissance (op.cit.), II, pp.121–130; and certain works of Owsei Temkin, in particular his Galenism: Rise and decline of a medical philosohy, Ithaca, New York, 1973 (cited as Temkin, Galenism).

49. Petrus de Abano, Concilator…., Venice, 1521; the Wellcome Library also has editions of 1483 and 1565, both printed in Venice. For further bibliographical information, see Thorndike, History, II, pp.917–926 and references below in Part III.
50. Symphorien Champier, Symphonia Platonis cum Aristotele: & Galeni cum Hippocrate, Paris, 1516.
51. John Henry Randall, The School of Padua and the Emergence of Modern Science, especially Chapter 1, pp.15–68, Padua, 1961. See also his The Career of Philosophy from the Middle Ages to the Enlightenment, New York, 1962, and "Paduan Aristotelianism Reconsidered" in Philosophy and Humanism: Renaissance Essays in Honor of Paul Oskar Kristeller, edited by Edward P. Mahoney, Leiden, 1976, pp.273–282. Randall's thesis has been criticised by, among others, Wightman, Gilbert, Edwards and Jardine; see below, Part II, Chapter 8 for detailed references.
52. R.R. Bolgar, The Classical Heritage and its Beneficiaries, Cambridge, 1954, p.281.
53. Vern Bullough, "Medieval Bologna and the Development of Medical Education", Bull.Hist.Med., 1958, 32:201–215 especially, pp.211–212.
54. A. Germain, La Renaissance à Montpellier: Étude historique d'après les documents originaux, Montpellier, 1871, pp.128–9.
55. A. Corlieu, L'Ancienne Faculté de Médecine de Paris, Paris, 1877, p.15. For university regulations at Paris and the various 'grades' through which the student passed, see Françoise Lehoux, Le cadre de vie des médecins Parisiens aux xvie et xviie siècles, Paris, 1976, pp.12–13 and her own sources, particularly Roger, Wickersheimer and Delaunay; see Part III, Chapter I, for detailed references to these and other studies of Renaissance and seventeenth century medical education.
56. Paul Oskar Kristeller, cited by Richard J. Durling, "A Chronological Census of Renaissance Editions and Translations of Galen", Journal of the Warburg and Courtauld Institutes, 1961, 24(3–4):230–305; p.244.
57. Randall, School of Padua, p.63.
58. See above, Note 51 and below, Part II, Chap. 8 and Note 2.
59. Temkin, Galenism, especially Chapter II, p.65 e.s.
60. Ibid., p.73.
61. Durling, Census, p.239.

62. Ibid., p.236.
63. Temkin, Galenism, Chapter IV, 134.
64. Durling, Census, p.242.
65. Ibid., p.245.
66. De sensu I, 436a 19-b1. Cited by Temkin, Galenism, p.65.
67. K I, 53–63; Durling, Census, pp.254 and 294. Quod optimus medicus sit idem philosophus, in Galen, Exhortatio ad bonas arteis … de optimo docendi genere, & qualem oporteat esse medicum, Basle, 1526 (and later editions).
68. De usu partium, Book XV. Translated by Margaret Tallmadge May, Galen on the Usefulness of the Parts of the Body, Ithaca, 1968, II, p.659.
69. For Cremonini see below, Part II, Chap. 11.
70. Durling, Census, p.243.
71. Germain, op.cit., pp.144–145.
72. Paul Oskar Kristeller, La Tradizione Aristotelica nel Rinascimento, Padua, 1962, pp.22–23. The English version is "Renaissance Aristotelianism", Greek, Roman and Byzantine Studies, 1965, 6:157–174.
73. Details of the editions summarised here are given below in Part I and in Appendix I.
74. For classic examples of the two positions, see Randall, op.cit., especially The School of Padua; and Ernest Renan, Averroès et l'Averroïsme, essai historique, Paris, 1866.
75. For a guide to Kristeller's large and sometimes confusing output, see "Bibliography of the Publications of Paul Oskar Kristeller for the years 1929–1974" in Philosophy and Humanism, pp.545–589. On this point I have used in particular his "Paduan Averroism and Alexandrism in the Light of Recent Studies" in Aristotelismo Padovano e Filosofia Aristotelica, Florence, 1960, 147–155 and La tradizione aristotelica nel Rinascimento. F. Edward Cranz, "Editions of the Latin Aristotle Accompanied by the Commentaries of Averroes" in Philosophy and Humanism, pp.116–128. Charles Schmitt, A Critical Survey and Bibliography of Studies on Renaissance Aristotelianism 1958–1969, Padua, 1971 and "Paduan Averroism studied through the Venetian editions of Aristotle-Averroes (with particular reference to the Giunta edition of 1550–2)"; Dr. Schmitt has very kindly allowed me to see a typescript version.
76. Cranz, "Editions", p.120.
77. Op.cit., Note 75.

Part I
The Sources: Classical Theories of Generation and Their Renaissance Editions

Chapter 2
Classical Theories of Generation in the Renaissance

Abstract Plato, Hippocrates, Galen and Aristotle – the four main classical examples – are considered as sources for the speculation surrounding the question of generation in the Renaissance They are important as the most accessible, and by far the most valued, of the classical authorities on generation known to Renaissance writers. Part I tries to make clear the answers of the four writers to such questions as: how is the semen formed? where, and of what substance? how does it transmit life, organised form and even detailed likeness to the offspring? how are the parts of the foetal body generated and distinguished? what does each parent contribute to this process? It also hopes to suggest the major sources, and clarify the major themes, of Renaissance theoretical discussions of generation. This is an attempt to look back at the major Greek medical writers not from the twentieth century but from the sixteenth.

Modern writers on topics in medical history often approach their subject from the point of view of modern medicine. This is true, as we have seen, of several notable contributors to the study of classical and Renaissance theories of generation. This approach, while natural, is even less appropriate in the case of generation than in many other cases. We have no cause to be smug when considering our success in 'solving' the ultimate questions with which Greek philosophers concerned themselves. And it will not help us to understand and make clear their viewpoint, complex and often obscure as it is, if we begin by considering it as obstructionist.

This chapter is planned with a different purpose in mind. It is very far from a comprehensive account of classical theories of generation. And its subjects are considered not as contributors to modern reproductive biology or as ends in themselves, as they might be treated by a classicist or a philosopher.[1] Instead, they are considered as sources for the speculation surrounding the question of generation in the Renaissance. Plato, Hippocrates, Galen and Aristotle – the four examples with whom I shall be concerned – would be central to any discussion of the subject but they are especially important as the most accessible, and by far the most valued, of the classical authorities on generation known to Renaissance writers. The table in Appendix I gives first and other major editions for the most important of their texts on generation, together with some information on their previous availability in

L. Deer Richardson, *Academic Theories of Generation in the Renaissance*, History, Philosophy and Theory of the Life Sciences 22,
https://doi.org/10.1007/978-3-319-69336-1_2

manuscript. Developments in scholarship were particularly important in making available more reliable Greek and Latin versions of Hippocrates and Galen; eventually the most popular works were translated into the vernacular as well. Aristotle's History, Parts and Generation of Animals also benefitted from a new translation by Theodore Gaza in the second half of the fifteenth century.

It is possible to distinguish classical theories of generation in several different ways. Adelmann makes a basic distinction between the 'Parmenidean' postion, emphasizing the primacy of being, and the 'Heraclitean' one, emphasizing becoming and the possibility that one thing can be transformed into another or arise out of another, essentially different, thing. He considers that this division marks a 'radical antithesis' between two classes of theorist which can be traced down to the present day. And he associates the former position with a preformationist approach, the latter with an epigenetic one.[2] This distinction, though sometimes useful, is certainly ahistorical, reading back from the controversies of eighteenth century biologists to a time when men did not see themselves, or one another, in these terms. Roger, on the other hand, divides his subjects into those who affirm and those who deny the existence of female semen.[3] This is a somewhat misleading distinction since, as we shall see, there is a whole range of possible opinion on this subject, rather than two distinct camps. Erna Lesky divides the topic in yet another way, according to the supposed origin of the semen in each theory. She distinguishes three types: the 'encephalo-myologenous' theory, in which the semen is assumed to be identical with the substance of the brain or spinal marrow; the 'pangenesis' theory, in which the semen is gathered together from all parts of the parents' bodies: and the 'blood origin' theory, in which the semen is considered to be elaborated from the blood of one or both parents. Plato, Hippocrates and Aristotle respectively can be taken to represent the three types, and Lesky considers that they represent successive – and progressive – modes of thought on the question.[4]

In what follows I hope to take account of these distinctions and to make clear the answers of the four writers to such questions as: how is the semen formed, where, and of what substance? how does it transmit life, organised form and even detailed likeness to the offspring? how are the parts of the foetal body generated and distinguished? what does each parent contribute to this process? In doing this I also hope to be able to suggest the major sources, and clarify the major themes, of Renaissance discussions of generation. This is, if you like, an attempt to look back at the major Greek medical writers, not from the twentieth century but from the sixteenth.

Notes

1. Classical authors, including those featured here, are surrounded by a formidable hedge of commentary by philologists, classicists and philosophers, daunting to a mere historian. Of the modern philological arguments (particularly fierce for the Hippocratic texts) I suspect that I have managed to remain almost as ignorant as my Renaissance subjects. But see I.M. Lonie, "A Structural Pattern in Greek

Dietetics and the Early History of Greek Medicine", Medical History, 1977, 21:235–260, especially p.244, Note 43, for his thoughts on the unprofitability (and indeed circularity) of distinguishing between 'Coan' and 'Cnidian' texts. Classicists and philosophers are alike in treating their subjects as contemporary thinkers (a point of view which Renaissance humanists would have understood). And their survey of historical development stops well short of the Renaissance: in the case of Richard Shute's classic essay On the history of the process by which the Aristotelian writings arrived at their present form, Oxford, 1888, with Alexander of Aphrodisias. Important sources for Plato's thought in the Timaeus are Francis Cornford, Plato's Cosmology: The Timaeus of Plato translated with a running commentary, London and New York, 1971 (reprint) and A.E. Taylor, A Commentary on Plato's Timaeus, Oxford, 1972 (first edition 1928). For Aristotle, I have consulted in particular G.E.R. Lloyd, Aristotle, the Growth and Structure of his Thought, Cambridge, 1968, especially Chapter iv; Stephen R.L. Clark, Aristotle's Man: Speculations upon Aristotelian Anthropology, Oxford, 1975, especially pp.28–47 (which deal chiefly with the philosophical problem of species) and Appendices A and B on pneuma and the sexes. Lloyd and Clark approach their subject as a classicist and philosopher respectively. Oxford, Peck, and Balme translations (see Chap. 6, Note 1 below) also provide useful notes and I have used Anthony Preus's commentary, "Science and Philosophy in Aristotle's Generation of Animals", Journal of the History of Biology, 1970, vol.3, Part 1, pp.1–52. Galen and Hippocrates have attracted the attention of medical historians as well as of translator-editors such as Daremberg, Littré and Kühn. Temkin, Galenism is particularly valuable, and Anthony Preus, "Galen's Criticism of Aristotle's Conception Theory" Journal of the History of Biology, 1977, 10(1):65–85, is a useful and relevant survey. More generally there is the often underrated Galen of Pergamon by George Sarton (1954) and the much less satisfactory Galen's System of Physiology and Medicine by Rudolph Siegel (1968). A brisk approach to classical medical and philosophical sources can be found in Thomas Hall, Ideas of Life and Matter, Volume I, Chicago, 1969, and in the secondary sources discussed in the Introduction, particularly Adelmann. Helmut Leitner, Bibliography to the Ancient Medical Authors, Bern, 1973, is a useful brief guide to modern editions.

2. Adelmann, Malpighi, v, II, pp.729–730.
3. Roger, op.cit., p.50.
4. Erna Lesky, Die Zeugungs- und Vererbungslehren der Antike und Ihr Nachwirken, Abhandlungen der Akademie der Wissenschaften und der Literatur in Mainz, Wiesbaden, 1950, p.1344. She adds a further, fourth stage in which a revived pangenesis is combined with atomism to produce new possibilities of explanation.

Chapter 3
Plato

Abstract The most important text for Plato's theories of generation is the *Timaeus*, though it is also discussed in the *Republic* (Book V) in the context of eugenics. Plato turns from the biological question, "where does human seed come from?" to the philosophical question, "how is the immortal soul confined in mortal life?" Plato's ideas on generation were quoted with respect long after observers had disproved his description of genitals as complete and separate animals. At the centre of the *Timaeus* is the idea that mortal existence involves 'trapping' a separable, immortal soul by binding it in an especially refined and prepared substance. And this substance is possessed of *pneuma*, breath. The association of ensoulment with *pneuma* and of both with the semen is a theme which will appear again, as is the notion of a special substance to bridge the gap between soul and body.

The most important text for Plato's theories of generation – unlike the others he wrote no work devoted specifically to this subject – is the creation-myth of the Timaeus, though it is also discussed in the Republic (Book V) in the context of Plato's programme of eugenics. It seems possible that the edition of the Timaeus available to scholars in the Middle Ages and Renaissance, with translation and commentary by Chalcidius, may have helped to make familiar the ideas of "lost" writers such as Chrysippus as well as those of Plato himself. This incomplete version was replaced from 1484 by Marsilio Ficino's elegant translation from the Greek. Ficino added a summary or "Compendium" to his version of the Timaeus, but deals only very briefly with the "medical" aspects of the treatise, which he considers are matters for the doctors to explain. He praises the elegance of Plato's description of the human body, and remarks on the amazing "fit" between the Platonic mind and the Galenic one.[1]

An animal is by definition a creature possessing soul, or anima, and as Lesky points out, Plato, in discussing the generation of animals, turns from the biological question, "where does the human seed come from?" to the philosophical question, "how is the immortal soul confined in mortal life?"[2] The special material substance in and by which the soul is 'chained down' and linked with the elemental substances of the body is the seed-stuff, and this is identified with the marrow in the brain, spinal cord and bone. This marrow, says Plato, was originally formed from the most

L. Deer Richardson, *Academic Theories of Generation in the Renaissance*, History, Philosophy and Theory of the Life Sciences 22,
https://doi.org/10.1007/978-3-319-69336-1_3

regular and perfect of the elemental triangles, those basic geometric shapes which can be used to construct the four regular solids with which the Platonic elements are identified. These triangles were separated from their fellows at the creation and mixed together in fixed proportions to form a range of seed-substances suitable for generating all the classes of living things. They varied, one can assume, according to the nature and perfection of the being to be generated, or the grade of soul with which that being was to be endowed. From these seed-mixtures, elemental though of the highest order, the Demiourgos produced the marrow. "In it he firmly implanted the different kinds of soul": the immortal or divine part of the soul in the sphere of marrow forming the brain, and the remaining mortal parts in the spinal and bone marrows "to which the whole soul was anchored. And round brain and marrow … he went on to frame our whole body", beginning with a protective layer of bone.[3] This account of the seat of the various parts of the soul is at variance with the more familiar one from the Republic in which Plato places the three parts of the human soul in the head (brain), chest (heart) and belly (liver). In the Timaeus, growth is explained as the assimilation of like 'triangles' from the food; and death occurs because "when the bonds of the triangles in the marrow fail and part under stress [of long existence], the bonds of the soul are also loosened; and when this happens in the course of nature the soul departs gladly."[4]

Sexual generation occurs, according to Timaeus' "likely story", in the second generation, when the souls of the timid and unjust are punished by returning to the world as women. Some anatomical adjustments are necessary to make sexual generation possible and to create a desire for it: Plato supposes that the excretory passage through which the urine is expelled "by air pressure" runs from the lungs via the kidneys to the bladder and is somehow joined with the passage in the condensed marrow, running from the head through the neck and spinal column. This marrow, he reminds us, is the seed:

> this marrow, being instinct with life, completed the process and finding an outlet caused there a vital appetite for emission, the desire for sexual reproduction. So a man's genitals are naturally disobedient and self-willed, like a creature that will not listen to reason, and do anything in their mad lust for possession.[5]

There are several points which can be made about this description. The first concerns the origin of the semen. It is neither elaborated from the blood nor gathered together from all parts of the body, but is a part of the soul-binding marrow implanted in all animals at their creation, and transmitted from generation to generation as semen. As such, it is possessed of soul: an animated, respiring substance. Proof of this is the effect which the seed has on the part where it seeks an exit: a compelling desire which is the mark of "a creature that will not listen to reason."

Plato specifically makes the point that this animated substance, and the desire which accompanies it, is common to both men and women. He makes no distinction between the contributions made by the two parents, although as we have seen he considers that all women, by their nature, possess an inferior grade of soul. Male and female produce equally fertile seed, though Plato does not make it clear how a single organised animal, or even several, are formed from the two 'seeds' sown separately into the womb. The womb itself is for Plato an even more obvious example of the activities of "a creature that will not listen to reason" than its male counterpart, and he ascribes a variety of female ailments to it, in words which were echoed by misogynists, and others, in the Renaissance:

> Much the same is true of the matrix or womb in women [i.e. as of the male genitals just described], which is a living creature within them which longs to bear children. And if it is left unfertilized long beyond the normal time, it causes extreme unrest, strays about the body, blocks the channels of the breath and causes in consequence acute distress and disorders of all kinds.[6]

Plato's description of the development of the embryo in the womb is very brief: sown in the womb as in a field are "seeds as yet unformed and too small to be seen which take shape and grow within until they are born into the light of day as a complete living creature."[7] Lesky argues that Plato's account is not based on 'preformationist' ideas such as those of the atomists,[8] but one should be wary, I think, of concluding too much from a brief account couched in the form of myth, not of medical fact.

However, Plato's ideas on generation were quoted with respect long after observers had disproved his curious descriptions of male and female anatomy as literal truths. At the centre of the story told in the Timaeus is the idea that mortal existence involves 'trapping' a separable, immortal soul by binding it in an especially refined and prepared substance. By means of this intermediary, it is possible to bring it into association with a mortal body: as Andrew Marvell wrote in his "A Dialogue of Soul and Body",

> A Soul hung up, as 'twere, in Chains.
> Of Nerves, and Arteries, and Veins.

This forced union of soul and body produces an unstable compound, but it is by means of this compound, an animated or ensouled matter, that life is maintained and transmitted from one generation to another. And this substance is possessed of pneuma, breath. The association of ensoulment with pneuma and of both with the semen is a theme which we shall meet again, as is the notion of a special substance to bridge the gap between soul and body.

Plato's idea that the male and female genitals are complete and separate animals is another aspect of his theory which was commented on by later writers, usually in the medical context of the causes of hysteria. Aristotle, in the short work De Motu Animalium (Book II 703b 5) which evidently found a place among Renaissance authorities on the question, suggests that the generative organ may be considered "alive" because "there flows from it the seminal potency, itself a kind of organism". Galen mentions Plato's ideas on the animal avidum generandum in several of his works, notably De semine, De locis affectis and De humoribus, as well as De naturalibus facultatibus and De usu partium.

Professor Screech, in his book The Rabelaisian Marriage, is one of the few Renaissance scholars to have explored this argument in its medical context. In his discussion of the passage in the Third Book of the Adventures of Pantagrúel, in which the doctor Rondibilis is advising Pantagruel on the question of marriage,

Screech suggests that Plato's idea of the uterus as an animal avidum generandum was widely accepted among the most advanced medical opinion of the time, and used to explain the inferiority, moral as well as physical, of women to men. He also suggests that this position represented a rejection of Galenic ideas of uterine pathology, and although I find it difficult to accept all the points of his thesis, for reasons which I have discussed elsewhere, the chapter is valuable in suggesting that a number of distinguished Renaissance medical writers were prepared to take Plato's theory seriously.[9]

Notes

1. "Replaced" is perhaps the wrong word, since Chalcidius's version was published in Rome in 1520: Plato Latinus edidit Raymundus Klibansky, vol.iv. Timaeus a Calcidio translatus Commentarioque Instructus, ed. J.H. Waszink, London and Leiden, 1962, p. clxviii. See Appendix I for early editions in Greek and Latin, including … Marsilii ficini florentini in libros Platonis…., Florence, [1484, 1485]. Compendium Marsilii Ficini in Timeum, 521^{r}-537^{v}, esp. 537^{r}.
2. Lesky, op.cit., p.1243.
3. Timaeus, 73 B–D. Cornford, op.cit., pp.292–295. Plato here puts forward a doctrine of the origin of the seed which comes from the Sicilian school of medicine, combining the theories attributed to Alcmaeon of Croton and Hippo of Rhegium.
4. Timaeus, 81 B-D.
5. Ibid., 91 A and B.
6. Ibid., 91 C.
7. Ibid., 91 D.
8. Lesky, op.cit., p.1235.
9. M.A. Screech, The Rabelaisian Marriage, London, 1958, esp. pp.90–96. Rondibilis's speech, which I take to be a skillfully constructed parody of the account in Timaeus, based on a number of sources – including Galen – comes from Rabelais, Tiers Livre, edited by Lefranc et al., Paris, 1912–13, Chapter xxxii, pp.245–7. Renaissance writers who mention the Animal avidum generandi, often in an anatomical or pathological context, include Vesalius, De fabrica, Book V, p.520, 533 (1543); Philipp Melanchthon, Liber de anima, p.44^{r} (1553); Nicholaus Rochaeus, De morbis mulierum curandis, 63^{r} (1542); Jean Fernel, Pathologia, Book VI, Chapter xvi, pp.201–206 (1554); Jean Riolan, Particularis methodis medendi, Sectionis Quartae, Tractatus Secundus, De morbis uteri: Chapter 1, pp.526–8, (Opera omnia, 1610); Ambroise Paré, Oeuvres, Book III, "De l'Anatomie", Chapters xxxiii, xxxiiii, pp.135–6, and Book XXIV, "Traictant de la Generation de l'homme, recueilly des Anciens & Modernes", esp. Chapters xlvi and lii, pp.982–990 (Quatriesme Edition, 1585). These authors are mentioned by Screech, and were therefore used as the basis of my own account, presented as a paper in the Tenth Renaissance Colloquium at University College, January 1–3, 1976, as "Plato, Rabelais and the Wandering Uterus", in

which I argued that many of Screech's Platonists seemed to take their accounts from Galen's description of hysteria, and that – as with the generation texts discussed in the Introduction to this book – the diversity among the texts which he uses makes it particularly difficult to classify them as 'Platonists' or 'Galenists'.

Chapter 4
Hippocrates

Abstract The first complete Latin edition of Hippocrates' *Opera* appeared in Rome in 1525, and in Greek in 1526. Separate translations of the most popular Hippocratic works were available earlier. The most important translations of the Hippocratic works on generation, those by Gorraeus and Cornarius, saw separate publication in 1545 and 1549. Both were extremely popular:

There are contradictions and inconsistencies in the 'Hippocratic' account. The works combine two distinct theories of the origin of the semen. In one, the semen derives from the brain and spinal cord, while in the other, the dominant one, it comes together from all, or the most important, parts of the bodies of the parents – the theory called pangenesis. Both parents contribute semen and there is no distinction made between the nature or importance of their contributions. The works most likely to have been taken by Renaissance doctors as epitomising Hippocrates' ideas on generation, *De genitura* and *De natura pueri*, together with their companion-treatise, *De morbis iv*, are probably by the same author, and they present a relatively consistent account. Hippocrates' explanation of generation is almost entirely mechanical. It requires no guidance by an intelligent, animated soul-substance.

Hippocrates' Opera were made available in printed form to Latin and Greek readers relatively late: the first complete edition in translation appeared in Rome in 1525, and in Greek the following year. Cornarius' translation of the Works went through numerous editions after its first publication in Venice in 1545. Separate translations of the most popular Hippocratic works were available before this, sometimes published with the works of Rhazes or as Articella, and the most important translations of the Hippocratic works on generation, those by Gorraeus and Cornarius, saw separate publication in 1545 and 1549 respectively. Both were extremely popular: Roger reports that there were 'at least' six editions of De genitura between 1542 and 1580, plus an Italian translation and two printings of Chrestien's French translation, De la geniture de l'homme. And he cites nine editions or translations of De natura pueri before 1580.[1]

Not surprisingly with a body of work which most modern scholars would agree to be by several authors, there are certain contradictions and inconsistencies in the 'Hippocratic' account. But the works most likely to have been taken by Renaissance doctors as epitomising Hippocrates' ideas on generation, De genitura and De natura

L. Deer Richardson, *Academic Theories of Generation in the Renaissance*,
History, Philosophy and Theory of the Life Sciences 22,
https://doi.org/10.1007/978-3-319-69336-1_4

pueri, together with their companion-treatise, De morbis iv, are probably by the same author, and they do present a relatively consistent account with which we may begin.

Lesky considers that the Hippocratic theory as we have it in these works combines and fuses two distinct theories of the origin of the semen: in one, which we have seen in one form in Plato, the semen derives from the brain and spinal cord, while in the other, which is the dominant one in Hippocrates, it comes together from all, or the most important parts of the body of the parents – the theory called pangenesis.[2] This position is characteristic of Empedocles and of the atomists, such as Democritus and Leucippus, but in De genitura it takes a somewhat different form. For the Hippocratic author, what 'comes together' is not a collection of atoms or of elemental particles from all parts of the body, but a part of the solid and fluid (humoral) constituents, notably the latter. He makes this point at several points in De genitura and in other works, but it is most fully expressed in a passage from the fourth book On Diseases:

> Semen, coming for the formation of the human being from all the parts of the body of the man and of the woman, and falling into the womb, coagulates: at the end of a certain time there results from it a human nature. The woman and the man have within themselves four types of fluid, which produce diseases … They are phlegm, blood, bile and water. It is neither the least nor the less active portion of these humours which goes to form the semen: and when the living being is formed, it has, conformable to its parents, just so many sorts of 'healthy' and 'disease-producing' humours.[3]

These humours, by their balance, help to determine heredity, resemblance, and may also explain why the child of a deformed parent is sometimes deformed and sometimes not, a point which strict pangeneticists found it difficult to explain. Since the defective part in the parent is of exactly the same constitution, humorally, as the sound, the child may be, and often is, born healthy.

> The children of deformed parents are usually sound. This is because although an animal may be deformed, it still has exactly the same components as what is sound. But when there is some disease involved, and the four innate species of the fluid from which the seed is derived form sperm which is not complete, but deficient in the deformed part, it is not in my opinion anomalous that the child should be deformed similarly to its parent.[4]

There are two, related, points to be noted in this argument. The first is that, as with Plato, both parents contribute semen and there is no distinction made between the nature or importance of their respective contributions. This equality is basic to any system of pangenesis, and is clearest in Hippocrates' discussion of heredity. Each sex contributes both 'male' (i.e. strong/thick/less abundant) and 'female' (weak/thin/more abundant) semen. The sex of the child is determined by the preponderance of semen provided by both parents. Resemblance in other features is similarly explained (as in the discussion of birth defects quoted above). The mother has a further important effect on the development of the child in utero: the size of the uterus and its available blood supply, as well as the maternal imagination, are all 'environmental influences' on the foetus.[5]

Further, Hippocrates' explanation of generation, most fully presented in these three works, is almost entirely mechanical. It requires no guidance by an intelligent, animated soul-substance. The proportions of the various humors contributed by the parents, and their 'strength', themselves determine the individual constitution of the child, subject to such outside factors as climate and season. This constitution, with its individual balance or imbalance of humors, is responsible in the same way for determining health or disease in the individual. In other works, for example, Airs, waters and places and On the sacred disease, the list of humors is somewhat different: black bile, yellow bile, blood and phlegm. But their role in causing disease and in determining inheritance – e.g. of epilepsy – appears to be identical.

In addition to the humors, contributions to the semen come from all the solid and fluid parts of the body. In De natura pueri Hippocrates explains more clearly just how the seminal substance is transformed. The "flesh" forming in the uterus respires, attracting both air, and blood which serves as nourishment for its growth. As it grows, differentiation of parts takes place:

> As the flesh grows it is formed into distinct members by breath. Each thing in it goes to its similar – the dense to the dense, the rare to the rare, and the fluid to the fluid. Each settles in its appropriate place, corresponding to the part from which it came and to which it is akin.[6]

This development occurs because of certain natural affinities: like goes to like, and this is natural, because like also came from like. The form of the parental body is thus recreated, as each part automatically 'goes back where it came from' and this of course helps to explain why the parts of the child's body resemble those of the parents'. There is a further detail, the part played by the air or breath drawn into the uterus:

> Now the formation of each of these parts occurs through respiration – that is to say, they become filled with air and separate, according to their various affinities. Suppose you were to tie a bladder onto the end of a pipe, and insert through the pipe earth, sand and fine filings of lead. Now pour in water and blow through the pipe. First of all the ingredients will be thoroughly mixed up with the water, but after you have blown for a time, the lead will move towards the lead, the sand towards the sand, and the earth towards the earth. Now allow the ingredients to dry out and examine them by cutting around the bladder: you will find that like ingredients have gone to join like. Now the seed, or rather the flesh, is separated into members by precisely the same process, with like going to join like.[7]

There is also a suggestion that the breath 'opens up' the vessels and hollow organs of the body, blowing them into shape as if they were inflated. But this breath, to judge from the description given here, has none of the psychic, intelligent attributes given it by Plato, by Diogenes (see below) or even by Aristotle and Galen. It is plain hot air, and its effects can be reproduced by a simple mechanical experiment.

Another kind of 'pneumatisation', also with its mechanical analogy, is to be found in the description, at the beginning of De genitura, of the way in which the semen is drawn from the parts of the body and transmitted to the genitals. It is here too, that the spinal cord reappears as a carrier of the semen, though not as the origin of it. Hippocrates tells us that 'veins' and 'nerves' containing fluid connect the reproductive organs with all parts of the body. During intercourse, these fluids in the body are heated, dilated, set into motion and "agitated by the movement it produces a foam, in the same way as all other fluids produce foam when they are agitated". The thickest and most active portion of this agitated liquid flows into the spinal marrow, which is connected with all parts of the body by afferent vessels and which in turn empties its contents into the kidney (by way of veins) and thence to the testes. The latter serve only as a passageway for the semen collected from all parts of the body. From the testes it passes to the genitals, not, Hippocrates is careful to point out, via the path taken by the urine.[8]

This complicated pathway employs nerves and blood vessels indiscriminately for the transport of the semen, passing through portions of what we would now consider as three separate organ-systems. But my purpose is not to condemn Hippocrates for anatomical inaccuracies, but to examine the implications of this viewpoint. It seems clear that it represents a conflation of two opinions: one similar to that held by Plato among others, in which the semen comes from the spinal cord and the other in which it comes from all parts of the body. The latter is the dominant position in Hippocratic physiology, but the spinal cord still appears as part of the route by which the semen reaches the genitals. A similar route is described in the Hippocratic text De natura hominis and quoted by Aristotle, who ascribes the work to Hippocrates' son-in-law, Polybus:

> The blood-vessels of largest calibre, of which there are four pairs in the body, are arranged in the following way … The second pair of blood-vessels runs from the head near the ears through the neck, where they are known as the jugular veins. Thence they continue deeply close to the spine on either side. They pass close to the muscles of the loins, entering the testicles and the thighs. …[9]

This too, though described as a pair of veins, is part of a multi-purpose pathway which serves the organs of generation in passing. And it makes clearer some of the more curious features of Hippocrates' discussion of sterility. In the case of castration, he attributes it to the fact that the 'numerous thin nerves' which carry the semen from the testes are cut by the knife. The resulting scar permanently closes the path and produces sterility. A similar, less drastic effect is produced by cutting the vein behind the ears: since, as in De natura hominis, "the greater part of the sperm travels from the head past the ears into the spinal marrow" and since "when the incision has formed a scar, this passage becomes obstructed" the result is usually infertility, though not the total impotence of the eunuch. In Chapter 22 of De aere aequis et locis Hippocrates attributes the sterility and effeminancy of many male Scythians to the effects of a similar operation, though later in the chapter he suggests that their un-Greek habits of wearing trousers and being constantly on horseback may have a contributory effect.[10]

The cause of sterility, and the question of eunuchism, is an important corollary to the cause of generation, and was discussed in this context by both Aristotle and Galen. Hippocrates' statement that sterility could be produced by an incision next to the ears was quoted with respect by Renaissance commentators, as was his argument that semen reached the testes via the spinal cord. Other points worth noting are the special emphasis he places on the rôle of four constitutive fluids or humors; the coming together of semen from all parts of the body of both parents; and the articulation and shaping of the embryo by the pneuma. It is worth emphasising that Hippocrates does not speak of the semen as having facultative and material parts or aspects. The humors composing it are material, ponderable fluids which by their very nature produce certain results when mixed in certain proportions. The mixing of the two semens, retained within the uterus, is all we mean by "conception":

> For when the womb has received the sperm it closes up and retains it, because the moisture causes the womb's orifice to contract. Then both what is provided by the man and what is provided by the woman is mixed together.[11]

The pneuma which directs development is similarly of simple construction: it is not a part of the semen itself but is 'breathed in' by the embryo from the outside air and is presumably identical with it. We shall see later how this position differs from those of Aristotle and Galen.

Notes

1. For Hippocratic texts cf. Littré, I, 373–9. For their history, cf. I, 502–10 (ancient texts and editions) 511–539 (manuscripts) and 540–554 (editions and translations). This is supplemented by the introduction to individual texts, especially VII, 467–9 for De genitura, De natura pueri and De morbis iv. Roger, op.cit., p.21. See also Appendix I, below.
2. Lesky, op.cit., pp.1237, 1300 e.s.
3. Hippocrates, De morbis iv: Littré, VII, 542, para.32. Translation mine.
4. De genitura: Littré, VII, 484, para.11.
5. Ibid., 478, 480, paras. 6–8; 482, para. 9 (blood supply); De mulieribus i: Littré, VIII, 64, para. 25; De natura pueri, 492, para. 14. Cited by Lesky, op.cit., 1327 e.s.
6. De genitura: Littré, VII, 470, para. 1; 474, para. 3; 496–8, para. 17.
7. Ibid., 498, para. 17.
8. Ibid., 470, para. 1.
9. De natura hominis: Littré, VI, 58, para. 11. Aristotle mentions this account among others in Historia animalium, Book III, iii, 512b 12-513a 7. The Platonic "spinal route" is given in Timaeus, 77 D.
10. De genitura: Littré, VII, 472, para. 2. De aere, aequis et locis: Littré, II, 78–82, para. 22.
11. De genitura: Littré, VII, 476, para. 5.

Chapter 5
Pneuma and the Pre-Socratics

Abstract To make clearer the range of associations which *pneuma* could have, we look at an extreme example of its importance in the work of the Pre-Socratic philosophers. They associated 'prime matter' with a single substance. Air, or breath – *pneuma* – has an observable association with life, so for Anaximenes *pneuma* or breath became identified with life and hence with soul, in the sense of intelligence, as well.

An important successor of Anaximenes was Diogenes of Apollonia. Fragments of one work, "On Nature", remain. From them we can construct the outlines of a system which considers air as the basic material principle. It also possesses intelligence. The air-soul of all living creatures is the same, since it all is breathed from the same source. Variations in density, temperature, motion and other qualities account for the differences among creatures. The association of air with life and with active intelligence has obvious implications for the activity of the *pneuma* of the semen.

For Aristotle and Galen, the *pneuma* has lost its central position as the single material and psychic constituent of the universe; but it retains a special importance as breath, present everywhere there is life, soul, intelligent, organised activity.

The rôle of the pneuma in classical theories of generation is, on the surface, something of a puzzle. Why should 'air' or 'breath' have a special function in embryonic development – as it does, though in different ways, for all four authors? To make clearer the range of associations which pneuma could have for these writers, it may be helpful to turn to an extreme example of its importance, in the work of the Pre-Socratic philosophers. A familiar feature of the cosmology of several of the Pre-Socratics was their association of "prime matter" with a single substance: e.g. water in the case of Thales, fire in that of Heraclitus and air for Anaximenes. Each substance has certain properties which makes its choice a reasonable one. Air, or breath – pneuma – in addition to its qualities of invisibility and rapid, powerful motion had obvious advantages as well in its observable association with life. Living things must breathe, and if they cease to breathe they die; and so for some philosophers pneuma or breath became identified with life and hence with soul, in the sense of intelligence, as well.[1]

L. Deer Richardson, *Academic Theories of Generation in the Renaissance*, History, Philosophy and Theory of the Life Sciences 22,
https://doi.org/10.1007/978-3-319-69336-1_5

An important successor of Anaximenes was Diogenes of Apollonia. Very little by him has survived, but fragments of one work, "On Nature" remain in the accounts given by Simplicius and by several other commentators, including Aristotle. From them we can construct the outlines of a system which considers air as the basic substance of animate and inanimate nature. Air or pneuma was for Diogenes the soul and substance of the world; of all living things; and of their seed. He believed, according to Simplicius, that

> the substance of the universe is infinite and eternal air, from which, when it is condensed and rarefied and changed in its dispositions, the form of other things comes into being.[2]

This air is the basic material principle of the universe, but it is also a principle possessing intelligence. This is particularly clear in the case of living things:

> Men and the other living creatures live by means of air, through breathing it. And this is for them both soul (i.e. life-principle) and intelligence... and if this is removed, then they die and intelligence fades.[3]

The air-soul of all living creatures is the same, since it all is breathed from the same source. But the air within living things is warmer than that outside, and variations in density, temperature, motion and other qualities account for the differences among creatures. The association of air with life and with active intelligence has obvious implications for the activity of the pneuma of the semen. As Simplicius again reports:

> He [Diogenes] shows also that the sperm of living creatures is aerated, and acts of intelligence take place when the air, with the blood, gains possession of the whole body through the veins.[4]

Two 'acts' of the aerated blood seem to be described here: one is generation and the other sensation or movement. For several of the commentators agree that Diogenes considered the semen to be aerated or pneumatised blood, or as "the foam of the blood". But they differ in their description of how the blood attains this foamy state. In one account it appears that the thicker blood is absorbed by the fleshy parts as it passes through seminal veins or vessels, leaving a surplus which "becomes fine and warm and foamlike".[5] This description of the passage of the blood through 'spermatic veins' to the testes or womb is quoted by Aristotle (History of Animals, III, ii, 511b, 31) and it resembles his own explanation in Generation of Animals (e.g. I. xix, 726b). In Simplicius' account, however, it appears that the pneuma mixed with the blood 'shakes it up' and so makes it thinner, finer, and more frothy. Clement of Alexandria gives a similar account of the aeration of the blood to form semen; both recall Hippocrates:

> But some suppose that the semen of living things is likewise by its nature the foam of the blood. Now owing to the native heat of men, it is shaken up in the embrace and expelled; foams up and is passed on to the seminal vessels. From this, Diogenes the Apolloniate will have it that the love-pleasure (ta aphrodisia) takes its name.[6]

Recalling a theory such as Diogenes' may help to make clearer some of the properties of the pneuma which seem to be assumed in its discussion by later writers. For Aristotle and Galen, the pneuma has lost its central position as the single material and psychic constituent of the universe; but it retains a special importance as breath, present everywhere there is life, soul, intelligent, organised activity. And it is this aspect which finds a place in the idea of a generation-pneuma. For none of our writers find it necessary to explain, or justify, the appearance of the pneuma in their theory of generation. It is a familiar figure which can represent everything from the physical power of wind to move objects to the intelligent power of soul to direct and organise movement. As we shall see, several of the Stoic philosophers revived the notion of pneuma as the substance of the soul (including the World-Soul) but it will be more useful to discuss their opinions when we come to Galen.

Notes

1. For classical doctines of pneuma, see Hall, Life and Matter, I and especially G. Verbeke, L'Evolution de la doctrine du Pneuma du Stoicisme à S. Augustin, Paris, 1945, pp.206–212, et passim. For the opinions of Aristotle and Galen, see the works cited below in Chap. 6, Note 20 and Chap. 7, Note 28.
2. G.S. Kirk and J.E. Raven, The Pre-Socratic Philosophers: A Critical History with a Selection of Texts, Cambridge, 1957, Chapter xviii, p.601; from Simplicius, Physics, 25.1.
3. Ibid., 605 (Fragment 4; Simplicius, Physics, 152, 18).
4. Ibid., 619 (Simplicius, Physics, 153, 13).
5. Kathleen Freeman, Ancilla to the Pre-Socratic Philosophers, Oxford (Blackwell), 1962, p.89. Translation of Diels-Kranz, Die Fragmente der Vorsokratiker griechisch und deutsch, Berlin, 1934–7. (Cited as VS). VS 64 B6. Also in Kirk and Raven, 618.
6. VS 64 A 24, cited by Lesky, op.cit., p.1348.

Chapter 6
Aristotle

Abstract The most important of the Aristotelian texts for generation is naturally *De generatione animalium*, his treatise on animal generation. Greek manuscripts of *De generatione* exist from throughout the medieval period; the earliest has been dated at c.1000 A.D. It was made available to Latin audiences in the translation of William of Morbeke (c.1260) and in that of Theodore Gaza (c.1450) which was printed, together with those of the *History* and *Parts of Animals*, in 1476. A printed Greek text became available in the 1497 (Aldine) edition of the *Opera*.

In briefly noting the distinctive features of Aristotelian generation theory, one would have to mention the separation of motive and material cause into male and female contributions; the origin of semen as a homogeneous residue of the blood; the insistence that parts are formed successively, beginning with the heart; and the importance of the *pneuma* – resembling the Stoic notion and yet quite distinct.

For our purposes, the most important of the Aristotelian texts is naturally De generatione animalium, his treatise on animal generation of all types. According to Peck it is the culmination of his series of works on "Zoology", combining the results of observation (as in the Historia Animalium); of a teleological discussion of the 'parts' and their form and function (De partibus and De incessu animalium) and of his treatment of the functions "common to body and soul" in De motu animalium and the Parva naturalia. These, together with the De anima, in which the form (or soul) of animates and its parts or functions are discussed, make up the 'animal works'.[1] Of these, Greek manuscripts of De generatione exist from throughout the medieval period; the earliest has been dated at c.1000 A.D. It was made available to Latin audiences in the translation of William of Morbeke (c.1260) and in that of Theodore Gaza (c.1450) which was printed, together with those of the History and Parts of Animals, in 1476. A printed Greek text became available in the 1497 (Aldine) edition of the Opera. Other important editions are those of the three 'Animal' translations of Gaza by the Colines press (1533) and the 1584 edition of the Historia Animalium by Scaliger.[2]

The basis of Aristotle's theory of generation – as of the rest of his philosophy – is his distinction of four kinds of cause, four senses in which we can say that one thing exists 'for the sake of' another. The first two, "the definition of the being" (or formal

L. Deer Richardson, *Academic Theories of Generation in the Renaissance*,
History, Philosophy and Theory of the Life Sciences 22,
https://doi.org/10.1007/978-3-319-69336-1_6

cause) and the cause "for the sake of which as end" (final cause) are the same thing in the case of animals, since for Aristotle the coming-to-be of any creature is determined by its logos or 'definition', the existence of the actual individual and species, which also serves as the end or goal of development: "for coming-to-be is for the sake of being, not being for the sake of coming-to-be". (De partibus animalium I, i, 640a 18). Logically at least, the chicken comes first. But these two causes play relatively little part in the detail of De generatione, and Aristotle could perhaps be accused of the same fault with which he chides his predecessors:

> The old physiologers, however, thought the opposite [i.e. that the process of formation determines the finished product] because they did not see that the causes were numerous; they recognized only the Material Cause and the Motive cause (and even these they did not clearly distinguish) whereas they paid no attention to the Formal Cause and the Final Cause.[3]

It is the material cause, and even more the motive or efficient cause, cause in the sense of "that from which comes the beginning of the movement" on which Aristotle focuses in this work: first because he feels it is the only one with which he has not already dealt in the other works; and secondly, because it is at the heart of the question: "to inquire about … [the motive cause] and about the generation of each animal is in a way the same inquiry".[4]

Aristotle's distinction between material and motive cause, and his emphasis on the importance of the latter, is likewise at the heart of his notion of sexuality. "Male is what we call an animal that generates into another, female that which generates into itself" (I, ii, 716a 14) but more importantly, in sexual generation, the male contributes "the source of the movement", the female the matter (716a 4). In other words, "the physical part, the body, comes from the female, and the soul from the male, since the Soul is the essence of a body."[5] By Aristotle's own definition of soul (e.g. in De anima 415b 7ff) and his own definition of cause, this implies that the male contribution incorporates the final, formal and efficient causes, the female only the material. This dichotomy runs as a motif through all aspects of his argument: his rejection of pangenesis; his discussion of anatomy; of the origin of the semen; of actual and potential; of the mechanism of heredity. But he is not always able to avoid contradicting this extreme statement of the case, as we shall see.

Aristotle provides a range of arguments, anatomical and philosophical, for his position. The philosophical argument at the beginning of Book II runs roughly as follows: since sexes do in fact exist, they must exist of necessity; or at least, since they are not universal, it must be better for the generation of those animals that have them to exist as male and female. Why? Because

> the proximate moving cause … is better and diviner in its nature than the matter; and it is better that the more excellent be separated from the worse. Because of this the male is separated from the female wherever possible and as far as possible.[6]

Anatomically, this difference is supported by the differences between the parts which serve generation in male and female. This in turn implies a difference in function, and this functional difference results from a difference in logos, in generative faculty or capacity, between the sexes. A number of the further arguments are linked to his discussion of the semen, and of the differences between it and the catamenia

(Book I, Chapter 17 e.s., and in Book II, Chapters 2 and 3). He begins Chapter 17 with a series of questions, which boil down to two: (1) Do both male and female discharge semen? (2) Is the semen drawn from the whole of the body? and begins by tackling the latter. But as we shall see, he considers that in discrediting the arguments for pangenesis, he is providing at the same time a series of reasons why the female need not be assumed to produce fertile semen:

> If the female does not discharge any semen, then it is consistent to say that semen is not drawn from the whole body either; or again, if it is not drawn from the whole body, there is nothing inconsistent in saying that it is not drawn from the female either, but that the female is responsible for generation in some other way than this.[7]

He begins the construction of this circular argument by reviewing four arguments for pangenesis: those based upon the intensity of sexual pleasure; the production of mutilated children by similarly mutilated parents; resemblance in individual features of child to parent; and the idea that 'if there is a special semen which gives rise to the whole' it is reasonable to assume special semens which give rise to each of the parts, so that similar causes can be adduced for whole and part. (721b 15–28).

Against this, Aristotle argues as follows. Concerning resemblance, he points out that children resemble their parents in ways (e.g. voice, way of moving) which are clearly not drawn from the semen, and that they may also resemble more remote (or collateral) ancestors who contributed nothing to it.[8] His second argument against pangenesis is more profound. One must assume that the semen comes together from the most basic possible components of the body, i.e. the elements (he accepts the Empedoclean four) or the 'uniform parts' such as blood, flesh and bone. If it is assumed to come from non-uniform parts as well, then a duplication is involved. But it is especially in the non-uniform parts, the "features" such as eye or nose, that we recognize resemblance. Further, what makes this resemblance is the assemblage of the parts, not their actual material composition:

> so if there is something which sets to work later on to bring this assemblage about, then surely this something, and not the drawing of the semen from the whole of the body, will be the cause of the resemblance.[9]

This notion of 'assemblage', together with a very literal-minded interpretation of the way in which the 'parts' are 'present' in the semen, dominates the rest of the argument. How do the disjointed parts of the embryo 'live' in the semen as imagined by Empedocles? If they are not disjointed, but joined together, why is the result not an animalculus: or rather two animalculi, since if both parents produce semen there will be 'two of everything'. Multiple fetations are no easier to explain by pangenesis; neither is the manner in which the embryo takes its growth (an important point) or the way in which the sex of the child is determined. (I, xviii, 722b–723b).

The point which Aristotle makes in the case of growth – that the aliment which is added and assimilated by the embryo cannot be of the same composition as the original conception, but must change into flesh and bone – leads him to his own definition of semen. In terms of the categories into which the substances of the body are divided, semen is a residue: that is, neither a part of the body nor a substance foreign to it (like a tumour). It is "the final residue of useful nutriment" or of "useful nutriment in its last stage." (I, xviii, 726a 26). But, "the last stage of nutriment is the blood in blooded animals" and so "the seed must be a residue of the nutriment which has become blood and is being distributed to the parts in its final stage." (I, xix, 726b 1;10–13). Those who are sick, too young or too old produce no semen because they have no surplus nutriment. Similarly, fat persons (in which the surplus residue is changed to fat) are less fertile than the thin. Weakness of the body following emission, and the occasional emission of blood, are further evidence of an origin in a vital portion of the blood (725b, 5-726a 7). Resemblance is explained as well: "for what has gone to the parts [as useful nutriment] resembles what is left over [as seed]." (726b 14–15).

Lesky believes that Aristotle's dissections (as described in Historia Animalium III, 510a 14f) in which he reported finding a blood-like fluid in the ductus deferens and a white, 'elaborated' semen on emission, contributed more to his theory that the semen has its origin in the blood than did the ideas of earlier writers such as Diogenes of Apollonia.[10] It seems likely that observation and theory reinforced one another. The presence of blood in the duct leaving the testicle might not have been considered significant by someone not already committed to the idea that the principle of generation was somehow derived from the blood. This observation, and the conclusions drawn from it, underline one of the characteristic points of Aristotelian anatomy: his refusal to admit that the testes (of their female equivalent, the ovaries, he was ignorant) play any role in semen production.

Testes are not necessary to generation, says Aristotle, because not all animals have them (he does not consider the seminal passages of reptiles and fish as equivalent structures) (I, iv, 717a 11ff). But since they exist in some animals, they must have a purpose, clearly subordinate, however, in these. The purpose he hits on is that they contribute to steadier and less rapid emission of the semen: in Vivipara, they act as weights to keep the winding seminal passages in place; in oviparous animals they act as temporary receptacles. The semen, he seems to think, is elaborated in the vessels leading from the testes to the pudenda: and it is these which correspond to the 'seminal passages' of the lower animals. But Aristotle was well aware of the complete change in the generative capacity, and indeed the nature of the entire animal, which accompanied the removal of these 'merely attached' weights (I, ii, 716b 5; I, iv, 717b 1; V, vii, 787b 20ff). He explains it by saying that "the principle from which the tautness of the body is derived is slackened" as a result of castration, and that the 'principle' accordingly undergoes a complete change from male to female. This is not due to any power inherent in the testes themselves: they are not "a ganglion of many principles" as some suppose. It just happens that when this "small alteration" takes place, "a principle changes at the same time." (V, vii, 788a).

The "principles" of male and female are basic, as Aristotle suggests, and involve much more basic differences than those of the generative organs alone. At times he seems to suggest that the female principle is the opposite of the male (as in the statement that the female provides matter, and the male soul, for the fetation). But in discussing the reasons why the female produces not true semen, but the catamenia, all that seems to be implied is that the female principle is inferior (I, xix, and xx). To return to the origin of the semen as a residue of nutrition. There are for Aristotle three generational residues, each with its natural collecting place within the body. These are respectively uterus (for the menses or catamenia); male pudenda (for the semen); and breasts (for the milk). The residual blood of which the catamenia are formed is exactly similar to that which goes to form the true semen. "The menses are a residue and are analogous in females to the semen in males" (I, 19, 727a 3); they are "seed that is not pure but needs working on"; only when they are "mixed" with the purer semen can they generate life (728a). The difference between true seed and menses is only in the degree of elaboration, or concoction, which the residue has received. It is in the capacity of the 'hotter' (and accordingly more perfect) male nature to elaborate the residue more fully, and therefore produce a more perfect, fertile product. This argument is based on two assumptions: that a 'hotter' nature is nearer perfection than a 'colder' one, and that the male nature is 'hotter' than the female.

But this line of reasoning, in which male is contrasted to female as perfect to imperfect, is almost immediately contradicted by another, in which the contributions have to be assumed to be opposite in nature, rather than simply different in quality:

> Now, that the female does not contribute semen to generation, but contributes something, and that this is the contribution of the menses ... is clear both from what has been said and according to reason on general considerations. For there must be that which generates and that out of which...If therefore the male exists as active and causing movement, and the female as passive, the female's contribution to the male's semen will be not semen but matter.[11]

How one, fully concocted, residue comes to be the agent, the other, less perfectly concocted the patient is not clear. At best, it suggests an imperfect fit between Aristotle's theory of generation, in terms of the blood-origin of the semen, and his explanation of the mechanism of generation in terms of proximate matter and proximate motive cause. This contradiction comes out as well in a passage in which Aristotle wonders why all women necessarily produce menses, but all men do not necessarily produce semen. Females necessarily produce a residue, because of their colder temperament: "The female system cannot effect concoction" (738a 34). But they also necessarily do so, because "the female always provides the material. ... hence, necessity requires that the female should provide the physical part, i.e. a quantity of material" (738b 20ff). But the male has no such necessity, since his contribution is an immaterial dynamis and a physical 'tool', the semen, is only secondary: "necessity does not require that the tools should reside in the product being made, nor that the agent which uses them should do so" (738b 24ff). As Aristotle explains it, neither the carpenter nor his saw are (or need to be) part of the table which he makes.

This distinction between the sexes as agent and patient creates practical difficulties when Aristotle comes to consider the question of heredity (IV, Ch.1–4). He can hardly deny that creatures are born having the female form, or that sons sometimes resemble their mothers as daughters do their fathers. He takes refuge in two arguments: (1) sex-determination is the responsibility of the heart, as the seat of vital heat and as the first organ to be formed: the difference in the 'principles' of the two sexes is a difference in temperature (765b, 36-766a 25, 30ff). (2) the power of the mother, or of the foetal material, to enforce resemblance (including sexual resemblance) is purely negative, coming not from any formative faculty but simply from a power of resistance (766a for sex determination; 768a for resemblance in general). In cases "wherein the generative and motive agent has failed to gain the mastery", says Aristotle, the opposite characteristics (which exist in potentia in the female matter) appear in the child; they are due to a 'relapse' on the part of the movements fashioning the embryo (IV, iii to 769a).

Again, and most importantly, it is clearly not a simple question of perfect v. imperfect when Aristotle turns from the question "out of what are the parts of the embryo produced?" to the question, "by what are they produced?" Is this "by what" external to the parts or semen, or is it in some way internal to it? And are the parts formed simultaneously or consecutively? In answering the first question, he explores in detail the notion of soul as it applies to generation; the second implies his discussion of potentiality and actuality (II, i, 734a).

The soul, as source of the movement, must be internal to the semen in some sense. First, because it is the soul of the semen: the activity or capacity which makes it possible to perform its function. In the same way, it is soul, or capacity, by which we distinguish the eye of a living man from that of a dead one (II, i, 735a). Secondly, it seemed to Aristotle, for whom motion of any kind implied contact, that it was unreasonable to assume that a mover outside the semen could form the parts (734a). But in what way does it make sense to say that the semen is possessed of soul? Plainly, it cannot be said to possess soul in the fullest sense, since its capacities are very limited. It possess soul only potentially (735a) and at two removes: for the semen is a potential embryo, and the latter is only potentially a fully-formed animal.

Aristotle approaches this concept of generation as the actualization of potentiality in three ways. First (a point discussed more fully in De Partibus Animalium, I, i, 640a) he reminds us that the seed is potentially that which already exists – the man or the horse or the oak tree which is the goal towards which its development tends. Secondly, 'potential' involves the capacity of the semen to transmit movements from the generating parent to the embryo. To do this, it must be regarded as somehow possessing, containing and conveying the three sorts of soul or faculty with which the embryo must be endowed if it is to reach maturity: nutritive, perceptive and intellective.

This is a complex and not entirely clear process.[12] The seed, as ensouled, is capable of self-movement; but it must also possess the capacity for transmitting the particular 'movements' initiated by the male parent to the embryo. It does so by initiating certain movements in the catamenia, which will cause the heart to be formed:

> Since the seed is residue, and is being moved in the same movement as that with which the body grows when the final nutriment is being particularized, when it comes into the uterus it constitutes and moves the female's residue in the same movement in which it itself is actually moving. For that too is residue and contains all the parts potentially, though none actually.[13]

Both here (735a 17) and in De anima (415a 25, 416a 19) the nutritive and generative soul/principle/faculty is the same. Aristotle's association of the heart with the seat of the nutritive soul, and his insistence that the heart was the first organ to be formed, seem to be based in part on his experiments on the development of the chick (H.A.

III, i). He argues the point several times in De generatione, and also in De partibus, (III, iv, 666a and b). But the nutritive soul is a logical necessity as well as (in the shape of the primordial heart) an observable fact:

> in the fetation, in a way all the parts are present potentially, but the first principle has made the most headway, and on that account the first to become distinct in actuality is the heart. This is plain not only to the senses (for after all it is a matter of fact) but also to the reason. Once the fetation which has been formed is separate and distinct from both the parents, it must manage for itself, just like a son who has set up a house of his own independently of his father. That is why it must have a first principle, from which also the subsequent ordering of the animal's body is derived. (II, iv, 740a 1–10).

This first principle cannot be the semen, since for Aristotle it is no part of the fetation. From the moment when the heart is formed, the foetus becomes an independent being with its own soul, developing successive capacities as it grows. Aristotle speaks of perceptive (or sentient) and intellective (or rational) faculties, but the organic seat of these faculties is less clear. Is the heart or brain the seat of sentient soul? Aristotle suggests the heart (II, vi, 743b 26: "As the source of sensation is in the heart, the heart is the first part of the whole animal to be formed".) And is the intellect to be associated with the brain, or is it a divine principle not requiring association with a body? (736a and b). The 'information' which is conveyed by the semen to the foetal material is described in terms of movements: and the analogy which Aristotle uses is that of the parts in an automaton, which possess a capacity (from their shape, material, relation, etc.) to transmit motion even when at rest (734b 9ff).

But how can the semen, which itself only possesses the capability for growth (i.e. nutritive soul) transmit to the human embryo the capacity to develop the higher faculties of perception and intellect? There are two problems here. "All principles whose actuality is bodily are clearly unable to be present without body" (736b 23). But the seed, itself only "a residue produced by a change in the nutriment" (763b 27) lacks differentiation and hence the sort of 'body' required to incorporate the 'bodily actuality' of perception. The intellect, on the other hand, is not a 'bodily actuality', for Aristotle, in the sense in which walking or even seeing or hearing are. But it too requires a body with which it can be associated, and it too must enter with the semen at conception, since there is no other point at which 'motion from outside' can plausibly be introduced. How is the physical transmission of these qualities possible, by a substance (the semen) which will never possess them in actuality, and to an animal (the embryo) in which they will be actualised only gradually, at birth or even later?

Aristotle seeks to solve both problems by associating "the capability of all soul" with a body "different from and diviner than the so-called elements":

> For within the seed of everything there is present that which makes the seed to be fertile, the so-called hot. This is not fire or that sort of capability, but the pneuma inclosed within the seed and within the foamy part [Aristotle had previously (736a) defined the seed as a mixture of pneuma and water] and more precisely the nature in the pneuma, being analogous to the element of the stars.[14]

At least one modern interpreter of Generation of Animals has argued that this passage meant that for Aristotle, generative heat was only a purer form of elemental fire, and not to be identified with the celestial element, aether or the quintessence.[15] A somewhat similar interpretation – that animal heat was not different in kind from either the elemental heat of fire or the celestial heat of the sun, but differed from them only in intensity – can be found in the writings of the unorthodox Renaissance natural philosopher, Bernadino Telesio.[16] In general, though, Renaissance interpreters seem to have taken Aristotle's meaning in this central but puzzling passage at its face value, although few do more than paraphrase it.[17] What Aristotle says elsewhere, in De anima and De partibus animalium, is not entirely clear on this matter, although it does clearly distinguish between the soul and the "substance of a fiery character" which is its vehicle.[18] But in De generatione itself he makes the distinction between generative or animal heat and fire quite clear. "The heat in animals neither is fire nor has its origin from fire" (737a 7). What distinguishes them is not simply a degree of purity, but that the former "contains a vital force", i.e. soul. This association of generative heat with soul is equally clear in the only other discussion of pneuma as generative: the passage (III, xi, 762a 18) in which Aristotle gives an account of spontaneous generation:

> Animals and plants are produced in earth and in wet because in earth water is present, and in water pneuma is present, and in all pneuma soul-heat is present, so that in a way all things are full of soul.[19]

Whether or not Aristotle is always entirely consistent in distinguishing animal heat from that of fire, it seems clear that he does so in these passages in the case of generative heat – which is not necessarily identical with the heat of animals. He is equally clear in associating this generative heat with pneuma and in comparing one or both with a divine element (the aether) and with the heat of the sun and stars.

In De generatione the pneuma is chiefly important as the 'body' associated with the "capability of all souls". But it has several minor, presumably associated roles as well. It is pneuma which is responsible for the 'swelling up' of the female breasts and of the testes at puberty: "a sign that the seed is in the menses" (I, xx, 728b 23ff) and as we have seen the pneuma is in the seed. Pneuma is responsible for the whiteness both of semen and of fair hair: this is not directly a cause of heat, but because the presence of heat (e.g. in the semen) implies the presence of pneuma, and it is "the air shining through" that gives the appearance of white, as in foam (II, ii, 735b 19f, V, vi, 786a 6f). Aristotle also says (II, vi, 741b 37f) that the parts of the embryo are articulated by pneuma, a statement which recalls Hippocrates. This is not the 'breath' of either mother or foetus, but the pneuma necessarily present in the hot and moist.[20]

As Balme points out, soon after Aristotle (and before him in the writings of philosophers such as Philistion and Diogenes of Apollonia) pneuma, in the works of the Stoic school, came to have the 'specialist' meaning of an indwelling divine spirit, in addition to its primary meanings of 'breath' or 'wind'.[21] Aristotle seems never to have systematized a pneumatology on Stoic lines, but there are certain resemblances, and differences, which are worth noting.

For Zeno (born c.332/3 B.C.) and his Stoic followers the soul was conceived as material, a kind of warm, rare pneuma penetrating the entire body. Judging by surviving reports, Zeno's account of generation resembled Aristotle's in some respects:

> Zeno said: the seed which men discharge is Pneuma combined with fluids, a separable part of the soul and a mingling from the semen of the ancestors; a mixture, which has come together from all parts of the soul: for this seed contains just the same rational power as the whole: when it reaches the womb it is met by another Pneuma, a part of the woman's soul, becomes one with this and grows in the hidden place, where it is moved and excited by that [Pneuma] and continually the moisture, out of which it increases itself, travels to it.[22]

But in Stoic theory the pneuma associated with the fluid is not simply a carrier of the generative heat: it is a part of the World-Pneuma and as such the source of the formative power (logos spermatikos) responsible for all generation; for the functioning of the parts of all organisms (including the Universe itself); and for integrating these parts into functioning wholes. The same logos is present in the semen, and explains its ability to produce all the bodily organs and integrate them into a working organism. This idea of the relation between whole and part, in which the smallest part "contains just the same rational power as the whole" reappears in Chrysippus (whose definition of the soul is quoted by Galen) and in other followers of Zeno such as Kleanthes.

It is not always clear, in Stoic writings, what physical substrate the pneuma has in the body. Usually it seems to be associated with semen, or with blood, and both appear to be considered as unified, homogeneous substances, despite their coming together from all parts of the individual. In the account by Zeno quoted above, at least, male and female contributions are evidently identical and equal. But his account differs from pangenesis: it is a mixture of pneuma, a 'part of the soul' of male and female which is separated and brought together in the uterus, and not differentiated representatives of the various parts of the body.[23]

In briefly noting the distinctive features of Aristotelian generation theory, one would have to mention the separation of motive and material cause into male and female contributions; the origin of semen as a homogeneous residue of the blood; the insistence that parts are formed successively, beginning with the heart; and the importance of the pneuma – resembling the Stoic notion and yet quite distinct. Galen put the distinction neatly when he said that for Aristotle pneuma was the work-material of the soul, whereas the Stoics made it the master of the work.[24]

Notes

1. A.L. Peck (editor and translator), Aristotle: Generation of Animals, Loeb Classical Library, 1953, p.vii. Translations are taken from this edition (cited as Peck), or from D.M. Balme (editor and translator), Aristotle's De partibus animalium I and De generatione animalium I (with passages from II 1–3), Oxford, 1973 (cited as Balme).
2. For the Renaissance grouping of the "animal works" see Part II, below. See Appendix I and Part II for further details of the Renaissance editions.
3. G.A. V, i, 778b 7, (Peck).
4. I, i, 715a 14f, (Balme).

5. G.A. II, iv, 738b 24, (Peck). See also P.A. 641a 25ff and De anima, loc.cit., for the soul as cause and principle of the body a) as the source of its movement; b) as its final cause c) as the essence of living bodies. On this point see also Anthony Preus, "Science and Biology", op.cit.
6. G.A., II, i, 732a 4f, (Balme).
7. G.A., I, xviii, 724a 8, (Balme).
8. An atomist, such as Leucippus, would probably have replied that the parents' bodies do contain, and transmit, atoms from an earlier generation, and that these same atoms, in a materialist world, could be assumed to determine 'acquired characteristics' as well as purely physical resemblances. But in fact, Aristotle's main target here seems to have been Empedocles. cf. Lesky, op.cit., 1316–7 and Balme's notes on this passage, pp.141–2.
9. I, 18, 722a 17ff; 722b 37–9. Peck's translation seems clearer than Balme's here.
10. Lesky, op.cit., pp.1352–6.
11. G.A., I, xx, 729a 21f, (Balme).
12. It is also an important theme in Renaissance discussions of generation, especially by natural philosophers: see especially the chapter on Cremonini in Part II and also the chapter on the animation of the foetus in Part III, below.
13. II, iii, 737a 18, (Balme).
14. II, iii, 736b 30, (Balme).
15. According to L.M. Balme, what Aristotle means to imply is as follows: "Every faculty of soul, whether or not it is associated with a bodily activity, is associated with a certain diviner sort of body (pneuma). Aristotle is not saying that soul is the activity of this body… The distinction lies rather between being associated with a bodily activity and being associated with a body. Even intellect has some association with body … It is not impossible, therefore, for Aristotle to recognize a physical association between intellect and body. Now he goes on to say that the heat in pneuma has the special property of being able to convey soul, including intellect. He does not explain this in physical terms, but judging from 737a 18–19 we may guess that he conceived it as a conveying of movements superimposed upon the heart's own movements – perhaps as a liquid conveys waves." (p.160).

    For Balme, it is not the pneuma itself which is the conveyor of soul, but the nature in the pneuma, i.e. the heat. And apart from the statement that this heat is 'analogous to the element of the stars' he finds the generative heat as presented here consistent in its attributes with the vital heat described by Aristotle in other works (for example, Parts of Animals, De respiratione, and De anima). But Balme considers that for Aristotle, in these other works and accordingly by implication here, vital or animal heat is equated with ordinary fire. Generative heat differs from fire only in what it can do (i.e. generate animals) and hence in being, perhaps, purer. From this comes the epithet 'diviner' which may only mean "less grossly material, purer, superior". "The generative heat is not identified with aither: if they are analogous, they must be different". Similarly, "pneuma is basically atmospheric air" (pp.161–4).

    Balme bases this argument for continuity on the gradations in the scala naturae which for Aristotle seem to correspond to variations in the nature of the hot (736b 32). "Since nature is continuous from non-living to living … animal heat need not be an altogether different element from other heat, nor generative heat from the rest of animal heat." But this argument surely makes too little of the clear distinction which Aristotle makes, both here and in the passages from Parts of Animals and De anima (p.163) between ordinary heat – the heat of elemental fire – and generative heat, which is identified with the vital heat of the sun and stars.

16. Bernardino Telesio, De rerum natura...., Naples, 1587; Book VI, Chapter 20, pp.253–4.
17. See, for example, Nifo, in Part II, below.
18. For example, De anima, II, 416a 9f (from the Oxford translation): "By some, the element of fire is held to be the cause of nutrition and growth, for it alone of the primary bodies or elements is observed to feed and increase itself. Hence the suggestion that in both plants and animals it is it which is the operative force. A concurrent cause in a sense it certainly is, but not the principal cause, that is rather the soul... Nutrition and reproduction are due to one and the same psychic power."

    And rather more fully, in De partibus animalium, II, vii, 652b 7: "For some [note Democritus] writers assert that the soul is fire or some such force. This, however, is but a rough and inaccurate assertion; and it would perhaps be better to say that the soul is incorporate in some substance of a fiery character. [Italics mine] The reason for this being so is that of all substances there is none so suitable for ministering to the operations of the soul as that which is possessed of heat. For nutrition and the imparting of motion are offices of the soul, and it is by heat that these are most readily effected. To say then that the soul is fire is much the same thing as to confound the auger or the saw with the carpenter or his craft, simply because the work is wrought by the two in conjunction."
19. As translated by Balme himself, (p.162).
20. For the question of pneuma in Aristotle, see also Arthur L. Peck, "The connate pneuma, an essential factor in Aristotle's solutions to the problems of reproduction and sensation" in Singer, ed., Science, Medicine and History, 1953, 1:111–121, and Appendix B of the Loeb edition of the Generation of Animals, pp.576–593. According to Peck, it is the pneuma, not the heat, which is the focus of Aristotle's concern, since he speaks of the embryo as formed by cold as well as heat: and the special property of the pneuma in accomplishing the actions of motion, generation and sense-perception, as he sees it, is its power of expanding and contracting (i.e. pushing and pulling). Peck takes the analogy with aither, the element of the stars, as implying a serious statement of similarity: (1) pneuma, and the aither of which the sun and the stars are made, is generative; (2) pneuma causes movement without itself undergoing qualitative change; the 'divine' (De caelo, 270b 10) aither likewise is not subject to change, and only to perfect, circular motion; (3) Both act as an intermediary between 'an immaterial mover and material objects' (Peck, Appendix B, esp. pp.586–9). See also H.A., VII, 7, 586a and Problemata, xxx, 953b 33 for another description of pneuma, as the 'blast of air' which causes the emission of sperm "by pneumatic pressure".
21. Balme, p.161.
22. Stoicorum veterum fragmenta, I, 128 (cited as SVF) coll. J. ab Arnim, Leipzig, 1903–5. Cited by Lesky, (p.1388).
23. Lesky, esp. p.1391.
24. SVF, II, 743; cited by Lesky (p.1388).

Chapter 7
Galen

Abstract Galen's contributions are *De semine* (*Works* 1490; separately 1533), *De foetuum formatione* (1535 and in French 1559) and *De usu partium* (1528 and in French 1566). Galen opposes Aristotle on four points especially: the existence of female semen; the material constitution of the embryo; the role of the testes (which for Aristotle had no part in forming semen); and the choice of the liver, rather than the heart, as the earliest centre of foetal development. These points of difference are central to academic debate in the Renaissance.

In *De semine* Galen speaks of the operations of heat and spirit. In part this recalls the Hippocratic *pneuma*, and the activities of the physical qualities – hot, cold, moist and dry – which Galen says form the simple parts. But the spirit appears to do more, intelligently arranging and composing the parts. If pressed on the nature of the essence lying behind this formative faculty, Galen is willing to say that it clearly reveals intelligence, but not to speculate on whether this essence, or soul, is mortal or immortal, corruptible or incorruptible, unitary or existing in parts. The result is a system which is far from logical.

It is to Galen himself that we must now turn. His most obvious contributions to the question of generation are in the works De semine (published 1490, in the Works, and separately in 1533) and De foetuum formatione (published in 1535, in the translation of Felicianus and in two other translations by Cornarius and Guinterius in the following year, and in 1559 translated into French). The short Galenic dissection manual, De uteri dissectione, which provides additional anatomical evidence for his acceptance of the 'female testes' and female semen, also appeared in 1490 and with other generation texts in 1535 and 1536. De gynaecǐs, a short work on womens' diseases, appeared in the 1490 Opera and in Champier's Speculum Galeni (1512, 1517). More generally, the important work De usu partium, only part of which was printed in the 1490 text, appeared separately in 1528 (Paris) and in a French translation in 1566.

Less obviously works on generation, but highly important for an understanding of Galen's ideas on the subject, are those which concern the workings of the soul within the body and the 'actions' of living things. Of major importance in this regard are De naturalibus facultatibus, first published separately in the Linacre translation

L. Deer Richardson, *Academic Theories of Generation in the Renaissance*, History, Philosophy and Theory of the Life Sciences 22,
https://doi.org/10.1007/978-3-319-69336-1_7

of 1523, which went through numerous sixteenth century editions, with translations into French and English; and De Hippocratis et Platonis decretis (or placitis) published in rival translations of 1534 and 1535 (and in the collected works from 1541). Less central, but still of interest, is Quod animi mores corporis temperamenta sequantur, which appeared in the collected works from 1490, in a Paris edition (with other works) in 1528 and in French translation in 1557. The pseudo-Galenic An animal sit quod in utero geritur also had considerable popularity in the Renaissance: it attracted five translators from its first appearance in 1540 (when two versions appeared) and was included in the collected works from the following year.[1]

It is probably easiest to approach Galen's work on generation as a reaction to Aristotle: much of Galen's writing is polemic in tone, and a considerable part of both De semine and De foetuum formatione is devoted to the refutation of the doctrines of other philosophers, Aristotle in particular. Galen opposes Aristotle on four points especially: the existence of female semen; the material constitution of the embryo; the role of the testes; and the function of the liver, rather than the heart, as the earliest centre of foetal development. These points of difference are central to academic debate in the Renaissance.

Galen begins Book I, Chapter i of De semine with a question: "What is the use and faculty of semen?" On this point, he says, there are two basic positions: one, represented by Hippocrates, in which the semens of both parents possess both material and efficient principles; the other, represented by Aristotle, in which efficient and material principles are represented by the male and female parent respectively. It is clear from the first that Galen sides with the Hippocratic position[2] and he goes on to make this clearer by taking the Hippocratic definition of conception (as retention of both semens by the uterus) as primary evidence that both semens provide material to constitute the foetus, and that both male and female contribute fertile semen. On both points, of course, he differs from Aristotle. Hippocrates' position, says Galen, is substantiated both by dissections and by the experience of pregnant women (De semine, I, ii, iv). This in turn leads to two further arguments: one in favour of the existence of female semen, the other suggesting that the material of which the foetus is formed is seed plus blood, not, as with Aristotle, blood alone. Galen does not develop one argument in sequence, but skips from one to another in ways which make clearer their interdependence. But he will be easier to follow if each of these threads is followed through separately.

Galen, like Aristotle, rejects the theory of panspermatic development as taught by Empedocles. His arguments are similar: (1) the necessity of "some third thing which will dispose and arrange" the scattered parts and the difficulty of imagining how complex organic parts – in particular the generative organs – could be formed by a 'tally' method; (2) the fact that the substance from which the embryo takes its growth is "only blood": "But if we say that each part is formed from altered, transformed blood, we shall have to assume the presence in the semen of an artificer to do this, and since we have such a one present from the beginning, it is ridiculous for us to assume parts that have been torn asunder."[3]

Galen also agrees with Aristotle (and Athenaeus) that the materials of generation, and of growth, are ultimately elaborated from the blood, rather than "drawn from all parts of the animal"[4] (De semine, II, v: Kühn IV 628). Semen is "blood completely (exacte) concocted by the vessels containing it"; blood, in turn, is "semen potentia" (Kühn IV 613–4). But he rejects the idea that blood can of itself do what semen does, or that it can transform itself into semen:

> Yet truly if blood has the power of simultaneously generating and forming such a humor [i.e. seminal] of and from itself, it does not need semen.[5]

If the blood cannot, of and from itself, generate the seminal humor, then a principle, the seat of some faculty, is necessary for this purpose. If the veins and arteries can themselves generate semen (as Aristotle says) why, asks Galen, don't they do it throughout the body? (I, xiii, Kühn 557–8). We have already seen that for Aristotle the testes could not claim to be "a ganglion of many principles" despite the major effects which followed their removal. For Galen, they are just that, responsible for semen-production and for masculinity and femininity in general, as the seat of a generative power or faculty.

His argument that the testes produce semen is backed – like Aristotle's argument that they do not – by the evidence from dissection (I, xii, Kühn 555–6). According to Galen, one can see that the winding passages leading to the testes contain blood, "whitening gradually" as it proceeds, whereas the body of the organ contains "the substance of semen itself". Moreover, you will find "abundant and thick" semen in the testes of an animal which has not been permitted to mate for a period; if dissected immediately after coition, the testes will be found to be emptied of their contents (I, xvi, Kühn 586). Galen suggests that the testes accomplish their function by means of a faculty of alteration: in effect, they assimilate blood to their own likeness.[6]

This evidence extends as well to the 'female testes' discovered by Herophilus and hence to the presumption for female semen. This argument is developed most fully at the beginning of Book II, where Galen quotes both Hippocrates, De natura pueri, and Herophilus, Book iii, De dissectione, as authorities. Galen's own short treatise on the dissection of the uterus[7] follows Herophilus' description of the position and appearance of the 'female testes' or ovaries in general, but he differs in insisting that the semen produced by the ovaries enters, not the bladder as in the male (making it useless for generation) but the uterus itself. The 'female testes' insists Galen, not only exist, but they are functioning organs, contrary to the opinion of Athenaeus, who argued that the female testes, like male breasts, were vestigal structures, formed, as it were, for the sake of symmetry, but not functional.[8]

Galen gives further arguments for the existence of female semen: from dissection experiments, similar to those earlier on male animals (II, i) and from heredity. If Aristotle is right, offspring should resemble either the father or the mother, but not both, as they do:

> for if resemblance is achieved by faculty, that is by semen, the offspring will turn out like the father only, none ever being produced like the mother; if from material, that is menstrua, not one will ever be like the father, but only the mother; but in fact, they clearly resemble both.[9]

To Galen it seems much more reasonable that both parents should produce a similar substance, as he argues syllogistically:

> If the offspring resemble both parents, they do so in consequence of a common cause existing in both; but in fact the offspring do resemble both parents; therefore they do so in consequence of a common cause existing in both. Then again: if offspring resemble their parents in consequence of a common cause, clearly they resemble [them] in consequence of the substance of semen or of menstrua; but the substance of menstrua is not common; therefore they are rendered similar in consequence of the substance of semen.[10]

Galen several times repeats that generation (as for his model, Hippocrates) involves the combination of male and female principles, each with a facultative and a material aspect. The conceptus "moves and disposes itself" (ipsum movet et concinnet): it cannot be distinguished into moved and mover. But he is not at all clear as to the identity of the 'female principle' involved. Is it the 'female semen' for existence of which he has fought so fiercely? Evidently not, since when he comes to speak directly of its function, it is given no chance to determine resemblance, since it apparently forms no part of the foetus proper. Female semen is the material of which the allantoid and other foetal membranes are made; it is the appropriate (proprium ac familiare alimentum) "food" of masculine semen, being nearest it in nature; and it has a part in stimulating female sexual desire. That is all (De semine, I, vii; II, iv).

In speaking of the problem of heredity or 'resemblances', Galen uses "semen" and "menstrua" interchangeably to mean the female contribution. He does not seem to be very clear on this point, although earlier he had made a firm distinction between the two substances (V, v). But it is clear that whichever term is used, the female contribution is not equal to that of the male:

> Indeed, that each and every animal is generated from material and faculty is generally acknowledged; truly both semen and menstrua have both principles, not however possessed of equal powers, but semen, indeed, [is] most powerful in faculty, but materially of moderate amount; the menstrua, however, plentiful indeed in matter, but very weak in faculty.[11]

As with Aristotle, the difference comes down to a difference in the degree of coction of the seminal humor, and this in turn to a difference in temperament – temperature, i.e. heat – between the two sexes. According to Galen, heredity involves three classes of 'resemblance': by sex, species and resemblance of features – each requiring a different type of explanation. Thus the sex of the child is the result of temperament, or temperature, both of the two semens and, importantly, of the uterus. Accordingly males are (usually) generated in the hotter and dryer right chamber of the uterus, females in the colder and moister left. Species, on the other hand, is determined by the nature of the subject-material (Galen, like Aristotle, believes that hybrids tend to follow the species of the mother) and resemblance of feature is due to the formative faculty of the semens. Since both sexes produce semen (or something like it) and since the semen of different 'emissions' can vary in strength, the particular combination in the conceptus will determine whether the child resembles the father or the mother in a given feature – an explanation which verges on pangenesis (II, v).

Semen, as we have seen, is the product of the testes. For Galen, no indwelling bodily faculty or capacity can exist without being suitably housed in an appropriate organ. For the faculty of generation, this organ is the testis (De semine, II, vi). But these organs serve not only – or even primarily – as semen-producers. They are responsible also for the masculine or feminine character of the entire animal body. Castrated animals change their nature because they are deprived of a principle of life – comparable with the heart – and of a source of vital heat. The heart, according to Galen, is the vital principle of the individual; the testes, of the species. They are responsible as well for maintaining the heat and strength of the entire animal body, for its full existence as an adult. Aristotle himself will admit, says Galen, that indwelling heat, or indwelling spirit, or good temperament are the cause of health and strength; and of these the testes are a major source. Galen devotes some space to demolishing Aristotle's description of the testes as 'mere weights'. But his positive statement: that the testes are a "principle", the seat of a faculty, and the source, with the heart, of vital heat in the body is perhaps more notable. In Book I, Chapter xvi, he compares the alteration caused by castration with the "alteration of the whole substance" produced by certain poisons.

If we turn to the process of development itself, as Galen describes it, we shall see that indwelling heat and spirit have a part to play here, as important as their function in the sexually mature adult.

Generation, together with nutrition and growth, is an effect of what Galen calls nature: i.e. the activities which animals have in common with plants, as distinct from the operations of soul, such as sensation and voluntary motion, which animals alone possess.[12] The development of a human embryo obviously involves more complex actions than that of a bean, but in the early stages, says Galen, the principles governing their development are identical. In particular, the basic parts of the animal foetus are formed, not from blood, but from seed – just as they are in plants. The spermatic parts of the animal are a reflection of the similarity of plant and animal development:

> Moreover, the first principle of all things is that of a plant, which produces artery and vein and nerve, bone also, not from blood, but from seed itself.[13]

The foetal membranes, bones, ligaments, membranes and all the 'fibres' of the body are formed from semen. But "all which are of the genus of flesh, are generated from the blood" (De semine, I, xi, 551). This division into sanguineous and seminal parts extends, as Galen explains it here, to the principles of growth and nutrition as well as generation. Seminal parts are nourished by seed, sanguineous by blood: a function which is proved by the weakness which follows the loss of either:

> From seed all the solid parts of animals are generated, from seed also they are nourished; just as on the other hand both the generation and the nutrition of all the fleshy [parts] derive from blood. For by the substance of which each part was first made, it is always nourished as well, for nourishment is the filling up (making good) of disappearing substance.[14]

It is possible to see this distinction of seed-formed and blood-formed parts as simply a refinement on Aristotle, since the seed itself is ultimately formed from blood, as we have seen. But it can also be seen as a late reflection of the idea we met in Plato, according to which the seed was formed of the 'special substance' of the brain, spinal cord and marrow, and this substance in turn became the organising core of the embryo, fleshed out by blood.

This analogy of plant and animal generation has the advantage of positing a common cause. One of the disadvantages of Aristotle's explanations, says Galen, is that an entirely different explanation is needed for plants (in which the seed clearly remains as part of the developing embryo) and for animals (in which, for Aristotle, the seed forms no part of the fetation). For Galen, the analogy is *functional*: seed = seed; uterus = soil; animal growth and nutrition = plant growth. These aspects can be abstracted from the (human or animal) individual and considered separately as the work of nature. It is also a descriptive analogy: the branching of the roots and branches of the tree and of the veins and arteries of the conceptus resemble one another visually.[15]

In *De semine* and *De foetuum formatione*, Galen contents himself with describing the successive stages of plant and animal development and their similarities. We must turn to other works to discover what he means by a faculty and which natural faculties are responsible for the work of generation. Galen gives a famous, or perhaps infamous, definition of faculty in I, iv of *De naturalibus facultatibus*: "so long as we are ignorant of the true essence of the cause which is operating, we call it a faculty". The example he gives is that of blood-making: a work (*ergon*) e.g. blood, is the result of an action (*energeia*), e.g. of the veins, made possible by a faculty (*dynamis*), the sanguiniferous.[16] A faculty is not a thing, but the capacity or power to perform some action or produce some effect, as evidenced by the effect or performance. A clear explanation on this point by Galen himself comes from Chapter i of his work, *Quod animi mores corporis temperamenta sequantur*:

> Many philosophers … have a poorly-defined notion of faculty; they imagine, it seems to me, that faculty is something which dwells in essences as we dwell in our houses, because they do not know that for each thing which is produced, there exists a formative cause, which is considered in the nature of a relation (Κατα τό πρόζ τι) and that there is a special and appropriate name for this cause, according as it produces this or that effect. The faculty of that which is produced resides in the relation of the cause to its effect; this is why one says that an essence has as many faculties as actions. [For example one says that aloe has a purgative, tonic, cicatrizing, drying, agglutinative, etc. faculty] … not indeed because there exists in aloes something in particular apart from the aloes itself, corresponding to each of its actions; for it is the substance of the aloes itself which produces all this.[17]

The natural actions, and the natural faculties responsible for generation, are of two types. Classified according to the kind of action involved, the basic natural faculties are the capacities for attraction, retention and expulsion. These capacities are specific for each organ: the kidneys attract, retain and expel urine, the stomach, food and so on. The importance of the specific attractive, retentive and expulsive capacities of the uterus to successful gestation are fairly obvious (*Nat.facs*., III, ii) and the uterus accordingly becomes a type of the retentive and expulsive faculty.

In addition to these specific organic capacities, Galen speaks in Book I, v of the specific activities responsible for the work of generation. They are genesis (subdivided into alteration and shaping), nutrition and growth. Both alteration and nutrition require assimilation, and these in turn can be subdivided into presentation and adhesion. Rather than following Galen in his attempt to classify processes by interminably resolving actions into simpler, mechanical sub-actions, let us return to his description of the process of generation (Nat.facs., I, vi). The form of any creature must be completed before it can begin to grow, and this is accomplished in two stages. In the first (alterative) stage, heat, cold and other physical qualities combine to form the homogeneous parts, such as bone, cartilage, nerve, vein and so on. These homogeneous or simple parts appear to correspond in number to the special alterative faculties possessed by the individual organs: particular faculties correspond to particular "simple parts". In the second stage these homogeneous parts are combined into organs by a formative faculty:

> the bringing of these together [synthesis] the combination therewith of the structures which are inserted into them, the outgrowth into the intestine, the shape of the inner cavities, and the like, have all been determined by a faculty which we call the shaping or formative faculty; this faculty we also state to be artistic – nay, the best and highest art – doing everything for some purpose, so that there is nothing ineffective or superfluous, or capable of being better disposed. This, however, I shall demonstrate in my work "On the Use of Parts".[18]

The organic seat of these earliest functions – as Galen describes them from the somewhat different perspective of De semine and De foetuum formatione – is the liver. Galen seems to have changed his mind on several points concerning priority of development in between writing the two works. De semine is evidently the earlier, and there (Bk. I, Ch.viii) he speaks as though the three primordial organs, heart, liver and brain, come into being simultaneously; the arteries, veins and nerves grow out from each organ respectively and the rest of the body is built up around them. In the latter work, he confesses to having changed his mind:

> When I was writing my commentary On the Semen and found it necessary to say something about the order of formation of the separate parts I said that the heart as well as the liver received the foundation, so to speak, of its generation in the first days of pregnancy, and I was led to this conclusion by considering the usefulness of the adult heart, which is most important. But when I found that all other physicians and philosophers too are pleased to think that the fetus is governed as plants are until it is clearly formed, it seemed more likely to me that the heart has no usefulness whatever at the beginning of generation and is certainly formed later than the liver.[19]

This position marks another dissent from that of Aristotle. It may have been based in part on observation: the blood vessels are initially the most prominent feature, in e.g. a foetal chick, and the organ especially associated with blood is the liver, which appears to be made up almost entirely of it. But it produces certain difficulties. At what stage is the embryo "clearly formed"? and what directs its development once the liver has lost this function? Galen is not entirely clear on these points. But he suggests that the second stage of development is directed by the heart, which is formed when the foetus begins to "breathe". And he implies that it must reach this stage very early in its development, since he notes that heart, brain and liver are all visible in a month-old abortus, though the liver and its attendant blood vessels are still the most prominent.[20]

As the nutritive blood is the hallmark of the liver, the innate heat and its attendant spirit is that of the heart. Hotter blood escapes from the liver system to form the arteries and from the centre of these the heart develops.[21] Because of this association, Galen's system is in some ways more logical when, as in De semine, he assumes that all three principles arise at once in the seed, than when, as in De foetuum formatione and De naturalibus facultatibus he suggests that the natural principle and its organic seat the liver arise first. For whenever the formative faculty of the seed is associated with a specific substance, this substance is the heat and spiritus drawn from the arteries – that is, from the heart-system – together with the spiritus in the semen itself:

> The semen, indeed, draws blood and spirit to itself through the vessels belonging to the uterus, into its own hollow space. Moreover, as was said before, a rarer and hotter blood than that in the veins is attracted through the arteries, together with spirit … The arteries indeed lead to another, hotter organ [the heart: the comparison appears to be with the liver and the 'thicker' venous blood] which, because of its excellent heat, never ceases to move, but is always alternately expanded and contracted. And the veins and arteries bringing materials to these organs are like their roots … Now their generation is from the substance of the semen, hollowed out like a tube. The third principle, from which all the nerves are produced, has its generation from the semen alone. For in the mixture with the female semen many bubbles are broken, from which spirit goes deep inside, desiring to preserve itself. For it was not like a vapour, but a self-moving principle of the animal. In the same way, it [spirit] also produced a ventricle or cavity containing a humour full of the spirit of the semen itself. Which, indeed, not easily drawn away from thence, builds its own strong home from the liquid substance of the semen which surrounds it, pushing to the outside that which was thicker and harder than itself, which, heated and dried in the passage of time, was the future bone. In the beginning, a faculty which shapes and forms the body operates; equally with the principle it does not yet appear because of its small size. When indeed they [the principles] first begin to appear, three of them [brain, heart and liver] are most important … With the passing of time the aforesaid three principles are more widely separated and transmit [the principles of] growth [germina; Galen's own word is, τά βλαστήματα] into the whole body of the animal, which they form. For the brain produces the spinal medulla like a kind of trunk, the heart the great artery … the liver the vena cava. And … around the medulla the spine hardens, over the brain the scalp [skull?] is drawn, and around the heart the thorax like the site of its strong and roomy dwelling….[22]

A number of points emerge from this passage. One is the dominance of the three primordial organ-principles, each building outwards to create the parts of the body. In De foetuum formatione the sequence of events is somewhat different[23] but this final stage is substantially the same. Another is the use made of the analogy with plant growth: the blood vessels grow outwards like roots, the medulla grows out of the brain like a tree-trunk. And finally there is the importance of the spirit which is drawn to the conceptus with the finer and hotter blood in the arteries, and the spirit in the semen itself, "which is a self-moving principle of the animal".

In Book II of De semine Galen speaks again of the operations of heat and spirit in hardening, drying out and inflating the cavities of the organs (II, v, 633). In part this recalls the Hippocratic pneuma, and the activities of the physical qualities – hot, cold, moist and dry – of which Galen speaks in the Natural faculties as forming the simple parts. But the spirit appears to do more, intelligently arranging and composing the parts of the organism. In the passage quoted above Galen suggests that it is to be identified with the "facultas opifex ac formans" which directs the operations of generation before the *internal* principle (or principles) of the embryo itself, appears. This faculty, which must in some way be identified with the semen, and surely with the self-moving, spirituous portion rather than the passive, liquid one, is the earliest principle of generation, and responsible for 'carrying over' the features of the parents' bodies. This same point is made at the end of *De foetuum formatione*: "For it seems that the semen itself is the creator [artifex], according to some, indeed, all of it, according to others the spirit contained in it".[24]

If pressed on the nature of the essence lying behind this formative faculty, Galen is willing to say that it clearly reveals intelligence – he makes it quite plain that he considers the opinion of the atomists absurd[25] – but not to speculate on whether this essence, or soul, is mortal or immortal, corruptible or incorruptible, unitary or existing in parts. He does suggest that the soul which forms the embryo may not be the same as that which governs it once formed: the person who builds the house is not necessarily the same as the one who lives in it.[26] And he refers the curious to another work, in which the opinions of Chrysippus and Plato on the soul are discussed. This might be the *De philosophica historia*, but is perhaps more likely *De Hippocratis et Platonis placitis*, which contains an important, though still cautious, statement of the Galenic theory of the soul. In the context of a discussion on the spirits responsible for vision, he admits that, if pressed, he would consider the substance of the soul, either as "a kind of lucid and aethereal body" (an opinion shared, he says, by Aristotle and the Stoics) or alternatively consider the soul itself as incorporeal, and this aethereal body as its vehicle, its medium in its operations in the rest of the body.[27] Earlier, in Book III, i (Kühn, V, 287–8) he quotes, and refutes, Chrysippus' notion that the *spiritus* (or *anima*) is one and continuous throughout the body, responsible for different effects in the parts according to their constitution, but with its principal seat in the heart. Elsewhere (V, 283–4) he also rejects the parallel idea that the substance of the soul is [spirit] nourished by blood, or can be identified with the blood itself. Cleanthes, Chrysippus and Zeno are cited for the former opinion, Diogenes, Empedocles and Critias for the latter.

Galen himself speaks of two spirits, *spiritus animalis* (πνεῦμα ψυχικον) and the *spiritus vitalis* (πνεὺμα ζωτικον). The former, associated with the brain and transmitted by the nerves, seems unlikely to be that involved in generation. A more likely candidate is the vital spirit, flowing out from the heart through the arteries. The 'natural spirit' with its seat in the liver seems to have been an elaboration of the system by Arabic commentators and later writers.[28] There is also, as we have seen, a spirit in the semen itself. But if this is to be identified with any organ-principle other than the heart, it should surely be the testes, rather than the liver: the former, as Galen describes them, are after all a principle of life coequal with the heart, a source of vital heat and a principle of generation. Galen himself never, to my knowledge (and evidently to Temkin's) makes this identification: perhaps because the testes (unlike the liver, heart and brain) have no attendant "pathways" by which their power can be transmitted throughout the body.

The result is a system which is far from logical in its implications. The vital *pneuma*, associated with the heart and the *pneuma* of the semen are *opifex* in generation, but it is the liver which Galen insists initiates and directs this process. Galen's position on the central questions of generation is often unclear, especially if several works are compared. This is not simply because he was sometimes confused (or so it seems) but also because often he approaches the same point from different angles, each time with a different opponent in mind. Temkin makes the point very well:

> Galen has pronounced views to which he clings persistently. But it was in the philosophical and dialectical nature of ancient science that details were often worked out in argumentation against views of others. Galen's emphasis shifted according to the point he wished to make or the view he wished to combat. This does not mean that the niceties of his opinions cannot possibly be reconciled. But it does mean that such reconciliations, if attempted, needs elaborate interpretation.[29]

If we try to distinguish the basic questions with which these ancient writers on generation are concerned, we find that the origin of the semen is of central importance, together with its role in the physiology of the parent individual and the manner in which it can be assumed to act in transmitting life, organisation and specific qualities to the offspring. Other more specialised questions grow out of this: what are the mechanisms of heredity? the role of each parent in determining resemblance? Of what is the body of the foetus constructed? on what pattern does it grow and develop? Mechanisms of heredity and descriptions of the process of development are of less concern to us, since they find their usual expression in the Renaissance in the midwifery manuals and "women's books", together with a mass of similar birth-lore, rather than in the questions of academic medicine in the sixteenth century.

In all these four accounts the central position of the semen is notable, together with the fact that it has a much wider range of functions than post-scientific medicine allows to its modern counterpart. It (or its female equivalent) is a vital constituent of the healthy animal body, whether identified with the material of the brain and spinal cord (Plato) or with essential fluids or humors of the body (Hippocrates, Aristotle, Galen). It is responsible for the existence of sexual characteristics and behaviour in the adult; for Hippocrates it is a part of the fluids, or humors, which are responsible for the constitution of the body and its propensity to health or disease; for Aristotle it is a vital, nutritive portion of that most vital substance, the blood; while for Galen the 'seminal humor' has overtones of both origins. Its importance arises partly through its association with extra-corporeal abilities, those of soul (Plato) or pneuma (Hippocrates) or both (Galen and Aristotle). And these same associations are used to explain its role in generation proper: a process which is conceived, not as a new creation, or even as an assemblage of disjointed parts, so much as the transfer of the capacity for life and organisation from parent to offspring.

For Plato, and also for Diogenes and the Stoics, what is transferred or transmitted is a part of soul – making the explanation of its effects correspondingly simpler to achieve. With Hippocrates, and successively with Aristotle and Galen, the explanation becomes more complicated. What is transmitted is, on the one hand, a vitally important sample of the fluids constituting (or nourishing) the parents' bodies. This seminal humor can be assumed to have certain capacities by virtue of "coming from" the body of the parent, and this is the aspect most emphasised by Hippocrates. But it obtains other capacities through its association with pneuma or breath, and through the further association of this pneuma with a vital force or essence, i.e. soul. What is involved is the transmission of certain specific capacities and powers which for all four authors (except perhaps Hippocrates) are necessarily associated with the possession of soul.

Viewed in this light, Plato, Aristotle and Galen – and Hippocrates with some reservations – can be seen as belonging to a developing tradition. The tradition can be seen most clearly from the vantage point of the physiology of Galen, whose dependence on and involvement with the ideas of his predecessors, and these in particular, is clear even when he disagrees with them. This tradition can further be distinguished from two varieties of ancient "mechanism": that of the atomists such as Democritus and Leucippus, with their "unworkmanlike" notion of generation as the result of a random collection of particles; and that of believers in pangenesis of the Empedoclean type. It can also be distinguished, though less firmly, from the pervasive animism and 'pneumaticism' of writers such as Diogenes and the Stoics.

So that Renaissance scholars who turned to these accounts – as in some sense all medical writers did – as the foundation and support of their own discussions on generation, chose their models from only a part of the spectrum of the range of ancient opinion. They chose, of course, from the part most readily available to them – the doctrines of the Stoics, of Diogenes and of Empedocles and the Greek atomists survive for the most part only in fragments. The success of the objections made by Aristotle and Galen in particular to the doctrines of their predecessors and rivals is not surprising when one considers that the works of the latter were often only available to Renaissance readers in the context of these refutations. How and when these objections came unstuck and mechanical physiology (together, it might be noted, with some extreme animistic doctrines) took on a new lease of life, belongs in general to the seventeenth century and not to the sixteenth. It is, in any case, another story.

Notes

1. Information about editions comes from Kühn and also from Durling. Also useful is the Bibliothèque Nationale, Catalogue des Ouvrages de Claude Galien conservés au Departement des Imprimés, Paris, 1913.
2. Though in fact he appears to be reading the terms of a later debate into Hippocrates' account: it will be remembered that Hippocrates himself does not distinguish the seed into material and facultative aspects. On Galen's attitude to Aristotle's theory in general, see Anthony Preus, "Galen's Criticism of Aristotle's Conception Theory", Journal of the History of Biology, 1977, 10(1):65–85.

3. De semine, II 3 (Kühn, IV, 618). Translated by M.T. May and quoted in Adelmann, Malpighi, II, 745–6.
4. At another point in De semine, however, Galen puts forward an explanation similar to Hippocrates, in which semen is formed from seminal moisture and vital spirit, coming together from all parts of the body and especially from the arteries: "It happens, indeed, that not only the seminal moisture is carried away from all parts of the animal at this time [i.e. coition] but vital spirit as well; for this is drawn out of the arteries together with the seminal moisture."

   ["Neque vero solam seminariam humiditatem ab omnibus animalis partibus auferri continget hoc tempore sed etiam spiritum vitalem; nam et hic ex arteriis una cum seminali humore evacuatur." (De semine, I, xvi, Kühn, IV, 588–9)]. This seminal humor in the parts also serves to nourish the parts framed by the semen: see below, and Note 14.
5. "At vero si sanguis ex se ipso ejusmodi humorem generandi simulque formandi facultatem habet, semine non indiget." De semine, I, v, Kühn, IV, 530.
6. De semine, I, xiv, Kühn IV, 563: "Nihil enim aliud est. permutatio et alteratio secundum qualitatem, quam assimilatio patientis ad ipsum efficiens."
7. Kühn, II, 887–908, De uteri dissectione; translation by Charles Mayo Goss, "On the Anatomy of the Uterus", Anatomical Record, v, 144, No.2, October 1962. More briefly in De semine, II, i (596–9, Kühn).
8. De semine, II, i, Kühn, IV, 599: female testes are "ipsa sola partium proportione in prima formatione facta, non etiam actione asservata" according to Athenaeus. Athenaeus's argument as quoted by Galen is taken up by Averroes and (Aristotelian) opponents of female semen in the Renaissance: see Part II. See also Part III and the illustrations in Appendix II for Renaissance ideas on the position and insertion of the "female testes".
9. Ibid., p.607: "Si namque a facultate, hoc est. a semine, similitudo perficitur, solis maribus foetus similes evadent, matri vero nullus unquam similis reddetur; si vero ex materia, hoc est. menstruo, foeminis solis, patri vero nullus unquam similis erit; atqui utrisque similes esse apparent."
10. Ibid., p.609: "Si utrisque parentibus foetus assimilantur, juxta communem causam in utrisque existentem assimilantur; atqui utrisque parentibus assimilantur foetus; igitur juxta communem causam in utrisque existentem assimilantur. Deinde rursus: Si juxta communem causam foetus parentibus similes fiunt, aut sane juxta seminis aut menstruorum substantiam similes fiunt; sed menstruorum substantia communis non est.; juxta seminis itaque substantiam similes redduntur."
11. Ibid., II, ii, p.614: "Quod quidem ex materia et facultate unumquodque animal gignitur, in communi confitemur; quod vero tum semen, tum menstruum habeat utraque principia, non tamen aequalibus viribus praedita, sed semen quidem effectivum validissimum, materiale vero modicae molis, menstruum autem materiale quidem plurimum, potentiale vero seu effectivum debilissimum."
12. Galen on the Natural Faculties, with an English translation by Arthur John Brock, Loeb Classical Library, 1928 (Kühn, II, 1–214) Book I, i, 1–2. "Nature" is compared by Galen with the lowest "souls" in the system of Aristotle and

Plato in several places, e.g. Kühn, IV, 665, 772–3, V, 521. (De foetuum formatione, Quod animi mores corporis temperamenta sequantur, De Hippocratis et Platonis placitis).

13. De semine, I, x, Kühn, IV, 546–7: "Principium autem plantae omnium primum habet, quod non ex sanguine, sed ex ipso semine arteriam et venam, et nervum, os item, et membranam procreet."
14. Ibid., I, xvi, Kühn, IV, 589: "Ex semine omnibus solidis animalis partibus generationem esse, ex semine etiam nutritionem; quemadmodum rursus carnosis omnibus ex sanguine tum generatio tum nutritione existit. Unaquaque enim pars, ex qua primum facta est. substantia, etiam semper nutritur, quod nutritio expletio sit substantiae defluentis."
15. Galen develops this descriptive analogy at several points: e.g. De semine, I, ix and De foetuum formatione, iii. The plant analogy is also used by Hippocrates (notably in De natura pueri) and by Aristotle himself (e.g. G.A., II, iv) to describe the development of the foetus, though not its origin. It is used also by Renaissance commentators such as Cremonini.
16. De naturalibus facultatibus, I, iv (translation Brock); see also Roger, op.cit., p.74.
17. Kühn, IV, 769–770.
18. Natural Faculties, (Brock), I, vi.
19. De foetuum formatione, iii, Kühn, IV, 663; translated by M.T. May and quoted by Adelmann, op.cit., II, 747.
20. Ibid., pp.670–1; 662–3.
21. Ibid., pp.671–2. This sequence (arteries to heart) is that of De foetuum: see Note 23.
22. De semine, I, viii, Kühn, IV, 539–541. "Semen equidem trahit ad se per vasa ad uterum pertingentia sanguinem et spiritum, utrumque ad suam propriam cavitatem. Attrahitur autem sanguis simul, velut etiam antea dictum est., cum spiritu per arterias tenuior et calidior, quam qui est. in venis sanguis, ... Arteriae vero ad aliud calidius viscus feruntur, quod ob excellentem calorem, velut flamma quaedam, moveri non cessat, sed semper vicissim distenditur et contrahitur. Et venae quidem ac arteriae materias in haec viscera deducentes velut radices quaedam ipsorum existunt ... Generatio autem his est. ex seminis substantia in fistulae modum cavata. Tertium vero principium, a quo omnes producti sunt nervi, generationem ex solo semine habet. In permixtione namque ad semen muliebre multae bullae ruptae sunt, ex quibus intro et ad profundum spiritus seipsum servare cupiens cessit; neque enim velut vapor erat, sed per se mobile principium animalis; quemadmodum et ventriculum sive cavitatem humorem continentem plenum spiritus ipsi semini produxit. Quo vero neque etiam inde prompte evacuaretur, firmam sibi domum facit, ex liquida seminis substantia, quae ipsum ambit, id, quod ipsi crassius et durius erat, ad externum lineamentum detrudendo, quod ipsum utique temporis progressu calefactum et resiccatum os futurum erat. Hoc in principio quidem facultas opifex ac animal formans operatur: verum juxta principium nondum apparet prae parvitate. Quando vero primum apparere incipiunt, maxima sunt tria haec ... Tempore

vero progrediente haec jam dicta tria principia longius disparantur, et in totum, quod eis afformatur, animalis corpus germina transmittunt. Spinalem namque medullam cerebrum velut truncum quendam producit, cor vero maximam arteriam … hepar vero venam cavam. Et sane apparet statim a principio simil cum horum generatione circa medullam spina ipsa eo modo, quem paulo ante dixi, firmata, circum cerebrum vero calva obducta, et thorax circum cor velut ampla quaedam ac firma domus situs."

23. The process described there (Chapter iii) is in three stages:
    1. the semen forms veins and arteries which draw blood from the uterine vessels to nourish the conceptus;
    2. the liver, and in succession, the heart, form as a kind of node or junction of these vessels, the brain being formed even later;
    3. Around these 'principles' the rest of the body is organised and built up.
24. De foetuum formatione, VI, Kühn, IV, 699: "Videtur enim ipsis artifex esse semen, nonnullis sane totum, aliis spiritus in ipso contentus."
25. e.g. Ibid, p.688, where he pours scorn on Epicurus' idea that "out of a kind of irrational and unskillful motion" (ex motu quodam irrationali et inartificioso) the end of the most perfect formation of all the parts, complex as they are, can be achieved.
26. Ibid., p.693.
27. De Hippocratis et Platonis placitis libri ix, Book vii, Chapter 7, p.643.
28. According to Owsei Temkin, "Galen's Pneumatology", Gesnerus, 1950, 8:180–9; cf. 188. See also L.G. Wilson, "Erasistratus, Galen and the pneuma", Bull. Hist.Med., 1959, 33: 293–314.
29. Ibid., (Temkin) p.189.

Part II
Philosophi Adversus Medicos: Renaissance Commentaries on Aristotle's Generation of Animals

Chapter 8
The Context of Debate and the Classic Questions

Abstract In Part II, generation is considered within the context of natural philosophy. The *Generation of Animals* is one of the 'animal works': the group of treatises in which Aristotle applied his philosophy to animate nature. These works are often grouped in a single volume in Renaissance editions of the *Opera.* As well as this logical framework, Renaissance editors sought to place the work within a didactic framework. The assumptions were carefully spelled out by translator/editors such as Gaza. In the Renaissance; certain specific questions are taken up as the focus for debate: questions concerning the place of the soul in generation; and a series of other questions which can be traced back to the "commentary" on *De generatione* by Galen.

These questions are prominent in the paraphrase-commentary on *De generatione* by Averroes, in the Renaissance translation of Jacob Mantinus. And they are taken up in an "Averroist" form by most of the commentators in this section. His Renaissance successors were in a position to use Averroes's authority as he had used those of Galen and Aristotle: not for slavish obedience, but as a guide to their own interpretations.

My attempt to understand Book VII of Jean Fernel's Physiologia, "On the Procreation of Man" has led backwards to the classical authorities on which the work was based. I now turn to a consideration of other, related attempts to explain human generation in the sixteenth century, and finally, most broadly, to an attempt to examine the contributions of natural philosophers and medical writers to the question, and to determine the relationship between the two groups.

In Part III, I approach this relationship as it appeared in discussions of a natural-philosophical question – the physical constitution of the natural world – which was at the same time basic to contemporary theoretical medicine and to medical theories of generation in particular. I have already suggested that it makes sense to approach the realms of natural philosophy and medicine as related and overlapping fields of knowledge – as well as, in some sense, conflicting ones.[1] They overlapped, I would argue, in education, in their method and their criteria of proof[2] and also in the questions they chose to tackle. And the problem of generation was one of the most important of the problems which they shared. Chapters on the role of spiritus and

L. Deer Richardson, *Academic Theories of Generation in the Renaissance*,
History, Philosophy and Theory of the Life Sciences 22,
https://doi.org/10.1007/978-3-319-69336-1_8

heat in generation, on the function of the soul, and on elements and temperaments provide examples of specific questions, basic to Renaissance theories of generation, on which medical and philosophical interest converged.

In Part III, I look at these problems from the point of view of medical writers who included such questions, originally from the realm of natural philosophy, in their medical treatises and compendia. In the section which follows, on Renaissance commentators on Aristotle's De generatione animalium, the position is reversed. Here, the question of generation is considered within the context of natural philosophy, rather than of medicine. And the authorities – Greek and Arabic commentators as well as those of the Renaissance – are men who were philosophi but not necessarily medici, although most of them wrote other works on medical topics.

The Generation of Animals has a special relation to two groups of Aristotelian texts. Most obviously, it is one of the "animal works", the group of treatises in which Aristotle applied his philosophy to animate nature. This distinct group of treatises, as it was recognised in the Renaissance, included Historia, De partibus and De generatione animalium, and, slightly distinct from the others, De anima and the group of shorter works known collectively as the Parva Naturalia. These works are often grouped in a single volume in Renaissance editions of the Opera: for example, they are Volume Six, "Libri omnes, ad animalium cognitionem attinentes" in the 1550 Junta edition.[3] Theodore Gaza's mid-fifteenth century translation of the History, Parts and Generation of Animals appeared in print early in the Renaissance (1476) and held the field almost unchallenged throughout the sixteenth century. It is the translation used in the major Renaissance editions, and printed with the commentaries on De generatione of Philoponus and Averroes. And it is the version quoted, or paraphrased, by the Renaissance commentators discussed below.[4]

The broader context of De generatione animalium and its Renaissance commentaries is to be found in the works central to Aristotelian natural philosophy, that is the Physica, De generatione et corruptione, De caelo and Meteorologica. Commentaries on these four works, plus De anima and the Parva naturalia, were a staple product of Renaissance natural philosophy: as the output of Nifo and Cremonini and the sample of Renaissance Aristotelian commentators provided by Lohr[5] suggests.

Taking the 349 commentators in Lohr's inventory for authors A-C as a sample, somewhat under half (150) wrote commentaries on at least one of the six Aristotelian works which I have described as natural-philosophical: Physica, De anima, De generatione et corruptione, De caelo, Meteorologica, and the Parva naturalia. Of those who produced multiple commentaries, one of the most impressive series is that of the Jesuit community at Coimbra; their commentaries on the six natural philosophical works were published in 1592, 1594, 1597 and 1598 and they also produced commentaries on dialectic and the Nicomachean Ethics.[6]

About half of the natural philosophy commentators (77) were Jesuits or members of other orders, and the proportion remains high in the later sections of the inventory; it seems likely that these commentaries, like those of Coimbra, were produced in connection with the teaching activities of the order.[7]

A second group of commentaries which is more central to our purpose is that produced by men who could call themselves philosophi et medici: that is, commentators who had taught, practised or studied medicine at some point in their careers. There were 29 of these in the A-C sample, and 20 of them wrote at least one commentary on a natural-philosophical work. Of the 29, most had their training and their careers in Italy, all but two or three in Italy or France, with Bologna, Padua, Pisa, Rome, and, in France, Paris, then Lyons, the most important centres. Professional examples such as these of the cross-fertilisation of the two disciplines – which has been studied especially with respect to the "Averroist tradition" at Padua[8] – seem to be confined almost entirely to French and Italian scholars. One might suggest – a suggestion which the published editions seem to confirm – that philosophical medicine was more highly developed there. In the A-C sample, the most prolific of these medical-philosophical commentators were Buccaferrea, with six commentaries, and Champier, with four. Overall, Physica received much the greatest number of commentaries, printed and manuscript (117). Next in order were De anima (71) De generatione et corruptione (62), De caelo (45), Meteorologica (42) and finally, well in last place, the Parva naturalia (14, with four of these on one book only). Later sections of the inventory broadly repeat this pattern, with Physica more than three times as popular, and De anima twice as popular, as the three other major works.[9]

These numbers contrast sharply with the single manuscript[10] commentary on the Generation of Animals which appears on Lohr's list to date, that of Cesare Cremonini. The contrast is indicative, I think, even though the census is incomplete. Cremonini himself was one of the most prolific commentators in the natural-philosophical tradition[11] and will be considered in more detail later in this section. Further, the developed tradition of natural-philosophical commentary which Lohr's census reveals, their numbers and their evident importance in teaching, suggest that Renaissance commentators and editors of the Aristotelian text De generatione animalium would expect to place their work within a certain context: that of the Aristotelian corpus as a whole, and of the successive commentaries which attempted to make of it a coherent system, particularly the works on method discussed by Gilbert, Jardine and Randall[12], the natural-philosophical texts applying this method to the physical world; and finally the group of "animal works" within which, according to this system or method, De generatione animalium held a specific, determined place.

As well as this logical framework, Renaissance editors of De generatione sought to place the work within a didactic framework. This appears especially in their discussions of the proper order of the "animal works" and the relationships between them: that is, their logical or methodological order, but also the order in which they were assumed to have been written, and the order in which they should be read or taught. In the medical tradition, there were similar debates on the correct order of the books of Galen and Hippocrates: i.e., the order in which they should be taught.[13]

In the case of Aristotle, editorial attention focussed both on the general question of the correct ordering of the "Opera ad animalia attinentia" and on the specific question of the place of Books VII and X of the Historia animalium within this sequence. And it was justified by appeals to Aristotle's own statements on the proper sequence of his works, together with an attempt to place them within a sequence which would deal with each of the four Aristotelian causes in a certain order, following the model of the Physics.

The order in which texts were printed, in the Junta edition of 1550, for example, or in other editions,[14] was not, then, as it might appear, haphazard, but based upon a reading of Aristotle's intentions in writing these treatises on animate nature. And the assumptions upon which the order was based were carefully spelled out by editors. This appears, for instance, in Gaza's Introduction, where he discusses the correct order of the three texts – History, Parts and Generation – which he translated.[15]

> History takes the first place, and as the name itself indicates, contains the exposition of a thing, as it is, or as it may be ... Then the books of parts, and of generation, declare why it should be so: the one especially the final [cause], the other the efficient.[16]

Before we consider the causes why an animal exists, we must have the knowledge that it exists, and the details of how it does so; hence History comes before the treatise on final cause (Parts) and Parts before that on efficient cause (Generation); this is accordingly the order in which the three texts are printed. In placing Generation of Animals last in the series, and associating it with the efficient or agent cause, Gaza is echoin Aristotle's introduction (G.A., I, i, 715a) in which he makes clear (1) that Generation of Animals is the last of the treatises dealing with the causes of animate beings and (2) that the questions which remain to be considered are, of the parts, those which subserve generation and, of the four causes, the motive or efficient cause.

In addition to De generatione, two books in the Historia animalium known to medieval and Renaissance commentators deal with the process of generation in man: they are Book VII and the pseudo-Aristotelian Book X. The place of these two books is given special consideration by Gaza in his introduction to Historia animalium, and again at the end of Book VI.[17] Once more, he emphasises the logic behind the sequence which he has adopted – and which differs from some of the manuscript examplars:

> But although this historical interpretation is contained in nine books, yet I have placed seventh what is included in the ninth [book] of the Greek exemplar, and I do not consider this to be done rashly [or by chance].[18]

Other editors have distorted the true order of the books, that is, their logical order. Since Book VII deals with the generation of man, it logically follows Aristotle's description of the mode of generation of the lower animals in Books V and VI, and should thus be placed as the third of the books on generation i.e. as Book VII, not Book IX. This also repeats the sequence followed by Aristotle in earlier books of the History, which deal in ordesr with the lower animals and with man, and it is consonant with Aristotle's intention as expressed in Book VI (Ch.18, 571b 5–7): having dealt with the pairing, conception and generation of animals of the air and water, "it now remains to treat of the same phenomena in connexion with viviparous land animals and with man".[19]

Gaza translates only nine books of Historia animalium, again leading modern editorial practice; though both the medieval versions and also a number of later

Renaissance editions of the Opera, including the 1550 Junta edition, add a tenth book. The omission was deliberate: Theodorus explains his rejection of "a certain fragment" which appears in both Greek and Latin exemplars by reference, once again, to the system of causes of each of the Aristotelian works: this fragment

> explains certain material and efficient causes of human generation, which are not part of history. For that reason I have not reckoned to include it among the books of the History, but if it is placed anywhere it belongs with the books on generation, in my opinion.[20]

According to Gaza, then, his texts are printed in the only correct order, and each is associated with a specific function within the Aristotelian system of causes: i.e. History is descriptive, Parts deals with the final, and Generation with the efficient and material causes.

Similar points concerning the logic behind the order of the "Animal Books" are further developed by Felicius Accorambonius in his Interpretatio obscuriorum locorum et sententiarum omnium operum Aristotelis.... (1590).[21] The order in which the "Animal Books" are considered in the Interpretatio is identical to their order in the Aldine (1497), Junta (1550) and other major editions[22] and follows Gaza's order for the three major works. As Accorambonius also explains, this is not simply due to inertia or editorial convention, but represents a conscious ordering of the texts according to the cause which they represent, and hence according to the rules of method:

> Treating of the knowledge [scientiam] of animals according to his usual method, Aristotle begins by dealing with the causes because-of-which the bodies of animals are formed.[23]

These causes should be dealt with in a certain order, according to the pattern established in the Physics for the study of the causes of inanimate nature; and also following the rules of method. Aristotle accordingly begins with a description of the nature of animals (i.e. History of Animals) then, turning to the question of the causes of animate bodies,

> he begins with the material cause, because it is a more universal cause ... [and] is more familiar to us, for the parts of animals are better known to us than is the soul, or the efficient cause of animals ... Then he deals with the formal cause in the book On the Soul because the formal cause is more noble than the efficient cause, and prior in the intent of nature [prius intenta a natura]. Next ... he deals with those accidents which are proper to animate bodies ... At length here in the books on the Generation of Animals Aristotle deals with the efficient cause of animate bodies.

Last of all come "those accidents which belong to animate nature by reason of the efficient cause."[24]

So the order of causes, as Accorambonius sets it out, is as follows:

Historia animalium	)	
De partibus animalium	)	MATERIAL CAUSE
De animalium incessu	)	
De anima	)	FORMAL CAUSE
De sensu & sensibilibus	)	
De memoria & reminiscentia	)	ACCIDENTS OF FORMAL CAUSE

De somno & vigilia	)	
De animalium motu	)	
De generatione animalium		EFFICIENT CAUSE
De longitudine & brevitate vitae	)	
De juventute & senio	)	ACCIDENTS OF EFFICIENT CAUSE
De respiratione	)	EFFICIENT CAUSE
De vita & morte	)	

Accorambonius points out that the correct place of De generatione in this sequence is, additionally, given by Aristotle in several places, notably at the end of De motu (which immediately precedes it in this ordering) and at the end of Book IV of De partibus.[25] So authority can be added to method in determining the order of the "Animal works".

As well as a clear idea of the sequence of his texts, Gaza had a clear idea of their significance to medicine and to natural philosophy. What he has to say on this point, again in his Introduction, is of interest to the question of the relationship between the two fields, and it may be useful to consider this before going on to commentaries on De generatione.

Aristotle's animal texts are, in the first place, a rich source of enlightenment for the doctor, who "admits himself that his own [subject] begins where the philosopher leaves off". If Galen was "Prince of Physicians", Aristotle was "First among Anatomists", and the doctor who studies these texts will be forced to admit Aristotle's outstanding authority in the field of anatomy.[26] But it is not only doctors who will find valuable lessons in these texts: "And moreover the Philosophers of our age, who learn the fourfold cause of things from instruction by that first book of natural observation [i.e. the Physics] do so, indeed, by a kind of reasoning in common".[27] The animal books, as we have seen, apply the methods of the Physics to the study of animate nature. Natural philosophers – "those who call themselves Physici" will accordingly find them especially worthwhile; since, like the Physics itself, they move from "the particular and sensible properties of a thing" to the search for universal causes.[28]

So "the books which pertain to the knowledge of animals" have a special role to play within the Aristotelian corpus, if we are to believe Gaza and Accorambonius. They provide examples of the relationship between the "particular and sensible properties" of animate nature, as described in Historia animalium, and the system of universal causes which explain these properties, as described in the other works, especially De partibus, De anima, and De generatione itself. With reference to causes, they provide, taken together, a comprehensive treatment of "the fourfold causes of things" as taught by Renaissance natural philosophers, for the realm of animate nature: an account which parallels, and supplements, the account of inanimate nature given in the Physics. Within this structure, De generatione represents the efficient cause, and corresponds directly to Book VIII of the Physics.[29]

If Renaissance editors and translators saw the "Animal Works", taken together, as providing a comprehensive review of the causes of animals or animates – in

Aristotle's definition, those physical beings which move themselves, or are moved from within – why, in the rest of this section, do I propose to concern myself only with the group of writings which can be considered as "commentaries" on De generatione animalium? Why not include other Aristotelian works which deal with the topic of generation, in particular Historia animalium Book VII and Book X, as well, perhaps, as some of the questions concerning generation in the Problemata – like Book X of Historia animalium a pseudo-Aristotelian work, but like it considered as part of the Aristotelian corpus by medieval and Renaissance commentators?

There are several reasons for my decision. The first is that, as we have seen, Renaissance writers distinguished carefully between the function of the various animal works according to the rules of method: History as descriptive, Generation as explaining the actions of the efficient cause. It makes sense, therefore, to respect this distinction as the possible basis of separate, parallel traditions of commentary based on the two texts. And this, I believe, is what we find.

Indeed, a brief survey of the contents of the sections of De generatione and of Historia which deal with human generation reveals that they are books of very different types. Book VII of Historia animalium consists of 12 short chapters, dealing with such topics as the signs of puberty and conception, the management of pregnancy and parturition, the choice of a midwife and postnatal care of the infant. Book X is also short and considers a similarly practical range of questions, centring on the causes of sterility or fertility, especially in women: the position and structure of the uterus, the consistency of the menstrua, affections of the uterus which may impede conception, uterine moles. It runs directly counter to Aristotelian doctrine in De generatione on the role of the female in generation, since the author of Book X argues that women, too, emit fertile semen.[30]

De generatione, as we can recall from the discussion in Part I of this thesis, is a book which concerns itself not at all with these practical questions of promoting fertility and safe management of childbirth; instead, as we have seen, it concerns itself with questions such as: what are the principles of generation? what is genital seed, and how is it formed? what are the causes of generation in general, and how do male and female parents contribute?

So we see that the distinction made by Renaissance editors between De generatione and Historia animalium was based on a genuine difference between the texts themselves, their subject and method of treatment. Book X of Historia animalium was rejected by Gaza and those who followed him, and even where it was printed in Opera omnia for completeness' sake it was placed at the end of Historia, not with De generatione, where Gaza suggested it might be better fit if indeed it belonged anywhere in the Aristotelian corpus.

There is a further practical reason for not including commentaries on Historia among the works considered below. The Renaissance does not appear to have produced many. Nifo's is the only one by a Renaissance commentator mentioned by Casaubon in his Index, although medieval commentaries by Arabic and Latin writers were printed in the Renaissance.[31] These included the section of the Canon Book III in which Avicenna deals with the practical problems of generation,[32] and the medieval commentaries on this 'golden chapter' by James of Forli, Dino of Garbi and others[33]; the commentaries "De animalibus" by Johannes Kenigale and Simon of Eversham[34] and Albertus Magnus' De animalibus, as well as the pseudo-Albertus Secreta mulierum.[35]

Pre-Renaissance commentaries on De generatione itself are even rarer than commentaries on De animalibus. Casaubon mentions only two, the commentary by Michael of Ephesus which was attributed to John Philoponus in the Renaissance and, indeed, much later[36]; and the Paraphrase of Averroes, printed in several Renaissance editions of Aristotle's Opera.[37]

According to Aristotle, the Generation of Animals is about two things: of parts, those which serve generation; of causes, the efficient cause. As we have seen, editors and annotators such as Gaza and Accorambonius agree. But the questions chosen for commentary in the Renaissance do not follow simply from this statement; rather, certain specific questions are taken up and become the focus for debate. What are these questions, and how do they arise? I think they can be divided into two categories: questions concerning the place of the soul in generation, in which the important authorities, to judge from Renaissance editions and citations, are Philoponus, Averroes, and certain works of Galen; and a series of other questions which, I think, can be traced back to the "commentary" on De generatione by Galen, especially as found in the two works De semine and De foetuum formatione.

The questions concerning the soul, as they appear in Renaissance editions of commentaries on De generatione by Philoponus and Averroes, are found especially in the context of the "element of the stars" passage. Two themes stand out, neither of which I want to pursue for the moment[38]: (1) The interpretation by Philoponus, and, following him, Averroes, of the "element of the stars" as vehicle only in the generation of the intellect; since other faculties can be seen to inhere in a physical body (as, in Averroes' own example, walking in feet) but the intellect has no such bodily vehicle.[39] (2) Averroes' description of the activity of the soul in generation as moving from potency to act by transforming materia, and his distinction of the form of the generated creature (the classic Aristotelian anima) from the virtus animata, which is assumed to be separate and immaterial, as the form of the craftsman is distinct from the form of his subject.[40]

Both these questions belong to a debate which is more nearly related, I think, to the developed Renaissance tradition of commentaries on De anima than to commentaries on De generatione[41]; at any rate, neither crops up significantly in Renaissance generation commentaries themselves. Much more significant in these are other questions, taken up by Galen, and pursued by Averroes, though not by Philoponus.

What are these questions? If we look again at Part I above we can distinguish four or five central questions raised by Galen in his response to the Aristotelian model of generation. What are the male and female contributions to generation, and related to this, what is the material constitution of the embryo? What is the function of the testes? In what order are the parts formed, and which is the principal part or parts? What are the operations of heat and spirit in generation?

These questions are prominent in the paraphrase-commentary on De generatione by Averroes, as it appears in the Renaissance translation of Jacob Mantinus. And I would argue that they are taken up in an "Averroist" form by most of the Renaissance commentators considered in this section. This argument rests on a new sense in which I propose to speak of Averroism. It is a use which has little to do with the interpretation of Averroist texts on the agent intellect and the immortality of individual souls, and less to do with the putative special excellence of "Averroist" Padua as a seedbed of the Scientific Revolution. At the same time, I do not use it, as Renan suggests, simply to denote an Aristotelian. I mean to imply neither orthodoxy nor heterodoxy, since few of the questionable opinions attributed to Averroes appear in the context of his commentaries on De animalibus, or of his medical works. I mean simply to suggest that Averroes was an important – perhaps uniquely important – interpreter of the works of Aristotle for Renaissance natural philosophers, especially in Italy.[42] There is another, special sense in which Averroes may have seemed especially appropriate as an authority. One of the themes of his medical work, Colliget – available in print from 1490 at least – appears to be the question of the relative claims of medicine and natural philosophy in the study of man.[43] As we shall see, this is an important thread in the work of many of Aristotle's Renaissance interpreters, in particular, perhaps, Cremonini.

The importance of Averroism in flavouring the natural-philosophical teaching of the Italian universities, and especially Padua, is a constant theme of recent historical accounts – however variously its importance is interpreted. Indeed, it does not seem too far-fetched to speak of an Averroes boom in the Renaissance. Charles Schmitt suggests that the high point of his popularity came not in the Middle Ages as one might expect but in the century between 1472 and 1575, when an impressive series of editions were produced. During this period, the number of works of Averroes made available more than doubled, with a notable spurt of new translations, many of them from Hebrew, and new editions, including the Junta Opera Omnia of Aristotle and Averroes appearing around 1550.[44]

Of particular interest to us are Averroes' Paraphrases or Epitomes on Parts and Generation of Animals, which were translated from Hebrew by Jacob Mantinus and published separately in 1521; they also appear in the Junta edition of 1550–2. In the Paraphrase on Generation of Animals, Averroes brings some striking arguments to the question, arising directly out of Aristotle, of whether the seed is animate, a living thing separate from the life of the parent, or not. He also addresses himself to several of the questions raised by Galen in his reply to Aristotle and attempts to mediate between the two authorities. The marginal headings provided in the Junta edition sharpen the impression that Averroes is setting out the debating points in terms which were to be followed by Renaissance natural philosophers in their commentaries on De generatione.

Book I, Chapter ii of the Gaza translation in this edition is headed "Principium generationis esse marem, et foeminam". And in his commentary on this chapter Averroes asks whether sexual generation results from a mixture of two seeds, or whether one seed contributes as form, the other as matter: the latter, of course, being the standard Aristotelian position. He goes on to make specific points about the difference between the sexes themselves, and their generative parts: male and female natures differ both in their definition, and ex sensu, i.e. in the visible differences between the sexual organs. These parts differ also in their actions: but though the organs of the two sexes are clearly distinguishable in form and in their powers, still there must be "a certain equality" in their actions, since both contribute to a common end.

In the last chapter of Book I, Averroes confronts the arguments for and against female semen more directly, though still keeping something of a delicate balance. On the one hand, menstrua is imperfectly concocted nutriment-residue, semen perfectly concocted. The male geniture is moreover a nobler, richer ("like butter" says Averroes) portion of the ultimate nutriment intended for the parts; its temperament is airy and fiery in contrast with the cold, moist temperament of the menstrua; and (unlike the menstrua) its major portion is natural spirit. But on the other hand, Averroes notes that semen and menstrua are parallel natures: as proved by their parallel appearance in the two sexes at puberty.[45]

Later in the commentary on Chapter 20, more direct arguments for and against female semen appear; they are marked by a marginal note, "Examination of the opinions of Aristotle and Galen on female seed, whether it has an active potency in generation".[46] Aristotle's arguments are given as follows. First, menstrual blood itself has a generative virtue. Secondly, "it is impossible for two things, having the same species of potency, namely either passive, or active, to contribute to the generation of one thing". If female geniture has an active virtue, it duplicates the action of the male geniture[47]; putting the two assumptions together, we are forced to the conclusion that the female virtue is passive, and is represented by the menstrua.

Galen, "et omnes Physici", concedes the first point, says Averroes; with regard to the second, he is disposed to disagree with Galen's solution that different parts of the body can have different causes (i.e. with the distinction between spermatic and sanguineous parts) and hence that male and female genitures can have differing active roles.[48] Instead, he sets out Galen's arguments for an active female seed, and their refutation. The first argument is from anatomy: the structural identity of the male and female sexual organs (Galen, we remember, argued that they represented the same structure turned inwards or outwards, and also described female 'testes', i.e. the ovaries) implies that they must have a similar action: so if the male organs produce fertile semen, so should the female ones. But, argues Averroes, (1) the relevant parts (e.g. the testes) differ in site, if not in figure; (2) the action of other Galenic homologues, such as vulva (uterus) and scrotum, or vagina and penis, is clearly not the same; (3) female testes may be vestigial, and hence non-functional, like breasts in the male. Secondly, Galen argues that unless useful, i.e. fertile, both the female structure and the substance produced would be 'in vain', which is contrary to nature. In reply, Averroes argues that the use of male geniture, or semen, is generation; but the use of female geniture as described by Galen (i.e. the fluid produced in the vagina, not the menstrua) is sexual pleasure, the use of which, in turn, is the conservation of the species. Thirdly, Averroes denies Galen's assertion that male and female contributions differ not in kind [specie] but in degree or amount [paucitate et multitudine] and that the production of infertile or imperfect eggs by birds represents a kind of 'falling short': "Indeed it is not imperfect, but differs in essence, which is the animated form, by which the chick is a chick". An unfertilized egg can never hatch, and that represents a, literally, essential difference.[49]

Galenic arguments from the resemblance of the child to its female parent are similarly dismissed, and Averroes ends the debate with Aristotle's own analogy, taken from artistic creation: "The male is like a workman, and his seed is like the tool of his art". And of course the female contribution, the menstrua, is like the material he "works".[50] On the question of the female seed, Averroes argues against Galen on largely Aristotelian lines: it is as though Aristotle were given right of reply. There are differences in detail; for instance, Averroes seems to accept that the male testes have an active role to play in generation. The same terms and the same handling of the debate appear again in Renaissance commentaries, medical and philosophical. I would argue that, especially in the centres where Averroism seems to have been an important philosophical influence, Averroes' formulation of these questions was important in establishing the form of the debate among Renaissance academics.

The commentary on Book II, Chapter iii is the setting for Averroes' consideration of the second 'classic question': whether the semen is physically part of the conceptus. Hippocrates, and following him Galen, suggested that it is: Hippocrates' story of the egg-like abortus produced by a dancing girl is cited as evidence. Avicenna admitted the male semen to be part of the form of the foetus, by virtue of its natural heat ("pars formae ... proprium calorem naturalem"). But Aristotle's view is different: the airy part of the semen mixes with the menstrual blood and forms bubbles from which arise the principal parts, but it separates once the work of formation is done. In Aristotelian terms, semen is an instrument: and therefore it acts according to three principles, the first two of which apply to artificial instruments:

1. An instrument is not part of the thing made: whether materially (as in Galen) or formally (as in Avicenna).
2. An instrument separates itself from the product once its work is completed.
3. An instrument touches all parts of the thing being made ("inside as well as out", in the case of semen).

It thus appears that "semen is the instrument (organum) of the formative power (virtus informativa) by reason of its airy part". This airy part or genital spiritus is not the same as the spirits which operate within the body once formed: for example, the vital spirit produced in the heart. For if it were, it would have to remain within the body, and could be formed in other ways than in the heart; both statements which are impossible by definition (the definitions of 'instrument' and of 'vital spirit'). So semen forms the foetus, says Averroes, but it forms no part of the foetus[51]: a position which he labels Aristotelian.

On the other hand, in his commentary on Chapter iv of Book II, ("Quaenam sint generationis causae, & quemadmodum partes quaeque constituantur in animalibus") Averroes follows Galen, incorporating the actions of Galen's four natural faculties into his model of generation. Thus, the male semen and the comparable female substance are released by the "expulsive virtue" of the generative parts; then the uterus attracts the male semen into itself, retaining the "part intended for generation" (presumably the airy, spirituous part, as we have seen) and expelling the useless portion.[52] Averroes goes on to discuss a third question which is prominent in the Renaissance debate: opinions on the priority of members, or the order in which the principal parts are formed. All physical philosophers, he says, believe that there is such a priority, that some parts are formed before others, but they differ in their order. Aristotelians believe that the heart is formed first among the major organs, and that internal parts are formed before external ones. Another group, which we can identify with the Galenists, considers that the three principal parts, heart, brain, and liver, are formed simultaneously. But the heart, says Averroes, is superior to the other principal parts by virtue of its heat, and indeed is their instrument, the principle of their formation; clearly, it must be formed first. And as the heart is the instrument or principle of the other principal parts, so they are the instrument, depending on the heart, by which the rest of the body is formed. As Averroes explains it, in a musical analogy, "the heart is to the brain and liver as a music teacher is to a flute, and the brain and liver are to the other members as the flute is to the pupil".[53] So the heart, as principle or origin of the natural heat, and also of the veins and blood, is superior to the brain and liver, and "works through" them to form the rest of the body.

The heart is superior as a principle of generation because of its heat; and the role of heat, as of spiritus, in generation is clearly important for Averroes, as it was to be for Renaissance writers. In a section on the generation of the similar parts, Averroes explains quite clearly how he understands the operation of the virtus informativa and its attendant heat and spirit:

> the first operation of the formative power, which together with the heat of the airy part of the semen, shapes [the parts], is the separating out of the material suitable for the substance of the organic parts themselves: then the separation of the material appropriate to each of the similar parts, which are the parts of these organic members. For it first distinguishes the material of the heart, then separates from it the similar members of the heart, which are the parts of the heart: and furthermore it first distinguishes the noble matter for the noble parts, and the ignoble for the ignoble ones.[54]

Following Aristotle, the action of heat and cold is sufficient to explain the formation of the similar, but not of the organic parts[55] (Meteorologica IV, xii, 390b); but Averroes seems unusual in assuming that the organic parts are discriminated first. Instead of building up from simple to complex as we might assume, the process of generation is assumed to be one of differentiation of an initially homogeneous substance, by means of an immaterial virtus informativa, "a power related to the soul" by which the semen is potentially animated.[56]

The major questions taken up by Averroes in his commentary, then, are: the respective roles of male and female geniture; whether the semen is materially or only formally part of the conceptus, which includes the question of what the parts are made of: seed, or blood alone[57]; and the order and relative importance of the parts. He also discusses the question of the animation of the seed and the activity of heat and spirit in generation. I would argue that these same questions are taken up, in similar form, by Renaissance natural philosophers, such as Agostino Nifo and Cesare Cremonini, when they come to consider the topic of animal generation.

Before I attempt to substantiate this claim with some examples of natural philosophers for whom Averroes seems to have been an important – perhaps uniquely important – interpreter of Aristotle, I should like to look at the medical side of the question. Averroes was not only "the Commentator" on the works of "the Philosopher". Like a number of his successors, he was himself doctor as well as philosopher, and in his medical works he considered the relative claims of medicine and natural philosophy to pre-eminence in the study of man. This is particularly true of the work known as Colliget, which was published in 1480 and in several subsequent editions and which provided a framework for, and distillation of, the concepts of theoretical medicine. In medicine as in natural philosophy, Averroes had a role as an interpreter, and in some sense conciliator, of the opinions on which Galen and Aristotle held conflicting views. Around 1499–1500 a volume of his shorter works appeared, including one with the title "Of the agreement [concordia] between Aristotle and Galen [on the question of the generation of the blood]."[58] This conciliatory spirit is evident in other medical compilations published in the Renaissance.

Averroes' medical commentaries have so far attracted less attention than his philosophical ones from modern scholars; one presumes that they, too, have been neglected by students of the Renaissance, as part of the "barbaric" Arab tradition which, it is assumed, was abandoned at the end of the Middle Ages. Mantinus, translator of the Epitome of Generation of Animals, argued that not Averroes but his medieval translators were responsible for the "rough and savage barbarisms" which had repelled the learned – a defeat which he hoped he had corrected in a translation, which, if not so very polished and elegant, was not too rough either.[59] In his preface to the 1550–2 edition, Thomas Junta expressed the hope that by publishing Aristotle and Averroes together he might reconcile opinions and end faction. And he describes the changes in fashion, both in philosophy and medicine, according to which first only the Arabs, and now only the Greeks, were thought to have produced works of value. According to Junta, both had an important contribution to make: the Greeks clarified the text and so Aristotle's meaning, but the Arabs went on to explore difficult questions which Aristotle had left unanswered.[60] Junta's edition contains not only Averroes' Aristotle commentaries, but also independent treatises, including the major medical works.

In the preface to his translation of Averroes' Collectanea, published in Lyons in 1537 and dedicated to two doctors, Johannes Galfredus and Hieronymus Mantuus Allobrogus, Joannes Campegius or Champier makes some of the same points about changing fashions in medical authority. He himself cannot agree with the many "juniores medici" who believe that nothing good has happened in medicine since the time of Galen and Hippocrates, "who transmitted to posterity everything which pertains to medicine, so learnedly, fully and accurately, that others could add nothing, which would contribute to the medical art".[61]

He reminds these enthusiasts for the classics that Averroes, "who often and with the greatest care sought to reconcile Aristotle with Galen" was one of the first to recommend Galen as an authority in medicine and natural philosophy. Indeed, according to Champier he was the first Galenist. And Champier reminds his readers that Averroes and other Arab scholars were important in transmitting classical texts to the Latin West.

But Averroes has in fact continued in popularity as a commentator with both philosophers and doctors. In France, lectures on Averroes' commentaries on the Aristotelian corpus are as common and popular as those on Aristotle, and "certain philosophers" flock to hear and praise. Translations of Averroes have been begun, at least, by men of the calibre of Lefèvre d'Etaples and Leoniceno.[62]

As for the doctors, Montuus, one of those to whom the edition is dedicated has confessed (although, says Champier, he is a true Galen addict) that he has gained more from reading Averroes than from any other author since the Greeks. Indeed, Averroes is "the greatest philosopher after Aristotle, and the greatest doctor after Galen" and so uniquely placed to interpret, and to reconcile, these two authorities. Like Mantinus and Junta, Champier hopes that in translating Averroes he has produced a work both elegant and useful.[63]

He goes on to consider some of the controversies which have arisen between doctors and philosophers, some of them, such as the debate on the seat of the soul, of very long standing. Among them, and equally ancient, are a series of controversies on the seed:

> from what it is made, what it is, produced by which [agents], that is, whether only from the male, or also from the female: which indeed by these two [i.e. Aristotle and Galen] and moreover by the most serious, eloquent and learned judges has been turned over for a thousand years, continuing without letup to the present day.[64]

Galen followed the ancient authorities, such as Pythagoras, Democritus and Epicurus, in arguing for female seed; and so abandoned the position taken up by Aristotle. Averroes' position is less clear cut, but he seems "not altogether to follow [Galen], but rather to stick to Aristotle. "Wherefore also all later schools contended on this question, and even now the case is still in the courts."[65]

Champier's Collectanea is a medical compendium or textbook in three sections: on the definition of health, parts, and temperaments; on the maintenance of health (regimen); and on disease. It is in fact a new translation of three parts of Colliget, which has seven books: anatomy or parts; health, including a discussion of temperaments; illness, both general and specific to each part; signs and prognostic; foods and medicines; regimen; and cure of disease. Thus the chapter on the instrumental parts of generation is Chapter 10 of Book I in Collectanea and of Book II in Colliget.[66]

Chapter 10 is concerned with the composite or instrumental parts (distinguished by both Galen and Aristotle from the simple parts such as flesh, nerve or bone) which operate in generation. Averroes clearly intended to give a brief summary of the nature and use of each of the generative organs, in the manner of Galen's Use of the Parts, but he is quickly drawn into two major controversies, one on the function of the testes and the other on the role of female seed. His technique in mediating between the rival authority of Galen and Aristotle is a kind of sidestep. Thus he admits, with Galen and against Aristotle, that the function of the testes is to make the seed, which they do by virtue of their substance, which is white and sponge-like. The testes assimilate the genital substance to themselves, making it white just as the liver assimilates blood to itself, turning it red. Elsewhere, in the Paraphrase on Generation of Animals, Averroes mentions Aristotle's story of the bull which fathered a calf after castration with some scepticism, saying that if the report were verified it would be a powerful argument against the importance of the testes in generation.

But Averroes follows Aristotle against Galen in arguing that the principal organ of generation is the heart, which transmits its powers to the testis, its instrument, via the heat and spirit with which it is endowed.[67]

Averroes' arguments on female seed are similar. Galen is correct to claim that the female has testes (i.e. the ovaries) and so must produce a seed: but Averroes supports Aristotle (and, he says, Galen) in claiming that this seed forms no part of the foetus, either materially or formally. Female testes, like male breasts, are vestigial, and cannot claim the same function as their counterparts. His arguments are drawn, on the one hand, from authority, primarily that of Aristotle; and on the other from an ingenious combination of logic and claims to observation.

The first prong of his argument is the observation that women can conceive without sexual pleasure; which, by analogy with the male, Averroes associates with the emission of seed. To test this claim, he has asked a number of local midwives to question their clients on this point, and their reports substantiate his statement. Averroes' midwives are cited frequently in the Renaissance debate.

Equally common in later arguments is Averroes' lady in the bath story: a woman "from a neighbouring region" whom her associates gave a blameless character, and who claimed to have become pregnant simply from using a public bath after "certain licentious men". She supports the same argument, for pregnancy without emission, in more clear-cut form; though Averroes admits that despite the lady's good reputation, the story "seems incredible".[68]

The second argument, which, like the first, is developed in parallel in his Epitome of De generatione, turns on the fact that in the human, parthenogenesis does not occur. But if the woman produced seed in generation, similar in its nature and powers to male seed, she could generate in and of herself, without need of a partner. Galen's answer to this is that the female seed forms some parts, the male others; but this only takes the argument back a step, says Averroes. For of these parts, one (it is in fact the heart) must be the principle of development on which the others depend; and if the female seed contributes this primordial part, "it will rightly be judged that our male seed contributes nothing to generation". And that, of course, is an unacceptable conclusion. The reply is the same to the argument that "a mixture and mutual tempering" of the two seeds is required for conception, just as the drug oxymel is neither honey nor vinegar alone, but a proper mixture of the two in a certain proportion. But the part which is responsible for this tempering has the generative faculty within it, and if this is the female contribution the argument is as before.[69]

Averroes is here making the Aristotelian assumption that the material and facultative parts are separate in the two sexes; and since, as he goes on to argue, the female contribution, i.e. menstrual blood, is clearly material, the male must be facultative.

What then is the function of female seed (which for Averroes, as for Galen and for writers up to the seventeenth century, was identified with the vaginal secretion)? According to Averroes it is expelled from the uterus, while the male seed is attracted into it by an occult sympathy. Its function, rather circularly, is to promote sexual pleasure and so the continuing of the species, just as saliva in the mouth promotes the desire for food.[70]

Having established that female seed has no role in the constitution of the foetus, Averroes returns to his interrupted discussion of the use of the testes, and then of the other generative parts, uterus and breasts. Here, he follows Galen's account – for example, of the oblique, lengthwise and crosswise fibres of the uterus which are responsible for its faculties of retention, attraction and expulsion.[71]

What part did these arguments, which emphasised certain clear areas of controversy, have on the theoretical discussion of generation in the Renaissance?

It might be argued that Averroes' paraphrase was almost bound to serve as a guide to later commentators, since there were remarkably few other commentaries on Generation of Animals – as distinct from History of Animals – to draw on. In the Index to his 1590 edition of Aristotle, Isaac Casaubon lists several medieval and sixteenth century commentaries on History, but only those of Philoponus and Averroes for Generation.[72]

***

In conclusion, I would like to suggest that the Commentator was important to the Renaissance debate on generation in two ways. First, by isolating and setting up certain "classic questions" as topics for debate, and marshalling the arguments on both sides. Secondly, by providing a model for commentary, and for attempts at conciliation between the positions; and by providing an example of the way in which logic and experientia could be used to supplement authority, and to draw out the implications of certain statements in the classical texts.

His Renaissance successors were in a position to use Averroes's authority as he had used those of Galen and Aristotle: not for slavish obedience, but as a guide to their own interpretations; as Agostino Nifo said, "when he was right". They had the opportunity for independent comparison of the classical authors he cited, helped by new translations and editions. J.B. Champier, for example, claims that he had read all of the known and published works of Aristotle and Galen in order to do his translation – a hefty number even in the 1530's. And, of course, they could use their own skills of reasoning and logic, and even their own observations, as a guide to what was correct: interviewing shepherds, to substantiate Aristotle's opinion on the testes (like Cesare Cremonini); or incorporating the results of their observations, in the field, the sickroom or even the anatomy theatre.

Notes

1. See Introduction and Part III for a development of this argument: also Temkin, Galenism; Roger, Les sciences de la vie, esp. Chapters 1 & 2, pp.72–3, for two works which emphasise the conflict between the authorities of medicine and natural philosophy.
2. On this point see the helpful accounts in Neil Ward Gilbert, Renaissance Concepts of Method, Columbia University Press, New York, 1960; N. Jardine, "Galileo's Road to Truth and the Demonstrative Regressus", Studies in History and Philosophy of Science, (1976), 7:277–318. Also Randall, The School of Padua, pp.15–68.
3. Sextum volumen, Aristotelis.. Libri Omnes. Ad animalium cognitionem attinentes, cum Averrois Cordubensis variis in eosdem commentariis, Venice (Juntas), 1550.
4. For the early thirteenth century version of De animalibus, translated from the Arabic by Michael Scot, which formed the basis of the commentaries by Petrus Hispanus, Petrus Gallego and Albertus Magnus among others and the later (c.1260) translation from the Greek by William of Moerbeke, see S.D. Wingate, The Mediaeval Latin Versions of the Aristotelian Scientific Corpus, with Special Reference to the Biological Works, London, 1931, esp. pp.72–92. For the Moerbeke version, see Aristoteles Latinus XVII, 2. v, De generatione animalium (Translatio Guillelmi), ed. H.J. Drossaart Lulofs, Leiden, 1966, and Aristoteles Latinus: Codices. Supplementa Altera, ed. L. Minio-Paluello, Bruges and Paris, 1961. Further information on the medieval versions and commentaries can be gleaned from Lynn Thorndike and Pearl Kibre, A Catalogue of incipits of mediaeval scientific writings in Latin, Cambridge, Massachusetts, 1937 (revised edition London 1963), and the supplements in Speculum, 1939, 14:93–105; 1942, 17:342–366; 1951, 26:673–695; 1965, 40:116–122 and 1968, 43:78–114. For sixteenth century editions, see F.E. Cranz, A Bibliography of Aristotle editions, 1501–1600, Baden-Baden 1971, cited as Cranz, Bibliography. See also Appendix I, below. Julius Caesar Scaliger produced a new translation of Historia animalium some time before his death in 1558, but this was not published until 1619 although his translation and commentary of Book X appeared in 1584 (Cranz, Bibliography, p.99). Major editions as late as Casaubon's dual language Opera omnia of 1590 continued to print Gaza for all three works, although Casaubon mentions the Scaliger translation and commentary of Book X of Historia animalium and says that the rest are "eagerly awaited by the learned" (Index, unpaginated). In the Fragmenta Praefationis ... in Aristotelis historiam de animalibus, printed with his Epistolae aliquot (Toulouse, 1620), Scaliger is critical of Gaza's translation, considering it elegant but filled with ignorant errors; but at the same time he says he respects him as a pioneer in the field and seems somewhat cautious of claiming to supercede such a standard version, (pp.73–77). Scaliger seems not to have translated De generatione, however, and if he did it was never published: see the introductory

letter in Αριστοτελους περι Ζοον ἱστοριας, Aristotelis historia de animalibus Julio Caesare Scaligero Interprete, cum eiusdem commentariis, Toulouse, 1619.

5. C.H. Lohr, "Renaissance Latin Aristotle Commentaries: Authors A-B", Studies in the Renaissance, 1973, 21: 228–289; "Renaissance Latin Aristotle Commentaries: Authors C", Renaissance Quarterly, 1975, 24 (4):689–741. "Authors D-F", ibid., 1976, 29:714–745; "Authors G-K", ibid., 1977, 30:681–741; remaining sections forthcoming. This is a continuation from 1500 to 1650 of the inventory of "Medieval Latin Aristotle Commentaries", which covers the period to 1500, published in Traditio, 1967, 23: 313–413; 1968, 24:149–245; 1970, 26:135–216; 1971, 27: 251–351; 1972, 28:281–396; 1973, 29:93–197. (Both series hereafter cited as Lohr, with volume number and year). Both are intended as preparation for the section on Aristotle commentaries in a forthcoming volume of the Catalogus translationum et commentarorium, edited by P. O. Kristeller and F.E. Cranz, of which volumes i (1960) and ii (1971) have so far appeared. (Lohr, 1973, 21:232).
6. Ibid., (Authors C) pp.717–9.
7. Ibid. For example Amaral (6) p.244; Alphonsus, Brutscher, Commoletus, Coscanus (5) p.243, pp.281–2; pp.716, 724; Anreiter, Arias, Bombinus, Borrassa, Bubalus, Carpentarius, Cerqueira (4) pp.245, 249, 271, 273, 282, 701, 710. Lohr, 1973, 21:229–232, deals with the didactic purposes of the various commentaries, including their use by the teaching orders to support Catholic orthodoxy.
8. See for example Gilbert, op.cit., especially Chapter 7, "Italian Aristotelians"; Randall and Jardine, op.cit.; on Randall and Gilbert with special reference to medical method, W.P.D. Wightman, "Quid sit methodus? 'Method' in Sixteenth Century Medical Teaching and Discovery", Journal of the History of Medicine, 1964, 19:360–376; and on the philosophical side, Bruno Nardi, Saggi sull' Aristotelismo Padovano dal secolo xiv al xvi (Università degli Studi di Padova, Studi sulla Tradizione Aristotelica nel Veneto, volume I) Firenze, 1958. For further references on the Averroist question, see below, note 42.
9. Lohr, 1973, 21:282; 28:710; et passim. For the D-K section of the census the numbers are: Physica, 109; De anima, 66; De generatione et corruptione, 33; De caelo, 39; Meteorologica, 30; and Parva naturalia, 15. Not all commentaries cover the entire work.
10. Lohr, 1973, 21:739, item 62, "Expositio super caput tertium libri secundi De generatione animalium". Lohr gives three versions of the manuscript, two in Padua and one in Venice. The printed commentary by Nifo is not mentioned in Cranz, Bibliography, who lists only those of Philoponus, Averroes and the Sententiae de animalibus of Cesare Odone, published in Bologna, 1563, (pp.78, 116, 125). Commentaries on H.A. and P.A. are listed also; pp.127, 145. See below, Chaps. 10, 11, 12, for Nifo, Cremonini and (among others) Odone respectively.

11. Ibid., pp.728–739. Cremonini's output as a natural-philosophical commentator can be judged from the commentaries, most in manuscript, which survive: seven on Physica, six on De anima, three on Parva naturalia as a whole, as well as others on specific aspects of all these works.
12. Op.cit., Note 2.
13. Work in this field is being done by Andrew Cunningham, Wellcome Unit, University of Cambridge. On Leoniceno and method, see William F. Edwards, "Niccolò Leoniceno and the Origins of Humanist Discussion of Method" in Philosophy and Humanism (op. cit., Introduction, Note 75).
14. In addition to the Gaza editions (see Note 15) the "Animal works" appear in the same order in Aristotle, Opera: De animalibus ... Venice, 1497 (Greek text), vol. iii; in the Junta edition of 1550 (Note 3) and in Aristotle, Libri omnes quibus historia, partes, incessus, motus generatioque animalium, etiam plantarum naturae brevis descriptio, pertractantur .../Opera/Tomus quartus, Lyons, 1560. See Aristoteles Latinus, (Note 4), for the medieval ordering of Aristotle's natural-philosophical works.
15. Aristotle, De natura animalium, De partibus animalium, De generatione animalium, [Translated by Theodorus Gaza], Venice, 1476, (Later editions in 1492, 1498, 1533).
16. In hoc volumine haec continentur. Aristoteles de historia animalium libri ix. De partibus animalium ... libri iiii. De generatione animalium libri iv. Theodoro Gaza interprete. De communi animalium gressu liber 1. De communi animalium motu liber 1. Petro Alcyonio interprete, Paris, (S. de Colines), 1533. "Historia primum obtinet locum, atque ut nomen ipsum significat, expositionem continet rei, quod est, sive ut sit ... Mox libri de partibus, ac de generatione causam cur ita sit, declarant: alteri finalem praecipue, alteri agentem." Praefatio, AA, iir-v.
17. Ibid., AA iiv; Opera, vol. VI, 1550, (Note 3), 38v: "Sequens hic Liber, quem Theodorus Septimum constituit, rationibus, quas in sua scripsit praefatione, Nonus in antiquis codicibus legebantur. Quod consyderatione etiam dignum est."
18. Ibid., "Sed quanquam libris novem interpretatio haec historiarum consummatur, tamen qui nonus in exemplari graeco continetur, huc ego septimum collocavi, nec id temere factum existimo."
19. Ibid., 59v: "Sequitur, ut pari modo de iis, quae in terrestrium genere animal pariunt, atque etiam de homine differamus." Quoted by Gaza in the Praefatio, AA iiv.
20. "Sunt etiam exemplaria tum graeca, tum verò latina, quae habeant fragmentum quoddam historiis additum: sed id causas quasdam materiales agentesque generationis humanae exponit, non historiam complectitur. Itaque non inter historiae libros hoc ego ponendum duxi: sed si collocandum uspiam est, libris de generatione, meo quidem judicio, conjugendum est." Ibid. Book X appears, however, in the 1550 Junta edition (De historia animalium liber decimus, Ioanne Bernardo Feliciano interprete), 59v-62r. It contains nine chapters, and concerns the factors which promote, or prevent, successful conception, almost

entirely with reference to the female. It is quite clearly not by Aristotle, however, since it argues that both sexes emit fertile semen (Chapters 5, 8), that conception requires simultaneous emission by both parents, and that the semen produced by both men and women varies in 'strength' (thus determining the sex and resemblance of the child). In short, it appears to be a commentary on the Hippocratic (or similar) works on generation which has been misplaced.

21. Felicis Accorambonii ... Interpretatio obscuriorum locorum & sententiarum omnium operum Aristotelis, & praecipuorum dubiorum, quae in singulis eius libris moveri solenit ... declaratio: ... & omnium ferme controversiarum, quae versantur inter Platonicos, Galenum, & Aristotelem examinatio, Romae, 1590.
22. Notes 14 and 15.
23. Op.cit., 644, "De generatione animalium annotationes". "Tractans Aristoteles scientiam animalium juxta eius consuetam methodum, prius incepit agere de causis, ob quas animalium corpora constituuntur."
24. Ibid., "incepit a causa materiali, quia est universalior causa ... [et] est nobis notior, partes enim animalium sunt nobis notiores quam sit anima, aut causa efficiens animalium. Postea tractat de formali causa in lib. de Anima, quia formalis causa est nobilior: & prius intenta a natura, quam sit causa efficiens. Postea ... agit de his accidentibus quae sunt propria animati corporis ... Tandem hic in libris de Generatione animalium agit Aristoteles de causa efficiente corporis animati...."; "his accidentibus quae insunt corpori animato ratione causae efficientis." The order he gives is identical with that in the 1497, 1550 and 1560 Opera cited above.
25. Ibid., (De motu, Ch. 11, 704a-704b; De partibus, IV, 14, 697b, 29–30).
26. Op.cit., (1533) AA r; (1550) iii v: "Medicus autem, quod fatetur se sua inde ordiri, ubi Philosophus desinit, nusquam plenius, quàm hic viderit. Quódque à Galeno Medicorum principe traditur, Aristotelem primum anatomen ... hic agnoscet; & quod idem & primus, & optimus fuerit author, percipiet."
27. Ibid., v. "Iam & philosophi nostri aetatis qui quadrifarium rerum causam ex prima illa librorum naturalis auscultationis institutione, accipiunt quidem communi ratione:"
28. Ibid. "& Physici, quos se appellant, dignius de caetero profitebuntur. rationes enim ideo quaerimus universales, ut demum res particulares & sensiles teneamus."
29. According to Accorambonius, op.cit., 644–5. "... Hoc consonum est cum tractatu de causis naturalibus in libris Phys. Ubi prius egit in primo libro de causa materiali, mox de formali & finali. Tandem in octavo tractat de causa movente."
30. De historia animalium Liber Decimus, ed.cit.; Ch. II and V, 60r-v; Ch. VIII, 61v.
31. See Lohr, Traditio, 1967, ff; Cranz, Bibliography, and Casaubon, "Index" for details. A later commentary on History, Book I by Christophorus Guarinonius was published in Frankfurt in 1601; it considers chiefly the differentiation of animals by their souls, i.e. the species question. The only important Renaissance commentary on De partibus animalium is that by Nicolaus Leonicus Thomaeus

(of Book I only) who also produced translations and commentaries of De animalium gressu and De animalium motu, as well as several of the Parva naturalia. Cranz, Bibliography, pp.145, 173.

32. Canon, Book III Fen 20 and 21.
33. Jacobi forliviensis ... expositio.... Papie 1479, Venice 1498 and 1502. The later editions are bound with additional commentaries of the same period. The 1502 edition has six sections: four expositiones or quaestiones on the aurem capitulum and two with text and commentary on Hippocrates' De natura puerorum. The contributions of James of Forli and Dino of Garbo, both north Italian physicians of the late fifteenth century, are considered further in Part III, Chap. 13 below, together with that of their predecessor, Peter of Abano.

Both are commentaries on Book I of Historia animalium only.

34. Mentioned by Casaubon, "Index". See Lohr, 1971, 27: 252 and 1973, 29:141–6, (Lohr's spellings are Joannes Kenighale and Simon de Favesham). Lohr also mentions a thirteenth century commentary on De generatione animalium by Gerardus de Brolio (1968, 24:160–1; Wingate, op.cit., p.92) as well as the works De animalibus of Petrus Gallego and Petrus Hispanus, 1972, 28:351–2 and 356–361; the possibly spurious commentary of Petrus de Alvernia (28:334–346); the Quaestiones de animalibus of Roger Bacon and T. de Hyondon and the commentary on Historia by Theophilus de Ferraris (29:115–121; 150; 157–8). Of these Casaubon mentions only Alvernia.
35. I have argued above that both De animalibus, which was based on the Michael Scot translation from the Arabic according to Wingate, op.cit., p.82, and the Secreta mulierum et virorum belong in the category of "women's books". De animalibus, indeed, seems to be a close commentary on the themes of Historia animalium, not De generatione. Its sections on generation are epitomised in the heading to Chapter VI, 3, ii, p.62 (Venice, 1519): "De regimine coitus et impregnatione et partus". Following the Renaissance editors discussed above, I should like to distinguish these issues from those which are discussed in De generatione and this thesis, though both De animalibus and De secretis were published, and popular, in the Renaissance. See Thorndike, History, II, pp.739–744 for a listing of the many fifteenth and sixteenth century editions of the latter. In some ways, De secretis comes closer to following the themes considered in this thesis than De animalibus itself. See, for example, Chapter 1, fol.A vi r (1542 edition) for the difference of opinion between Aristotle and medical accounts of the role of the male seed; A vii r for the use of the testes; B ii r for the sources of the seed and menstrua; vo. for the nature of the heat active in generation; and B vii vo. for the question of whether the seed, once separated from the body of the parent, is animated or not. The discussion draws on a range of Aristotelian natural-philosophical texts, in particular De generatione et corruptione and Metaphysica, but is very confused. Two other important medieval authorities whom I have chosen not to consider in detail – preferring to concentrate on the impact of the classical tradition in the Renaissance – are Thomas Aquinas and Aegidius Romanus or Giles of Rome. St. Thomas's ideas on generation were based on Aristotle and expressed in the Summa Theologica,

Part I, Questions 81, 119 and 128; Part III, Questions 32 and 33 especially; they seem to have focused on the problem of ensoulment (Needham, op.cit., p.93 and Note 1). For Giles of Rome, see the detailed recent study by M. Anthony Hewson, Giles of Rome and the Medieval Theory of Conception: a Study of the De formatione corporis humani in utero, London, 1975. Giles's approach is somewhat metaphysical; he is cited as an authority by Renaissance commentators, but the focus of his work is again not on the questions considered here.

36. Aristotelis quinque de Animalium generatione libri ex interpretatione Theodori Gazae cum Philoponi:, Venice, 1526. A Greek edition appeared in the same year.
37. Michael Hayduck points out in his Praefatio to the standard modern edition of this work that the false attribution to Philoponus was corrected only in 1854, by Valentine Rose in his commentary on the order and authority of Aristotle's works. Rose pointed out "quae ut [the G.A. commentary] plane eadem cum scholiorum in 1. de partibus animalium auctore Ephesio … ita a prolixis Io. Philoponi commentariis plane aliena est, et … multo inferior", Commentaria in Aristotelem Graeca (C.A.G.), vol. 14, pt.3, Joannis Philoponi (Michaelis Ephesii) in Libros De Generatione Animalium Commentaria, edited by Michael Hayduck, Berlin, 1903. See also, C.A.G., vol. 22, i, Michaelis Ephesii in Parva Naturalia Commentaria, ed. Paul Wendland, Berlin, 1903, in which the order of his scholia is given by Michael of Ephesus (Praefatio & p.149; C.A.G., 22, 2, Michaelis Ephesii in Libros De Partibus Animalium. De Animalium Motione. De Animalium Incessu Commentaria, ed. Michael Hayduck, Berlin, 1904. Michael's commentaries were also published in the Renaissance, that on the Parva naturalia in Florence in 1548 and on De partibus and other works in Venice by the Aldine Press, 1527. Both of these appear to have been reasonably well known, and both are cited by Renaissance commentators, in particular Nifo, who also follows "Philoponus" quite closely on some points – see below, Chap. 10.
38. I hope to consider them in more depth, especially as they relate to Galen and Averroes, in the section on the soul in generation in Part III.
39. Ed.cit., 221v (1550), "Utrum sperma sit animatum", (Commentary on G.A., II, iii, 736b).
40. Ibid., "Haec enim virtus animata est similis arti, & continetur in genere naturae coelestis; & id quod ipsam generat, est de necessitate quid separatum (sive immateriale) cum videatur agere in aliud absque instrumento corporeo." (Here at least, he still seems to be considering the generation of the intellect).
41. Ed.cit., (1550), fol.206r e.s.: "Aristotelis Stagiritae, Peripateticorum Principis De Generatione Animalium Liber Primus, Theodoro Gaza interprete: Cum Averrois Cordubensis Paraphrasi, Jacob Mantino Hispano Hebraeo, Medico interprete." Volume six includes commentaries or paraphrases on De partibus, De anima and six of the Parva naturalia as well as De generatione. Cranz, Bibliography, notes a translation of Averroes, De spermate in an edition of Aristotle, Omnia opera, Venice, 1560, (pp.75, 123). This I have not seen. Averroes' Epitomes on P.A. and G.A. were first available in Latin translations

from the Hebrew in the Renaissance; Mantinus's seems to have been the only version. On Averroes editions see also Renan, op.cit., pp.377–383; Cranz, "Editions" (op. cit., Introduction, Note 75) and H.A. Wolfson, "The Twice Revealed Averroes", Speculum, 1961, 36:373–392 (especially 382–385) and "Plan for the Publication of a Corpus Commentariorum Averrois in Aristotolem", ibid., 1963, 38:88–104 (especially 89–94). Mantinus's translation of the Epitome of P.A. and G.A. from the Hebrew was first published separately in 1521, with a dedication to Pope Leo X: Cranz, "Editions", p.123.

42. Cranz, "Editions". See also the works by Schmitt and Kristeller cited in the Introduction, Note 75, and Schmitt, "Towards a Reassessment of Renaissance Aristotelianism", History of Science, 1973, 11:159–193. Other relevent recent studies are those by Nardi, op.cit., Note 8 above; A. Poppi, Introduzione all'aristotelismo padavano, Padua, 1970; and for the earlier period, N.G. Siraisi, Arts and Sciences at Padua: The Studium of Padua before 1350, Toronto, 1973. And see the "Randall thesis" (op.cit., Note 51, Introduction) and the response to it by the authors cited in Notes 2 and 8 above.
43. I am grateful to Katherine Park Dyer for bringing this point to my attention. For her illuminating study of another area on which medical and natural-philosophical interest converged, see Katherine S. Park, "The Imagination in Renaissance Psychology", unpublished M.Phil. thesis, London, 1974.
44. Schmitt, typescript, "Paduan Averroism", op.cit. He makes a similar point about the continued popularity of both Aristotle and his commentator in "Renaissance Aristotelianism"; see especially the table of sixteenth century Aristotle editions in Note 14 and pp.170–1 and Notes 75 and 76.
45. Ed.cit., 207r, 215r-v.
46. Ibid., 215v.
47. Ibid.
48. Ibid., 215v-216r.
49. Ibid., 216v, "Non enim est imperfectus, sed differentia essentiali, quae est forma animata, qua pullus est pullus."
50. Ibid., 216v-217r, "... mas penes ipsum est veluti faber, & semen eius ut instrumentum artis...." And to continue the simile, "virtus animata est similis arti..." (Note 34). The artistic image appears again in his flute-lesson analogy (below and Note 53) and in his development of the notion of the semen as instrument.
51. Ibid., 221v-222r. I am not quite clear whether virtus animata = virtus informativa.
52. Ibid., 226r.
53. Ibid., 226v.
54. Ibid., 227r, "initium operationis virtutis informativae, quae informat cum calore seminali aereo, est discretio materiae convenientis ipsi substantiae membrorum organicorum; mox discretio materiae aptae unicuique membrorum similarium, quae sunt partes huius membri organici. ipsa enim prius distinguit materiam cordis: postea vero secernit ab ea membra cordis similaria, quae sunt partes cordis: denique primo distinguit materiam nobilem pro membro nobili, et ignobilem pro ignobili."

55. Ibid., 227v.
56. Ibid., 221v, "non possumus ullo pacto dicere animam inesse semini: neque etiam possumus dicere ipsum nullum habere animam: sed bene possumus dicere ipsum habere potentiam animatam. Intelligo autem per potentiam animatam illam virtutem, qua semen efficit figuram, & procreationem … Hanc equidem virtutem appellat Galenus informativam, & non est anima nutritiva, … nec est etiam intellectus separatus, … sed est virtus ad anima relata:" To be considered further in the section on the soul in generation.
57. In addition to the arguments given above, Averroes takes this up at the end of his commentary on II, iv (227v) where he says that differences in heat and cold of the organs formed (i.e. differences in temperament, which he also uses to distinguish the forms of individual and species in the case of the semen: (213r)) and not the supposed origin of the parts from the seed or blood explain the differences in 'replaceability' of some organs from others.
58. Cranz, "Editions", p.121.
59. Quoted in ibid, p.123.
60. Ed.cit. (1550), Vol.I, fol. 2r-v. The point is made by Cranz, ibid., pp.125–6 and quoted also by Renan, op.cit., p.387. Junta goes on to suggest that this "generation gap" is partly responsible for the quarrels which exist between philosophers and doctors, and, in effect, suggests the use of Averroes, and Aristotle, as a guide to Galen's medical philosophy and as a cure for this malady: "Et, ut illi doctissimis Galeni operibus adjuti, corpora vel aegrota morbis levant, vel sana conservant, ita isti Averrois simul & Aristotelis eruditione locupletati, ingenia vel nutrire possint vel expolire…." (ibid., 3r-4r).
61. Collectaneorum de Re Medica Averrhoi philosophi, … Sectiones tres … A Joannes Bruyerino Campegio … nunc primum Latinate donate, Lyons, 1537. Praefatio, A, 2r: "… qui universa, quae ad medicinam pertinent, tam doctè, tam copiose, tamque accuratè posteritatè tradiderunt, ut à caeteris nihil addi possit, quod rem medicam adjuvet. "Campegius is better known as Jean-Baptiste Bruyerin Champier, nephew of Symphorien, in whose house the Preface, dated January, 1537, was evidently written. I am grateful to Dr. R.G. Lewis for confirming this for me, although the details also appear in Renan, op.cit., p.382.
62. Ibid., [A, 4r]. Wolfson (op.cit., 1961, p.383) reminds us of Leoniceno's statement that Averroes was the most admirable commentator of Aristotle, excluding the Greeks: a statement strikingly similar to that which Champier attributes to his dedicatee, Montuus.
63. Ibid., [A, 4r to B, 1r]. An 'unlettered' translation is a bar to understanding as well as to acceptance of Averroes as an authority, according to Champier: "Cum igitur medicinae candidati à lectione horum collectaneorum, ob tam incultum, tamque horridum sermonem abhorrerent, unde Averrhoi ingenium eruditionemque intelligere minime poterant, officium meum esse putavi, ut si quid virium in nobis esset, quibus efficere, ac praestare haec meliora valerem, hunc materiae campum ingrederer, & pro dignitate excultiorem faceremus: in quo gloriari non verebimur, nos primos esse, qui huiuscemodi barbarorum interpre-

tationem susceperimus: quae (ut speramus) ad eloquendum aptior, & ad medicandum commodior utiliorque judicabitur." (Praefatio, B, 1r).

64. Ibid., A, 4v-B, 1r: "Afferrem hoc loco mille alias controversias, inter philosophos simul, ac medicos coortas, ut est de somno, ac vigilia, de sensu: denique de semine, ex quibus fiat, quid nam sit, à quonam depromatur, itemque num ab uno tantum mare, an etiam à foemina: quae sane lis duobus, & eo amplius, annorum millibus, sub doctissimis, eloquentissimis, gravissimisque censoribus versata, hactenus sine fine perseverat."
65. Ibid.: "Quamobrem omnis quoque posteriorum schola super hac re decertavit, & adhuc sub judice lis est."
66. Collectanea is indeed, as the Junta edition makes clear, "sectiones tres, tribus colliget libris, Secundo, scilicet Sexto, et Septimo respondentes" and Champier is there praised for his learning and for the polished elegance of his translation. Two other translations are mentioned, the "antiqua" or vulgate translation still used in the Junta edition itself, and a new translation by Mantinus of Book V, Chapters 57–59 printed in parallel with it. Ed.cit., Vol. X, Averrois Cordubensis Colliget Libri VII...., fol. 13r. See also Renan, op. cit., p.382. Cranz, Bibliography, p.183, attributes the Book V translation to Champier. The earliest Opera containing Colliget which he lists is the 1550–2 edition; earlier editions of 1472–5,1483,1495–6 and 1497 were certainly incomplete and for the most part contained only translations of (some of) the commentaries on Aristotle. Colliget was, however, published earlier in the Renaissance: apart from the partial Champier translation, the Wellcome Library has it in medical compendia of 1490 and later, and published separately with dubia et solutiones by M.A. Zimara, Venice, 1542; I have used both this and the Junta versions.
67. Collectanea, Chapter 10, "De partibus dissimilaribus quae sunt genitivae potentiae instrumenta", fol. [E, 4v] – F, 1r.
68. Ibid., F, 1r-v. "... quod etsi incredibile videretur, fecit nihilo minus vitae illius anteactae integra, atque inculpata castimonia, ut fidem verbis suis adhiberem."
69. Ibid., F, 1v-F, 2r. In the Paraphrase a similar argument appears, particularly in the commentary on G.A. I, Ch.20; see above and ed.cit., 215r–217r.
70. Ibid., F, 2r.
71. Ibid., F, 2v-F, 3r.
72. See above, Note 4.

Chapter 9
Renaissance Commentators on "De Generatione": Felix Accorambonius (fl. 1540–90)

Abstract Similar points concerning the logic behind the order of the "Animal Books" are developed by Felix Accorambonius in his *Interpretatio obscuriorum locorum et sententiarum omnium operum Aristotelis…* (1590). The section "De Generatione Animalium Annotationes" of his *Interpretatio* is concerned with a discussion of the major controversies and questions arising from the text and from commentaries, in particular those of Galen in the work *De semine*. Accorambonius concentrates on four linked problems which are important to the philosophical, as distinct from the medical, debate on *De generatione*: the problems of spontaneous generation and of species; the mind and its relation to the semen and to the *virtus formativa*; and the relation of elemental, animal and celestial heat. Accorambonius also takes up the difficult question of the origin of the mind. His *Interpretation* is an attempt, in the manner of Averroes, to discuss controversial questions and reconcile Peripatetic doctrine with Catholic faith; Aristotle with Plato; and even, often, Aristotle with Galen. He seems particularly interested in examining the controversies which divide Aristotle and Galen.

Accorambonius' contribution to the Renaissance debate on the purpose of Aristotle's animal books and their correct order has already been considered in the first chapter of this part. The remainder of the section "De Generatione Animalium Annotationes" of his Interpretatio is concerned with a discussion of the major controversies and questions arising from the text and from commentaries, in particular those of Galen in the work De semine. In discussing Book I of G.A., for example, he takes up the question of the role of the testes; the arguments against female semen; whether the male parent contributes materia; and the broader philosophical question of the role of the generans.[1] His attitude to Galen is by no means simply hostile. He accepts the Hippocratic and Galenic notion of the attractive uterus. In discussing the role of the male testes, he argues that there is no real quarrel between the two authorities. For Aristotle, the testes are not merely weights, but receptacles, places of seminal concoction, and (from the convolutions of their passages) organs which have the effect of slowing the process of coition. His objections to Galenic doctrine on the existence of female seed and the materiality of the male contribution are based on a combination of theoretical arguments and practical ones. Women cannot produce

L. Deer Richardson, *Academic Theories of Generation in the Renaissance*,
History, Philosophy and Theory of the Life Sciences 22,
https://doi.org/10.1007/978-3-319-69336-1_9

seed because they cannot be assumed to have two "excrements", seed and menstrua, fulfilling the same function; furthermore, in reply to Galen's analogy of male and female pleasure, Accorambonius argues that since women do not feel pleasure at the same place (i.e. the clitoris) at which Galenic 'semen' is produced (i.e. within the vagina), this humor is not semen.[2] Similarly, he argues against the existence of seminal parts both on the grounds that they are of a different nature (earthy and dry) from that of semen itself (airy and watery, as described by Aristotle in Meteorologica IV 10 389a 19), and because in an aborted foetus the 'seminal parts', e.g. bones, are observed to be still red and blood-like. In general, he takes the line that while Galen claims to be a great philosopher, he just has not understood Aristotle.[3]

In the remainder of his discussion, and especially in his commentary on De generatione II iii 736b 30ff, III xi 762a 19–22 and IV x 777 b 17–31, Accorambonius concentrates on four linked problems which are important to the philosophical, as distinct from the medical, debate on De generatione: the problems of spontaneous generation and of species; the mind and its relation to the semen and to the virtus formativa; and the relation of elemental, animal and celestial heat.

He begins his treatment of Book II with a development of the mechanical analogy which Aristotle uses to explain the way in which the external mover or efficient cause-i.e. the male parent-transfers the motion of generation to the semen and thence to the menstrua, to form the heart and other parts in a certain, unalterable order.

> Which happens, because just as in trains of wheels or other instruments, when first the showman or other external mover allows the first wheel to move, it moves others, and so finally the whole statue moves. This is most evident in clocks: for we see that when the first wheel is moved, it moves all the other wheels, and so also moves the hand, which shows the hours, which from outside seems to move of itself, when, however, it is moved by the wheels within, which are moved by the first wheel, which is itself moved by some external mover, either a hanging weight or some other weight which fulfills the same function, because the resting parts in it are so constructed as to be able to move easily, when the first is moved by some external force.[4]

Apart from a rather charming description of the working of a weight-driven clock, Accorambonius is making the point that one must, as a philosopher, look beyond the obvious facts of observed experience: the clock hand, like the developing foetus, appears to 'move' and develop of itself. It is only when one looks 'inside' the question that one concludes that it is in fact moved by the motion of the male parent, transferred in the semen.

As well as transferring motion from parent to child, the act of generation must in some sense transfer its form or anima. How does this transfer take place? The question is most difficult in the case of creatures which are generated spontaneously, since, as Accorambonius explains at the end of his discussion of Book I, we cannot say that they are generated in potentia from another nature which has the form of their species in actuality. Must we then introduce Platonic Ideas as models for such creatures? Aristotle, says Accorambonius, deals with this problem in the first book of De generatione et corruptione [I v 320b 17–25] where he distinguishes three ways in which one thing can be said unqualifiedly to come-to-be out of another. The generans can be (i) of the same species (fire generates fire, man generates man); (ii) of the same genus; (iii) of the same act or actuality. Spontaneous generation is an example of the generans idem actu, where the generating power is that of the stars (calor stellarum).[5]

Accorambonius pursues this problem, and the relation between the Aristotelian anima and the Platonic Idea, in his discussion of the passage in Book III [xi 762a 19–22 … "so that in a way all things are full of Soul"] in which Aristotle explains more fully the mechanism of spontaneous generation. Spiritus, the instrument of the soul in the sexual generation of animals, is air, and in this air is animal heat. But this air or spiritus is also present in earth, water and all the substances in the natural world from which animals generate spontaneously. So within elemental heat – the heat of fire – there is animal heat, and this in turn takes its powers from the heat of the stars. The stars, by means of their light, infuse vital heat into elemental fire, and through fire to the other, inferior, elements. And hence it is no wonder if animals generate spontaneously from decayed matter in earth or in water; since they, like animals generated sexually, are generated "by the heat of the stars."[6]

Accorambonius' attempt to reconcile the doctrines of Plato and Aristotle comes through most clearly in his suggestion that the Platonic Idea may fill the same function as the Aristotelian form and his linking of the Platonic anima mundi with the Aristotelian phrase that "everything is full of soul". Similarly, he compares the Platonic planetary Intelligences with the celestial bodies whose motions are, for Aristotle, cause of all generations, corruptions, alterations.[7]

The action of the heavens in generation is further considered when Accorambonius looks at the passage in Book IV [x777b] in which Aristotle says that it is heating and cooling which up to a certain proportion bring about the generation of substances, and that "the limits of these processes" are controlled by the movements of the heavenly bodies, in particular the sun and moon. What Aristotle seems to be saying here is that there is a relationship between astronomical periods and the lifespan of man and other animals; and he seems, as well, to be contrasting elemental actions of heat and cold with the supra-elemental actions of the stars. For Accorambonius, however, the meaning of this passage is that the heavens act on the natural world as universal causes, through the operation of their light and motion which themselves produce heat and cold and the natural actions of heat and cold:

> light and motion operate in these inferior bodies so as to produce heat, and cold, and they are the universal causes of the generation of natural bodies, and the decay of inferior things, especially animals and plants.[8]

If the universal causes of generation are the sun, moon and stars, the particular causes of plant and animal generation are male and female[9]: the female menstrua and the innate heat "like the heat of the stars" which is in the male semen and which the anima sensitiva uses to produce a formed animal from the menstrua. These particular causes too are important, and the relationship between the two levels of action is taken up most fully by Accorambonius in his commentary on the "element of the stars" passage, and his consideration of the question "is the seed animated?"

Here, Accorambonius also takes up the difficult question of the origin of the mind, and its relation to the means by which the material parts of the body are formed. As we have seen, both Philoponus and Averroes identified the virtus or power which is in the seed and which "corresponds by analogy to the element of the stars" with the formation of the mind, since they assumed that the formation of the body was the responsibility of terrestrial elements. It is this 'Arabist' and theologically worrying position which Accorambonius opposes, though his own is by no means clear of its traces. Aristotle believes that "the intellectual soul is immortal, and that it has its own action, not in any way linked to the body", according to Accorambonius.[10] At the same time, he believes that the animae or virtutes materiales: the vegetative, sensitive and motor powers of the anima, cannot exist before the body, or enter from outside, but must be present in potentia in the matter of the foetus.[11] Mens, the intellective part of the soul, is thus not the virtus formativa (as Scotus claims), for the mind has no action communicating with the action of the body. The formative power has all its action in the body, which it forms. And it is this virtus formativa of which Aristotle speaks when he posits a certain heat which is in the semen, renders it fertile, and "corresponds by analogy to the element of the stars."

> Wherefore in Aristotle's opinion the formative power is not mind, but is the spirit and heat innate in the seed, or the heat of the stars in those which arise not from seed but from putrid matter ... this power is not outside the seed, but in the seed itself.[12]

There are, then, in effect, two levels or grades of soul involved in the process of generation. On the one hand, the intellect: immaterial, incorporeal, of the same nature as the Angels and planetary intelligences; divine and having no connection with the actions of the body; not potentially present in the embryo but entering the foetus from outside, "together with the seed". The sensitive and vegetative souls, on the other hand, are linked to matter. This means, as Accorambonius explains it, that in the body of the geniture (that is, the mixture of male semen and female menstrua in the uterus) there is a double soul-principle (principium duplex animae): (i) the mind or intellect, which is separable from the body and (ii) an inseparable principle, which is the spirit and innate heat in the semen. This second principle of generation is a body, and corporeal, because it is in and inseparable from the semen and evaporates together with it once it has set the sequence of developmental motions going. The mind, on the other hand, since it is separate from the body of the semen, remains when the latter evaporates.

The exact nature of the relationship between the mind and the male semen which acts as its vehicle is not entirely clear. On the one hand, the mind is separate from bodily organs and functions and enters deforis, from outside; on the other hand, it enters with the semen, at the moment of conception ("simul cum semine") and, according to Accorambonius, it is present or contained within it by virtue of the heat and spirit which the seed contains:

> Truly, the reason why the mind is in this way present and contained within the seed, is that in the seed, and especially human seed, there is a power and heat corresponding by analogy to the element of the stars, that is, of the nature of the celestial bodies, wherefore just as the

intelligences are present in the celestial bodies, which they move and foster, so the divine and incorporeal mind can be present in the seed.[13]

This analogy, which takes the mind as equivalent to a planetary intelligence and the 'power and heat' within the seed as corresponding to its heavenly body, maintains the distinction between mind and virtus formativa which Accorambonius has already argued. But when he goes on to argue that the mind is "all in all, and all in each part" because semen is concocted from the "residue of the total nutriment" he is on slipperier ground.[14]

The argument which distinguishes the virtus formativa and the mens, and identifies the former with the innate heat of the semen, is directed, says Accorambonius, against two opinions. First, against the opinion of Avicenna and of Scotus that the power within the seed which forms the animal is to be identified with an image or Idea of the animal (and hence with the form or anima, which would contradict Christian teaching of the divine character of the immortal part or mens). Secondly, and curiously, against the alchemists, who attempt to generate gold by fire; since "neither from material fire, which they use, nor from fiery things ... can any animal, or mineral, or anything else be formed."[15]

Unless Accorambonius is being entirely inconsistent, it seems clear from his attack on the pretensions of the alchemists that he is making a distinction between "material fire", i.e. the ordinary fire which one can see burning in a grate or furnace and elemental fire or heat, "the heat in elements and mixtures." It is the latter which is identified with animal heat and with the innate heat which is in the male semen and which acts, together with the vital spirit, as instruments of the soul in the formation of the foetus.[16] Also involved in the process is the celestial motion and heat which warms (or fosters) the innate heat of the seed, and without which no generation could take place. It is celestial heat, which, in Aristotelian terms, brings potency to act.[17]

Accorambonius' interpretation of the process of generation, then, leaves us with four tiers of "action" working on the unformed matter of the menstrua. At the lowest level is the mechanical transfer of motion from the male parent via the semen, which he compares to the working of a clock. Directing this action are the two soul-principles, the higher representing the intellectual and immortal parts of the soul, the lower its corporeal actions, identified with the heat and spirit in the seed. Both have a role in generation, the lower acting as the instrument of the higher in forming the foetus. Part of this description seems to be based on Galen's association of the formative faculty with the heat and spiritus drawn from the arteries together with the spiritus in the semen itself.[18] And above and beyond the individual act of generation is the action of the heavens, without whose heat and motion, nothing would come to exist.

It is clear both from his title and from what he says in the text that Accorambonius intends to provide something more than a mere paraphrase. His Interpretatio is an attempt, in the manner of Conciliator, to discuss controversial questions and reconcile Peripatetic doctrine with Catholic faith; Aristotle with Plato; and even, often, Aristotle with Galen. At the same time, he rejects the "Arabist" and Scotist association of the mens with the "substance analogous to the element of the stars", identifying it instead with the innate heat and spiritus which Galen describes as working in the geniture.

He seems particularly interested in examining the controversies which divide Aristotle and Galen. Annotationes to the three books of Galen on Temperaments follow his exposition of the Aristotelian "animal books", and he explains the reason for this:

> Because in the books on animals by Aristotle we have dealt with many controversies between Aristotle and Galen, it seems relevant to what follows if we add, as a kind of appendix following these works of Aristotle, annotations on the three books on temperaments by Galen, and in these examine briefly, in what ways he differs from Aristotle and for what reasons, since in this also natural truth will shine forth.[19]

Despite its relatively late date (1590), I have chosen to look at the Interpretatio before the more specialised commentaries of Nifo, Cremonini and other authors. This is because it is a general, encyclopaedic work, which considers De generatione in the broader context of Aristotle's writings in general, and his works on natural philosophy in particular. It thus makes explicit certain themes which are present, but less clearly evident, in earlier Renaissance commentaries.

Subsequent interpretation of the "animal books" of Aristotle left Renaissance scholars with a series of unsolved problems, or points of debate. For many of them, as for Accorambonius, "natural truth" was to be found by considering the points on which Aristotle and Galen disagreed, comparing their statements and trying to conciliate them or mediate between them. Averroes gave an example of one way in which this could be done, and in his commentary on Book I of De generatione Accorambonius follows his lead in the questions he draws out for comment, while in Books II – IV he cites many of the same passages as appear in the Paraphrase.[20]

His discussion of the levels of ensoulment present in the foetus is a useful introduction to the work of Nifo and Cremonini, which is considered in Chaps. 10 and 11, below. And his treatment of the problems of spontaneous generation; of species; of the role of mind, the formative power and heat in generation is helpful in drawing out the questions which were important to natural philosophers. In other respects – for instance his four-tier explanation of the powers active in generation – he resembles some of the medical writers considered below in Part III, particularly Fernel and his followers.

Notes

1. Felix Accorambonius, Interpretatio obscuriorum locorum pp. 646–7; 651–6; 665–6; 657 ff.
2. Ibid. pp. 651, 654, based on G.A. I xx 728a 32–35. Later, however, (p.666) he argues for an association between pleasure and the female emission – while still denying it is semen.

3. Ibid. p. 656 for seminal parts; p. 657: "Non possum certe non summopere mirari Galenum, qui cum maximum se esse philosophum et dialecticum profiteatur, non modo illum lateat ratio, sed ne verba quidem Arist. intelligat."
4. Ibid. p. 660: "Quod evenit, quia cum sint intus compositae rotae, vel alia instrumenta, quae unum aliud movet, cum primum a praestigiatore, vel ab alio extrinseco motore latentur movetur prima rota… illud alia movet, et sic denique tota statua movetur, quod in Horologiis appertissime apparet: videmus enim prima rota mota, moveri cunctas alias rotas, et ita extrinsecus quoque moveri indicem illum, qui horas ostendit, qui exterius ex seipso moveri videtur, cum tamen moveatur ab interioribus rotas, quem a prima rota moventur, mota et ispa ab extrinseco motore, seu perpendiculo, seu alio pondere quod perpendiculi vicem gerat, quia partes quiescentes ita sunt constructae in his rebus, ut valeant facile moveri, cum primus inest motus ab aliquo extrinseco motore." This theme of the order of formation of the parts is taken up by Cremonini in his De origine et principatu membrorum: see below,Chap. 11.
5. Ibid. p. 658.
6. Ibid. p. 677: "Spiritus enim est aer qui reperitur in aqua, et in terra, and in aere est animalis calor, ergo in aere, in aqua et in terra est calor animalis ideo omnia plena sunt animae, quod enim in igne sit calor animalis patet quia ignis est calidus, primus inter elementa, calor enim animalis est calor ille in elementis et mixtis, qui proportione respondet elemento stellarum, quia sydera mediante lumine infundunt calorem vitalem in ignem, and aerem… postea in elementa inferiora." For another passage in which spiritus is identified with air, see the commentary on G.A. II ii (p. 661): ".. Ex quo clare docet, quod superius per spiritum intellexit aerem, and flatum, seu flatuositatem." Accorambonius here seems to dissent from Aristotle's own position, that "the heat [which is] in elements and mixtures" is not the same as animal heat, and that the latter only is generative, and analogous with the element of the stars. (See G.A. II iii 736a 28 ff, and my comments in Part I, above). He comments at length on this passage, and his conclusion that elemental heat = animal heat = sidereal heat seems to be influenced by Platonic ideas. In this he resembles Contarini (see below, part III, Chap. 14) who in his De elementis produces a similar model of the stars working through the elements and particularly through fire as the highest element. The problem of the operation of celestial influences in the sublunar world is one which also concerns Cremonini, as we shall see.
7. Ibid. p. 677. For Aristotle on the celestial causes of motion, see especially De generatione et corruptione, II 10 and 11. There are similar passages in De caelo, Physica and Metaphysica.
8. Ibid. 685. "..Lumine et motu tantum operantur in his inferioribus producentes calorem, et frigus et sunt causae universales naturalium generationum, and interituum rerum inferiorum, praesertim animalium, ut plantarum."
9. Ibid. p. 685.
10. Ibid. p. 662: "quod anima intellectiva sit immortalis, et quod habet propriam actionem, non communicantem corpori ullo pacto."

11. Ibid. "..partes animae materiales, quae operantur simul cum corpore, ut est vegetativa, sensitiva et motiva secundum locum, non possunt praeexistere, neque deforis ingredi, sed opus est ut insit potentia in materia embrionis."
12. Ibid. 662–3: "Quapropter ex sententia Arist. virtus formativa non est mens, sed est spiritus, et calor innatus seminis, vel calor stellarum in his, quae non ex semine, sed ex putredine oriuntur... haec virtus non est extra semen, sed in ipso semine."
13. Ibid. 663. "Ratio vero cur eiusmodi mens insit, et adsit semini, est, quia in semine, and praesertim humano est virtus et calor proportione respondens elemento stellarum, hoc est naturae corporis caelestis, quare sicut corpori caelesti assisstit intelligentia, quae movet illud corpus, et fovet, ita semini potest adesse mens divina et incorporea."
14. Ibid. "Unde habemus quod anima intellectiva est tota in toto, et tota in qualibet parte, quia si est in semine, quod est residuum totius nutrimenti, est in toto nutrimento, et est in toto corpore, unde semen excernitur, quia totum est intelligens, et est tota in toto, quia est indivisibilis, et tota in qualibet parte, cum sit in semine residuo totarum partium." This seems to come dangeriously close to identifying mind with its seed-vehicle.
15. Ibid. "neque ex igne materiali, quod utimur, neque ex rebus ignitis.. non possunt neque animal, neque mineralia, neque quid aliud constitui."
16. Ibid. 665. "quod animalia in utero formantur a motione, seu motu maris... ex menstruo mulieris.. ad quem motum concurrit calor innatus, and spiritus vitalis, quibus anima utitur, ut instrumentis aptissimis ad movendum, et distribuendum singulis membris materiam propriam."
17. Ibid. p. 677.
18. Cf. De semine I viii (Kuhn IV 539–541). Accorambonius several times refers to the spirit in the seed as 'vital spirit'; see for example the passage quoted in note 16, above, and Part I, Chap. 7 note 22.
19. Interpretatio, p. 694. "Quia multas controversias inter Aristot. and Galenum in libris de animalibus Arist. tractavimus, non ab re futurum mihi visum est, si post illos Arist. libros, velut appendicem, addemus annotationes in tres libros Galeni de temperamentis, et in eis breviter, in quibus differat ab Aristotele et quibus rationibus, examinemus, ut in his quoque naturalis veritas elucescat."
20. For example, Book II iii 736b 30ff (on the element of the stars); III xi 762a 19–21 (on spontaneous generation); Book IV x 777b 17–31 (which is also taken up by Nifo). Other important texts for Accorambonius's interpretation of generation are De generatione et corruptione, I v 320b 17–25 (different ways of coming to be) and Metaphysica VI vii 1032a 12ff, which, together with Averroes's commentary on it, is also an important passage for Nifo.

Chapter 10
Agostino Nifo (1470? –1538)

Abstract Nifo produced commentaries and translations for major Aristotelian treatises. His medical activities as a practising doctor and university professor of medicine bring to his commentary on *De generatione animalium* the perspective of the medical tradition at Padua in the late fifteenth century. The *Expositio* of *de generatione animalium* is largely a section-by-section paraphrase of the text, with occasional points of difference raised by later commentators. Nifo begins his commentary with a summary of the themes of each of the five books of *De generatione*. He touches on a number of questions which we have already met, including the role of *spiritus* and heat in generation, the sense in which the seed or conceptus is ensouled, the generation of the *mens*. Nifo argues that sexual and spontaneous generation, though the mechanism by which they operate is different, are parallel and equivalent processes. All he could say was that similar effects (i.e. generation) must have had similar causes (similar measures of heat). It may well be that these differences in the kinds of explanation which were acceptable to Renaissance doctors and natural philosophers had more profound effects in determining their approach than the simple division into 'Aristotelians' and 'Galenists'.

It might be said that Agostino Nifo is the only Renaissance author of a printed commentary on De generatione animalium which was known to his contemporaries, and even then the recognition is equivocal. He is the only one whom Isaac Casaubon lists in the Index to his edition of the Opera (1590), but Casaubon mentions only Nifo's commentaries on History and Parts of Animals, omitting the work on Generation of Animals which was published with them.[1] Of other interpretations of De generatione available in 1590, Casaubon lists only the Gaza translation and the commentaries of Philoponus and Averroes, which were sometimes printed with it.[2] He also mentions that commentaries on the remaining books, to go with his translation-commentary of De historia animalium Book X were "eagerly awaited' from J.C. Scaliger. An edition appeared in 1619 at Toulouse.[3]

Among earlier interpreters of the other "animal books", Casaubon mentions the medieval commentaries on Book I of De historia animalium by Johannes de Kenighale[4] and Simon of Eversham[5] and on De motu by Peter of Alvernia.[6] Among Renaissance interpreters, he includes, together with Nifo and Scaliger, Nicolaus

L. Deer Richardson, *Academic Theories of Generation in the Renaissance*, History, Philosophy and Theory of the Life Sciences 22,
https://doi.org/10.1007/978-3-319-69336-1_10

Leonicenus' work on Book I of De partibus and Paul Belmisserus's translation of the first two books of De historia into Latin elegiac verse.[7]

Nifo is the only writer Casaubon mentions who wrote commentaries on all three of the major works De Animalibus. And he is the only commentator in this category whose works I have been able to find in print (excluding an "omnibus" commentary such as Accorambonius's Interpretatio). His is also the only straightforward Aristotle commentary by a Renaissance author included in this section, apart from the manuscript Expositio of De generatione II iii by Cremonini.[8]

Agostino Nifo (or Augustinus Niphus) has been the subject of several studies[9] but these have tended to concentrate on one portion of his work: the writings on psychology, and in particular his dispute with Pietro Pomponazzi on the immortality of the soul.[10] In the present state of scholarship, neither the basic facts of Nifo's life nor the state of his reputation are entirely beyond dispute. Thorndike gives his dates at 1473–1546; more recent scholars have suggested, however, that he was probably born in Sessa in late 1469 or 1470[11] and that he died on 18 January, 1538.[12] Nifo studied at Padua as a student of Nicoletto Vernia and took his degree in philosophy and medicine in or before 1492, the year in which he was appointed extraordinary professor of philosophy in second place. In 1495 he became ordinary philosopher in second place, rising to first place in 1496; by 1498 he was on a level with his old professor Vernia. He left Padua in 1498 or 1499 for a teaching career that was highly successful but genuinely peripatetic, teaching successively in the short-lived new university of Salerno (c.1507–10); at Naples (c.1510–13); at Rome; at Pisa (1519–21), where he was again ordinary professor in first place, with a handsome salary; and finally at Naples again (c. 1531) where he taught both medicine and philosophy. In the intervals he seems to have practised medicine, and at various times numbered Pope Leo X and the Spanish Grand Captain among his patients.[13]

His conspicuous success has perhaps contributed to the difference of opinion as to Nifo's merits as a scholar. One of the earliest evaluations, the Judicium de Augustino Nipho of Gabriel Naudé (1645), is highly laudatory and attempts to defend Nifo against certain damaging charges, including a continuing interest in astrology. It contains certain clear inaccuracies, however, and has generally been discredited by modern scholars.[14] For the French historian of Averroism, Ernest Renan, Nifo was an intellectual lightweight, a trimmer with no great gifts except an ability to move smoothly from one position to another and write books which were fashionable. He was the epitome of that "inoffensive Averroism" which Renan saw as dominating Paduan teaching in the sixteenth century.[15] But Edward Mahoney, who has made something of a speciality of Nifo, clearly considers him as a distinguished representative of Italian Aristotelianism – however fashionable he may have been – and suggests that the criticisms made by his contemporaries may have arisen from jealousy and professional rivalry.[16]

Whatever his ranking among the philosophers of his own day, Nifo was remarkably prolific by any standards: both as a commentator on Aristotle and Averroes and as the writer of independent treatises on a wide range of subjects.[17] Three groups of these writings concern us particularly: his Aristotle commentaries, in particular on the natural philosophical works and De Animalibus; editions and commentaries on

Averroes; and a small group of medical works. These provide a context for understanding the Expositio de Generatione Animalium, which we can then examine in greater detail.

Nifo's earliest published work was a two volume edition of the works of Aristotle and Averroes in 1495–6, followed by an edition and commentary of Averroes' Destructio destructionum in 1497.[18] Mahoney has argued that Nifo's early acceptance of Averroes as a faithful interpreter of Aristotle's doctrine of the soul in this work and the early De Anima commentary of 1503 changed to criticism after he learned Greek, and that this more critical attitude to Averroes appeared in Nifo's De intellectu, also published in 1503.[19] The Destructio went through several later editions (1517, 1529, 1542) however, and whatever his dissatisfaction with Averroes may have been, Nifo published a defence of the Averroist doctrine of mixtures in 1505 and commentaries on two other Averroes works, de substantia orbis and de animae beatitudine, in 1508. These also were reprinted at least to the midcentury.[20]

Apart from the 1503 De Anima commentary and a commentary on De generatione et corruptione which Naudé lists as being published with Nifo's treatise on the infinity of the prime mover in 1504, his earliest commentary on a work of Aristotle seems to have been a translation with commentary of the Physics in 1508.[21] There followed a Disputatio on the Metaphysics (1511); commentaries on Book XII of the Metaphysics of Aristotle and Averroes (1518); a second commentary on De Anima (1522); and commentaries on the Parva Naturalia and Physiognomia (1523); translation and commentary of De caelo (1525); and of De generatione et corruptione (1526); and a commentary on Meteorologica (1531). The last of this group to be published were the commentaries De Animalibus (1546) and a late exposition of the Metaphysica (1547).[22] Nifo also wrote commentaries on Aristotle's logical works, books in the Organon and the Topics; these appeared in the 1520s and 1530s.[23]

Nifo's medical works seem to have been produced either early or late in his career. His work on critical days appeared in 1504 and continued in popularity at least until 1614.[24] It was one of a series of works published by Nifo which had an astrological aspect.[25] His other medical treatise, De ratione medendi, was published in 1551; this, together with his commentaries De animalibus, may have been a product of his teaching of medicine in Naples at the end of his career.

Each of these groups of works provides a different aspect of the background for the work in which we are primarily interested, Nifo's commentary on de generatione animalium. The importance of the natural philosophical works of Aristotle and commentaries on them, as a context for Renaissance interpretations of De generatione, has already been discussed. Nifo produced commentaries, and in three cases translations, for the four major natural philosophical treatises, Physica, De caelo, De generatione et corruptione and Meteorologica. He commented on the De Anima and the works of the Parva Naturalia; and also wrote a series of commentaries on the Metaphysica. His commentary on Book XII of this last work is of special relevance because of references to it (and his commentary on De Anima) in the De generatione animalium commentary itself.[26]

In addition to his position as interpreter of the natural philosophical and logical works of Aristotle, Nifo is of interest to students of Renaissance Aristotelianism as an interpreter of the works of Averroes.[27] As already mentioned, recent discussions of Nifo's Averroism have centred on the question of his opinions on the immortality of the soul and on the apparent shift in his opinion of Averroes as a reliable guide to Aristotle. For our purposes, the chief interest will be to see whether Nifo's commentary on *De generatione* relies heavily on Averroes as an authority and the effect which this reliance might have on the questions considered: that is, whether Nifo takes up the questions which, as we have seen, were highlighted by Averroes in his own *Paraphrase* of *De generatione*.

The third group of texts – the small number of medical works produced by Nifo – together with his activities as a practising doctor and university professor of medicine, suggest that he would bring to his commentary on *De generatione animalium* another set of interests and authorities: those of the medical tradition as developed at Padua in the late fifteenth century.[28]

Agostino Nifo called his commentaries on the animal works *expositiones*; that is, line by line analyses of the text, similar in structure to the paraphrase and gloss, and ultimately based on the teaching device of the lecture.[29] The *Expositio* on *de generatione* is printed in sections, following the portion of the text to which it refers, in the same manner as the Renaissance editions of the commentaries of Averroes and Philoponus. The work is dedicated to Pope Paul III, and in his *Praefatio* to all three commentaries Nifo places his treatise within the context of the "teaching books' of Aristotelian natural philosophy, and argues for the importance of the "animal works" within this context in terms very similar to those used by Gaza, sometimes lifting actual phrases from the translator's *Introduction* of 1476.

Nifo begins by noting certain controversies among both ancient and modern writers regarding the "fruitful exposition" of the text, and by setting out the order to be followed in the preface itself. He will begin with the matter or subject of the "animal books"; then consider the authorities who have written on this topic; the place of *De animalibus* within the natural philosophical writings of Aristotle; the divisions and order of parts within *De animalibus* itself; their *inscriptione*; and finally, their usefulness.[30]

With regard to the subject matter of *de animalibus*, Nifo notes that there is a controversy. Is the true subject of these works man, the exemplar through which we can come to understand the rest of the natural world, as certain "juniores" or recent writers have claimed?[31] Or is their true subject animals in general? Nifo holds the latter position: the title of these books is after all *peri ton Zoon*, and they concern animals in general – that is, in Aristotle's definition, bodies which are self-moving – and not man alone. Indeed, man is not a reliable guide to the internal arrangement and workings of the other animals, which must be studied in and of themselves, for their own sakes, as Aristotle himself did.[32]

Nifo next considers the authorities, listing Latin (*ex nostris*) and Greek and Arabic (*ex externis*) authors who have written on terrestrial, aquatic and flying animals, on lesser animals and reptiles and on animals in general. For the most part he names ancient authors, listing, for example, Democritus, Aristophanes, Appollodorus Lemnius, Polybius and Celsus for terrestrial animals; Empedocles, Hippocrates, Herophilus, Erasistratus, Asclepius and Heraclides for lesser animals; and for animals in general, the major authorities ex nostris he gives are Pliny and Albertus Magnus (of whom he says that his treatise has "as many errors as words") and ex barbaris, Avicenna and Averroes. In a class by himself, however, is Aristotle, together with his disciple Theophrastus.[33]

In the next section, in which he discusses the method to be followed in his expositio (whether "by chapters, or by the method of instruction, which Aristotle teaches in the book On Parts") he further considers the relative merit of the earlier interpreters. His prime authority is Aristotle himself, because of the breadth of his knowledge, and because he is the only one to treat the question of animate bodies methodically, that is, according to a proper teaching method.[34] Of the commentators, he mentions the exposition of John of Philoponus and the scholia of Michael of Ephesus[35] with greatest approval. He considers that Averroes' Paraphrase gives an inaccurate interpretation; but he will follow Averroes "when he is right" and Avicenna "when he agrees with Aristotle". Albertus Magnus receives the harshest judgement of all: Nifo will follow him in some things, "but generally we contemn him, since he is almost always on the wrong track".[36] All four of these writers – in particular the two earlier commentators on De generatione, Philoponus and Averroes – are cited frequently in Nifo's Expositio.[37]

Nifo – like Gaza, Accorambonius and Cremonini – also considers the place of the works de animalibus and their order within the teaching works on natural philosophy. This is a difficult question, which again provokes debate. The initial order which Nifo gives is:

1. Physics, which deals with mobile bodies, or with natural bodies in general.
2. De caelo, which deals with the "where" or place of moving bodies.
3. De generatione et corruptione, which deals with the simple form of bodies.
4. Meteorologica, which deals with the form of imperfect mixtures.
5. De lapidibus, which deals with the form of natural bodies in general.
6. Liber de animatis, that is
    (a) On plants, by Theophrastus or Eudemus
    (b) on animals, by Aristotle himself: that is History, Parts and Generation of Animals
    (c) on man, in works by Avicenna and Albertus Magnus (presumably the Canon and De animalibus).

But this sequence has raised certain objections and suggestions for alterations. The first is that the (pseudo-Aristotelian) book on mixtures belongs in the series in fifth place, following meteorologica, and possibly replacing de lapidibus. The second objection is that the so-called de homine is not a "book" at all, and that its subjects belong either to de animalibus or to de anima: "what is animal belongs to the book on animals, what is truly rational, to do with understanding or the intellect belongs to the book on the soul."[38] So we might have a sequence 5) de mixtis 6) de lapidipus 7) de stirpibus 8) de animalibus, moving in order from lower and simpler natural bodies to higher and more complex ones[39].

Does the work de anima belong in this series, and if so, where? As we have seen, Renaissance editors and interpreters of the animal works, such as the Junta press and Accorambonius, included De anima among the "libri ad animalium cognitionem attinentes", placing it between De partibus and De generatione. Aristotle himself considers the place of the soul in the field of study of the natural philosopher at the beginning of De partibus, and suggests that, apart from the intellect, the remaining parts of the soul should be included as "something which constitutes the form of the living being", and as the final cause of the body which serves it. Nifo tells us that he considers this question more thoroughly in his commentary on De Anima itself.

By referring to this commentary, we do obtain a clearer idea of his opinion. In the Praefatio to the work, he begins by listing the "Partes physicae" of the Aristotelian corpus in the same order as in De animalibus but omitting De mixtis. Item seven on the list is thus De animalibus, which according to Nifo concerns the "motions" determined by the sensitive soul. Should not De anima follow as part eight, with motions controlled by the intellective soul as its subject?[40] The commentators disagree on this point. Themistius, says Nifo, classes De anima with the mathematical sciences and Simplicius and Philoponus do not include it in their lists of the natural philosophical works of Aristotle; but Alexander of Aphrodisias, Averroes, Avicenna and Aristotle himself, in Metaphysica book XII, De Sensu and De partibus animalium, all include it among the physical works. Aristotle, in Nifo's view, considers De anima as a mean between straight physics and straight metaphysics. And Nifo opposes Aristotle's argument, (De partibus I i 641a 33–641b^{10}) that if natural philosophy includes the intellectual soul, it must include everything knowable by the intellect as well. But if we assume that De anima is one of the works of natural philosophy, where in the series does it belong? There are two possibilities: before de animalibus, as greater in dignity; or after it, as being more general in its scope.[41] This is an argument we have already seen in the commentaries of Gaza and Accorambonius.

The position of other works in the series is threatened by their doubtful authenticity. According to Nifo, de lapidibus is not accepted as a genuine work of Aristotle by the Greek Peripatetics, but only by the Latins, such as Pliny and Albert, who wrote commentaries on it. As for de stirpibus, Nifo points out that Aristotle promised a work on plants in Book I of De generatione animalium (he is probably referring to the passage at the end of chapter i (716a 1) in which Aristotle says "Plants, however, must be investigated separately") and he is inclined to belive, in opposition to Averroes, that the work on plants in 16 books is by Aristotle himself. But he notes that this work, too, is not commented on by the Greeks, and that Gaza attributes it to Theophrastus.[42]

How many books are there in De animalibus itself, what is their order, and why? Aristotle claims fifty books, but Gaza, "who is expert in both languages, and is no less learned than eloquent in Aristotelian philosophy", counts only 18. The discrepancy is resolved by Pliny, who includes De anima and the Parva naturalia among the 'animal works', bringing the number up to the required fifty. For the order of the various treatises, and the reasons for this order, Nifo follows Gaza.[43]

He follows "Theodorus" again, word for word in some cases, on the question of the utility of the "animal books" to the two groups with whom we are most concerned, the doctor and the natural philosopher. In reply to the objection that Aristotle wrote a great deal about flies and worms, but very little about God, Nifo points out the importance of the study of the Creation as a way to the knowledge of its Creator, and the close relationship between the highest and lowest levels of living things – both familiar arguments. He also suggests that, in the manner of a bestiary, animal behaviour illustrates moral precepts: we can look to the bee for an example of industry, to the elephant for a model of chastity, and so on. Furthermore, animals serve as a guide in the study of healing, since they instinctively eat, sleep, drink and exercise the correct amounts for health, and so illustrate all the principles of regimen. He quotes Gaza's Introduction on the usefulness of these texts to the natural philosopher and doctor, and concludes that they are the best possible guide to nature herself, and to her methods in the generation and constituting of animals.[44]

The Expositio of de generatione animalium itself is largely a section-by-section paraphrase of the text, with occasional points of difference raised by the opinions of later commentators. Modern scholars have emphasised the philological interests of Nifo's late commentaries. What is perhaps more evident is the careful, rather pedantic, concern for order and structure, and the reliance on authorities, particularly the commentaries of Philoponus and Averroes, both of which were made available in the 1520s.[45]

Nifo begins his commentary with a summary of the themes of each of the five books of De generatione. The first two, which deal with the definition of male and female, and with the causes of generation and the respective role of the two sexes[46] are most important for our purposes. The other three books deal with oviparous generation, with the problems of inheritance and sex-determination and with secondary characteristics such as eye and hair colour.

Nifo begins his exposition of Book I of De generatione on a philosophical level, with a consideration of the four Aristotelian causes as they apply to generation. Aristotle told us that De partibus deals with the material causes of animals in general, De generatione with the parts serving generation in particular and with the efficient cause. But the topic of animal generation raises the question: do the formal and final causes of generation always coincide? 'Philoponus' in his commentary on De generatione and Averroes in his commentary on Book XII of Aristotle's Metaphysics appear to believe that they do, says Nifo. But he himself is not so sure: there are certain cases in which the form and the end of the thing generated are not the same, or coincide only in their subject.[47]

For a fuller treatment of this problem, especially as it applies to the case of spontaneous generation, Nifo refers the reader to his own commentary on the Metaphysics Book XII. This commentary was originally published much earlier in his career (1518) at a time when he had recently written commentaries on several of Averroes' treatises and was writing his reply to Pomponazzi on the immortality of the soul. It was reprinted in a more modern edition in 1558, together with Nifo's commentary on the other books.[48] Here, Nifo follows Averroes very closely on the question of whether spontaneous generation is different in kind and in cause from sexual generation, virtually lifting lengthy passages from the earlier commentator. In Book XII, Aristotle considers the coming-to-be of substance, and concludes that neither matter nor form per se come to be, since this would imply infinite regress. He goes on to list the kinds of coming-to-be, in a passage which both Averroes and Nifo consider important:

> ... each substance comes into being out of something that shares its name... For things come into being either by art or by nature or by luck or spontaneity. Now art is a principle of movement in something other than the thing moved, nature is a principle in the thing itself (for man begets man) and the other causes are privations of these two. (XII iii 1070 4 ff)

The key phrase for Nifo is the parenthetic one, 'for man begets man'. Human generation is an example of the coming-to-be of substance by nature, with a generator of the same name and species.[49] But is spontaneous generation, in which, as it was understood, not frog but slime generates frog, also an example of "generation by nature"? If not, to which of the other types does it belong? This question was raised according to Nifo, by the Greek commentator Alexander of Aphrodisias, who opposed Aristotle's argument that generation by nature necessarily implied a generator of the same name and species. Alexander's objections were discussed by Averroes in his commentary on Book XII, which is Nifo's major source throughout.

Alexander used two examples: animals generated spontaneously, and the mule, to make his point. Both are examples of natural generation, not because their cause is of the same name and species as they are, but because it is "similar in operation or power" to those active in sexual generation. This similar cause is the heat of the stars; and Alexander reports that this solution appears in Aristotle himself, in "one of the books of physics."[50]

The most important contribution made by Averroes to this argument is his addition of virtus formativa to the generative power of the stars, so that there are two "similar causes" at work in spontaneous generation:

> He [Averroes, reported by Nifo] says that they come from 'something which is the same' in two ways: first, as Alexander says, from the heat of the sun and other stars. Secondly, from the formative power found in the decayed particle or corpuscle. For he conceives that just as two seeds come from the father and mother, from which arises a certain body which is called an embryo, so from the heat of the sun and stars is generated a certain corpuscle similar to the embryo, and just as in the embryo there is a subtle spirituous part which is called the formative spirit, so in the corpuscle there is present a subtle part, airy and spirituous: therefore the identity is double, both universal, which is that between celestial and animal heat, and particular, found in the individual seed.[51]

The only difference is in the source of the power-- seed or sun-- which activates the embryo or corpuscle, not its nature. This is an interpretation with which Nifo appears to agree, although his much later commentary on the "book of the physics" to which Alexander referred, De generatione animalium, is much less forthcoming on this point.

Nifo calls on Averroes again, towards the end of his commentary on chapter iii of Book XII, for support against Plato's argument that the univocal causes of all generation, whether from seed or from slime, are the universal forms or Ideas. His follower Themistius used Aristotle's definitions to argue that spontaneous generation was an example of coming-to-be by chance, not by nature, and so required the mediation of immaterial Ideas. Averroes opposed this notion, and Nifo follows him closely, with the smugly anachronistic comment, "Averroes agrees with us in expounding the text of Aristotle."[52]

In the introduction to his exposition of De generatione, as we have seen, Nifo gave Averroes only guarded praise:

> Averroes who wrote paraphrases of the books of parts and of the generation of animals admittedly gives a faulty interpretation, in my opinion. Nevertheless, we have followed him where he speaks correctly.[53]

But in the body of the commentary, Nifo cites Averroes relatively frequently, and despite his opinion in the preface, always, I think, with approval. In particular, on the questions of the role of the testes and of female seed, he follows the lines of Averroes' argument very closely. This may be partly because, as a commentator, he was expected to take account of earlier interpretations of the text and to cite them – whether or not he approved; and, as I have mentioned, apart from Averroes, Philoponus (whom he also cites relatively frequently, especially on certain 'philosophical' points) and of course Galen, commentaries on De generatione were relatively rare. More positively, it may have to do with Nifo's evident approval of Averroes as an interpreter of the more straightforwardly philosophical works of Aristotle, such as De Anima and Metaphysica, on which Nifo, too, had commented.

Some interpreters have argued that Nifo's "Averroism" was abandoned very early in his career, but, as I have already suggested, these interpretations are generally based on a narrower definition of Averroism than the one I am using here. It may well be, for instance, that Nifo left behind his support for unorthodox opinions on the immortality of the soul in the early 1500s. In any case, his commentary of De generatione is much less metaphysical in flavour than these early works, and sticks to an uncontroversial and rather literal paragraph-by-paragraph study of the text. His discussion of Books I and II touches on a number of questions which we have already met in the Averroes paraphrase: are testes necessary? what is the nature of semen, and its origin? how do male and female contribute, and are their contributions similar or different? how do heat and spirit contribute to the causes of generation?

On the questions of the role of the testes and of female seed, he follows the lines of Averroes' argument very closely. Indeed, in setting out and evaluating opposing views on the use of the testes, he takes up where the Commentator left off. He notes Averroes' doubts concerning the reliability of the story of the bull who fathered a calf after castration. It would be important evidence, he had said, against the importance of the testes to generation – 'if it could be verified.' Nifo reports proudly that it has been verified: he has himself interviewed local shepherds on this matter, and their testimony confirms Aristotle' story. And he uses Averroes to prove that on this point, in his words, "Galen's reasons are frivolous."[54]

In discussing the question of female seed, he again turns to Averroes; this time to the "testimony of honest women" – Averroes' sample of midwives – which proved that conception without pleasure is possible, and thus that the female need not produce a fertile seed. Averroes argued, further, that sexual pleasure and production of a genital moisture do not necessarily coincide, since fair women, for instance, produce more of this humour than do dark ones. Thus this humour does not meet the requirements of a true female seed, as Galen had argued. Nifo follows him in this, and also in his argument that female "testes", though they exist, cannot be assumed to have the same function as in the male. What counts is not the similar figure of the two organs, but the differences in heat between the two sexes, which means that the male alone is able to produce fertile seed. Nifo concludes the discussion of these two major questions, in Book I, with the words,

> These are what can be collected together from what Averroes says on these points. There could be much more to say, which is discussed in the following books, and in other works.[55]

Averroes' evidence is valuable, not only because he agrees with Nifo – at least sometimes – but also because from his statements one can 'collect together' the important points relating to these disputes.

Averroes's importance as a guide is less clearly evident in Nifo's commentary on Book II of De generatione. Here, he tackles the major remaining theoretical questions: the role of spiritus and heat in generation, the sense in which the seed or conceptus is ensouled, the generation of the mens. On these questions, he cites the commentary of John Philoponus (Michael of Ephesus) fairly frequently, though in several places he clearly disagrees with it. It may be, too, that here, as in Book I, he is viewing earlier commentators through Averroist spectacles – as he warned at the end of the first book. In any case, his commentary here is much less adventurous than his treatment of similar questions in the much earlier Metaphysics commentary, and amounts to little more than a paraphrase of the Gaza translation at several points.

Nifo begins his treatment of Book II by asking what Aristotle meant when he asked by what the geniture was 'first moved' ("quo nam primo movente, et quali materia.. oportet explicare.") According to Philoponus, 'primo movente' referred to the 'corpora divina', ie. the heavenly bodies which were said to cause all earthly movements. Nifo suggested, more piously, that God, ultimate cause of both celestial and sublunar movements, was meant. From the context of the passage, it seems likely that Aristotle himself meant simply the efficient power of the male parent.[56]

Indeed, it is the male parent as cause of motion or agent, with the seed as his instrument, that Nifo goes on to discuss. In what sense can it be said that the seed is ensouled, or alive? Nifo suggests that of itself, the seed does not exist as an animal, possessing soul, but only as an instrument. It is the instrument of the faculty proceeding from the father alone, since, as we have seen, Nifo does not accept that the female parent produces fertile seed. And it is this faculty alone – and not, as some medical writers have argued, the heat in the mother's womb – which is responsible for triggering off the process of actualisation of the foetus. The male parent's faculty in the seed simply makes use of this heat, as a blacksmith uses a forge.[57]

In what sense, then, if at all, is the seed animated? Nifo reviews the major arguments of ancient and Arabic authorities. Aristotle made use of the simile of the sleeping geometer to suggest that the seed was animated only in potentia: it had the capacity to become a living thing, but this was not yet actualised. This fairly negative position seems to come closest to Nifo's own opinion. Philoponus, on the other hand, argued that the conceptus possessed a vegatative soul, like that of a plant, and that even an egg had a soul of sorts. Actualisation began as soon as seed met menstrua, and Nifo suggests that for Philoponus seed itself was animate: a position which, he argued, created important difficulties.

Philoponus also seems to have suggested that all the faculties of the soul were present in the embryo at the same time (ie. from the beginning). Most theories, though, made an exception in the case of the mind or rational soul, and suggested a sequence of ensoulment as the embryo developed.[58] What of the origin of the mind, the (rational) soul "by which man is man"? How – Plato and Avicenna had asked – is mind received into the conceptus for the animation of a creature-which-will-be-human, but has as yet no human attributes or capabilities? Various solutions had been suggested by the ancients, but according to Nifo "astral" theories (like those of Plato, Calippus and Eudoxus) and the transmigration theory of the Pythagoreans were both rejected by Aristotle. But mind needed an appropriate vehicle for its activities (just as, in Aristotle's own example, the power of walking required feet). Nifo's suggestion – an interpretation which we shall also find in Cremonini – was that when Aristotle spoke of the element of the stars, he was referring to the generation of the intellect, not of the child as a whole. This element, or the spirit which was likened to it, provided a vehicle for the rational soul or mind.

The same passage also suggested an answer to the question "what makes semen fertile?" According to Nifo, the correct answer is "the generating power" (virtus genitalis) and this power is transmitted from father to child by means of the spiritus seminalis in the seed. Diffused through the menstrua like wine in water, the spiritus seminalis "heats, alters and directs it, and at length brings it to be just such a body as the parent animal." In the same way, "heat from the stars diffused in air or water is called 'the element of the stars' because by it all things which there are in this inferior world are generated and composed."[59]

In Book I, Nifo argued that what counted in determining the differing roles of male and female in generation was not the 'figure' of their parts (which could be seen to be similar) but the 'measure of heat' in these parts (which he argues was unequal).[60] Here, Nifo argues that sexual and spontaneous generation, though the mechanism by which they operate is different, are parallel and equivalent processes. For in each case an equivalent 'measure of heat' is diffused through an appropriate material – menstrua, water, air – and the result is the generation of new life.

For the medical explanation of the causes of generation, epitomised by Galen's On the Use of the Parts, anatomical observation of the parts serving generation was essential. Anatomical details could be compared to establish analogies or homologues, and these interpreted to give the cause of the process which their structure or figure made possible. This is an approach which would still be understood by a physiologist, though his notion of the structure of a part, and of its use, would differ from Galen's. But an explanation which depends on comparing measures of animal heat is much more difficult to check or to establish by independent means. All Nifo and his fellow natural philosophers could say was that similar effects (ie. generation) must have had similar causes (ie. similar measures of heat). It may well be that these differences in the kinds of explanation which seem to have been acceptable to Renaissance doctors and natural philosophers had more profound effects in determining their approach than the simple division into 'Aristotelians' and 'Galenists' would suggest.

Whether one shares Mahoney's belief in Nifo's distinction and importance in the history of ideas, or agrees with Nardi's harsher evaluation of him as "a man of vast erudition, but confused and pretentious",[61] his background as a medical practitioner, Aristotle commentator and Averroes scholar meant that he brought a particularly rich breadth of knowledge to his commentary on De generatione animalium. His relatively balanced, even conventional brand of Aristotelianism serves as a useful background to the more polemical figure who is the subject of the next chapter.

Notes

1. ʼΑριστοτελους···τα σωζομενα Operum Aristotelis… nova editio, Graece & Latine… adscriptus ad oram libri… emendationibus..Latinae interpretationes adjecta sunt..Accesserunt…fragmenta quaedam, etc. 2 vols, [ed. I. Casaubon] Lyons, 1590. "Index alter [following an alphabetical list of interpreters' names] quo declaratur quid sit a quoque ex superioribus interpretibus in singulas Aristotelicae philosophiae partes scriptum" (unpaginated). Casaubon's entry for Nifo reads, "Augustinus Niphus Philotheus Suessanus Arist. libros de Hist. animal. partibus earumque causis ex Th. Gazae translatione ad Graeca exemplaria recognovit, & scholiis illustravit."
2. Ibid.,
3. Ibid.; "Julius Caesar Scaliger librum decimum De historia animalium transtulit, & doctissimis commentariis illustravit. Eiusdem in reliquos eius operis libros commentarii avidissime a doctis viris expectantur."
4. Ibid., This must be 'Joannes Kenigale' of Lohr, 1971, 27:252; and probably also "Johannes Tydenshale", Wingate, op. cit., p. 83.
5. Ibid., See Lohr, 1973, 29:141–6.
6. Ibid., See Lohr, 1972, 28:334–346.
7. Ibid., For Belmisserus, see Thorndike, History VI, p.
8. Nifo's is the only commentary on De generatione animalium mentioned by Thorndike (History, V, 254). Neither Lohr's list of commentaries from 1500, now complete through K, nor the subject indices of Lipenius (1679) add any further titles. As Casaubon's and Lohr's lists indicate, medieval commentaries De Animalibus were printed in the Renaissance, but most of these relate to History of Animals, especially Book I.

9. Monographs include E.P. Mahoney, "Nicoletto Vernia and Agostino Nifo on Alexander of Aphrodisias: an unnoticed dispute", Rivista critica di storia della filosofia, 1968, 23:268–296;"Pier Nicola Castellani and Agostino Nifo on Averroes' doctrine of the Agent Intellect", Ibid., 1970, 25:387–409; "Agostino Nifo's early views on immortality", Journal of the History of Philosophy, 1970, 8:451–460; "Antonio Trombetta and Agostino Nifo on Averroes and intelligible species. A philosophical dispute at the University of Padua" in Storia e cultura al Santo fra il XIII a il XX secolo, Vicenza, 1976; pp.289–301; "A note on Agostino Nifo", Philological Quarterly, 1971, 1:125–132. As already mentioned, none of these deal directly with the question of Nifo as a commentator on the biological works. But they supply some useful details in passing.
10. For a sample of the considerable literature on this debate, see Mahoney, op. cit. (J.H.P.) p.451, especially notes 2 and 4. Paolo Zambelli, "I problemi metodologici del necromante Agostino Nifo", Medioevo, 1975, 1:129–171, is largely concerned with Nifo's commentary on Averroes' Destructio destructionum and especially his De daemonibus, as well as the immortality dispute. For earlier mentions of Nifo in more general works on Latin Aristotelianism, see especially Renan, Averroisme 367–371; Lynn Thorndike, History, V, chapter 5, especially 71–82 (the remainder of the chapter is an exposition of De daemonibus) who gives a useful summary of biographical and bibliographical details to date (1941). More recently, there is Nardi, Saggi (op. cit. II; note 8) who is, however, largely concerned with the psychological writings, and for a different but equally unbiological approach, Randall's School and Career (1961 and 1962; op. cit.)
11. Thorndike, op. cit. p.71. For the earlier suggested date of 1469–70, cf. Nardi, Saggi, p.284 n. 8; he finds evidence for the earlier date in a manuscript family tree, for the latter by inference from a passage in De intellectu.
12. Mahoney, J.H.P. (1970) p.451 note 1; Rivista, (1968) 282–3 and references.
13. Thorndike, op. cit. pp.71–2; for details of the Paduan and Pisan periods see also Mahoney, Rivista (1968) 282–3 and (1970) p.391 n. 9. ; Nardi, Saggi, p.101.
14. Gabriel Naudé, in Augustini Niphi sua Tempestate Philosophi Omnium Celeberrimi Opuscula Moralia et Politica: cum Gabrielis Naudaei de eodem Auctore Judicio. Paris, 1645.
15. Renan, Averroisme, especially p.369: "Niphus semble avoir été un de ces chevaliers d'industrie littéraires si communs en Italie au xvie siècle. … Ses traités politiques et moraux avaient de la vogue..Cette legèreté de caractère ne permet pas de prendre bien au sérieux la doctrine philosophique de Niphus.."
16. Mahoney, op. cit.; he makes these points in both Rivista articles.
17. Bibliographical details are given by Naudé in his edition of the Opuscula (op. cit. note 14), "Librorum Omnium Editorum Catalogus," and by Thorndike. I have also used the printed catalogues of the British Library and Wellcome Institute Library.
18. Venice, 1495 and 1496. See Cranz, "Editions", pp.116–128, especially 120, for other versions, and for Nifo's role in these early editions. The first single edition of the collected works of Aristotle and Averroes was published by Nifo's

teacher Nicoletto Vernia in 1483. Even earlier (c.1472–1475) the natural philosophical works had appeared in editions by Laurentius Canozius.
19. Mahoney, op. cit. (J.H.P.) especially p. 452, note 8; Nardi, Saggi, p.101.
20. W.I.H.M. Printed Books, vol. I, lists an edition of the de substantia commentary of 1559, bound with an earlier (1524) edition of Nifo on de animae beatitudine.
21. Naudé, op. cit; B.L. Printed Books.
22. Ibid.; W.I.H.M. Printed Books, vol. I;Book XII of the Metaphysics is a reprint of the 1518 translation.
23. Naudé, op. cit.; B.L. Printed Books.
24. Ibid.,
25. Those in the British Library include his commentary on Quadripartitum (1513); De nostrarum calamitatum causis (1505); De figura stellarum (1520); De falsa diluvii prognosticatione (1519); De auguriis (1534, 1614 and 1546 in French translation); De temporum signis (1540). See Thorndike, V chapter 5 and Zambelli, op. cit. for further details of this aspect of Nifo's work, especially his treatise De daemonibus.
26. The Preface to Book Lambda of the Metaphysics was translated by Elias Cretensis, according to him at the request of Giovanni Pico della Mirandola, and published in 1488 with a dedication to Cardinal Grimani; Nifo dedicated his 1497 edition-commentary to him as well (Cranz, "Editions", 119–20).
27. In addition to the works already cited in this section by Mahoney, Thorndike, Renan, Nardi and Cranz, cf. H.A. Wolfson, "The Twice-Revealed Averroes" and "Revised plan for the Publication of a Corpus commentariorum Averrois in Aristotelem" Speculum 1963, 37:88–104, both reprinted in Wolfson, Studies in the History of Philosophy and Religion, Cambridge, Massachusetts, 1973; Kristeller, "Renaissance Aristotelianism" and "Paduan Averroism and Alexandrism" and C.B. Schmitt, Critical Survey and "Paduan Averroism" (op. cit. Chap. 1, notes 72 and 75); Poppi, Introduzione and Siraisi, Arts and Sciences, (Part II, Chap. 8, note 42).
28. For the development of the Paduan link between medicine and natural philosophy, see especially Siraisi, Arts and Sciences.
29. Augustini Niphi …Expositiones in omnes Aristotelis Libros [de Animalibus] Venice, 1546. See Lohr, 1967, 23 and especially 1973, 21:228–232 for a list of types of printed commentary in the Renaissance and for the continuity between medieval and Renaissance teaching methods and, therefore, types of treatise.
30. Ibid., Augustini Niphi…Praefatio in Libros de animalibus Aristotelis. Ad Paulum III, Pont. Max. Fol iir.
31. Ibid.,
32. Ibid., iir–v.
33. Ibid., iiv: "Inter hos autem diligentissime scripsit Aristotelis, quem Theophrastus fere in omnibus est."
34. Ibid., iiir: "solus Aristotelis per Paediam de eis tradidit, per quem modum libro Physicorum docuit agere de rebis naturalibus."

35. For Renaissance writers these were, of course, distinct personages; cf. C.A.G. 14, part 3 and note 37 to Chap. 8, above.
36. Ibid., iiv. "Aristotelem sectemur, et ex interpretibus, Philoponum…et Ephesium… Averroem etiam, qui paraphrastice libros de partibus, et de generatione animalium, licet meo judicio perperam interpretetur. Tamen in quibus recte dixit, secuti sumus. Avicennam quoque saltem in quibus Aristotelem secutus est. Albertum vero in quibusdam, nam generatim contempsimus, cum quaesi semper sit delirus."
37. Ibid., iiir. "Est enim primus… liber Physicorum, ubi agitur de ente mobili, aut de corpore naturali.
38. Ibid.,
39. Ibid.,
40. Augustini Niphi…Expositio..et..Collectanea Commentariaque in Tres Libros Aristotelis De Anima …Venice, 1559. Praefatio (unpaginated): "Tertio dubitatur, quia non videtur negocium physicum hic completum, quia adhuc considerari potest corpus, ut mobile motu terminatio ad animam intellectivam."
41. Ibid.,
42. Expositiones in.. [De Animalibus] Praefatio, iii r–v.
43. Ibid., iiiv–ivr and "De historia animalium.. expositione," p.1.
44. Ibid., Praefatio, ivr–vr.
45. Mantinus' translation of Averroes's paraphrase of P.A. and G.A. was published in 1521 (Cranz, "Editions", p.123) and editions of the commentary by Michael of Ephesus attributed to Philoponus appeared in 1526 in Greek and a Latin translation by Nicolaus Petrus Corcyraeus. See note 36 to Chap. 8, above.
46. "Aristotelis De Generatione Animalium Liber Primus, cum Augustini Niphi Medicis Philosophi Suessani Expositione", p.1. "Primo, declaravit quid mas, et foemina sint… In secundo… ostendit qua de causa mas, et foemina sint in rerum nature, et quid emolumenti ex eis afferatur rebus."
47. Ibid., p.2. Aristotle speaks of the subject matter of P.A. and G.A. in a passage at the beginning of the latter (G.A. I i 715a, 7–18) which was often quoted by commentators.
48. The two editions are: … Augustini Niphi..In duodecimum Metaphysices Aristotelis et Averrois.. Venice, 1518; Augustini Niphi..Expositiones, in Aristotelis Libros Metaphysices, Venice, 1558. Spelling and punctuation, together with page references, are taken from the 1558, posthumous, edition; but this follows the text of the much earlier commentary on Book XII alone.
49. Metaphysica XII (Lambda) iii 1070a 4ff; Nifo, Expositiones in…Libros Metaphysices (1558) p.527.
50. Ibid., pp.528–9.
51. Ibid., p. 529. "Ipse dicit quod fiunt dupliciter ab univoco, uno modo, ut dicit Alexander, a calore Solis et aliarum Stellarum. Secundo modo a virtute formativa reperta in particula seu corpusculo putrefacto. Imaginatur enim quod sicut a patre et matre deciduntur duo spermata, e quibus fit quoddam corpus quod embryo dicitur, sic a calore Solis et Stellarum generatur corpusculum quoddam similis embryoni, et sicut in embryone est pars subtilis spirituosa quae spiritus

formativus dicitur sicut in corpusculo illo adest pars subtilis aerea spirituosa: ergo duplex est univocatio, et universalis qualis est inter calorem coelestem et animal id, et particularis reperta in particulari semine." For Nifo's own position on the "two seeds" and for further detail on the "formative spirit" see below.

52. Ibid., pp. 536–7. For Averroes, see commentaries 13 and 18 in Metaphysics XII iii; Ed. cit. VIII, 142v–143r. "Averroes convenit nobiscum in exponendo Arist. textum.."
53. See note 36, above.
54. "De Generatione Animalium ..Expositione", pp. 11–14. Nifo's account of the role of the testes is based heavily on Aristotle, G.A. Bk I iv and Bk V vii 788a 5ff, and also on the arguments of Philoponus, ed. cit. (1526) 4r–5v and of Averroes, ed. cit. vol VI, chapter vii, especially pp. 207v–208r.
55. Ibid., pp. 25–41 (chapter XVII) and 45–60 (chapter XIX). For Philoponus's arguments against a fertile female seed, see ed. cit. 22r ; for Averroes's, see especially ed. cit. p. 216, and above, Chap. 8 for arguments that (i) conception without pleasure is possible; (ii) pleasure without flow of the uterine moisture (identified by Galen with female seed) is possible; (iii) amount of moisture varies with constitutional type. For Nifo's identification of his own argument with that of Averroes, cf. p. 60: "Hae sunt quae ex eis, quae Averroes dicit, colligi possunt. essent alia complura dicenda, quae tum in sequentibus, tum in aliis libris sunt disputata."
56. Ibid., p. 61. For Philoponus's arguments, see ed. cit. 27v. The context for this statement is the first chapter of Book II of G.A., following a passage which describes male and female as the "principles" of generation.
57. Ibid., p. 71. "..semen non sit actu tale univoce, quale animal, quod ab eo fit, tum est actu aequivoco tale, tum instrumentum patris, qui est tale actu..."
58. Ibid., pp. 71–76 (commentary on G.A. II iii); for Philoponus on the animation of the seed, cf. ed. cit. 32v–33r, 34v–35v.
59. Ibid., p. 79. "Appelavit autem huiusmodi caliditatem, naturalem, quia est quasi naturalis forma spiritus seminalis, per quam agit in menstruo.. et caliditatem a stellis diffusam in aere, vel aqua vocavit elementum stellarum, ex eo omnia quae in hoc mundo inferiore sunt, generatur et constant."
60. Ibid., p. 60: "Non enim figura est principium agendi. sed si esserit eiusdem mensurae caloris, tunc essent eiusdem operationis. est enim calidum membrorum seminalium principium, atque causa seminis." The medical theory of temperament, which had been elaborated by Galen and applied to the action of parts of the body, the body as a whole and also to the action of drugs, did, however, relate function to the level of heat; see below, Part III.
61. Nardi, Saggi p. 451: "Agostino Nifo, professore al Padova dal 1492 al 1499, uomo di vasta erudizione, ma confusario e pretenzioso.."

Chapter 11
Cesare Cremonini (1550–1631)

Abstract The Italian natural philosopher Cesare Cremonini takes up the "many controversies between Aristotle and Galen in the books on animals" in a spirit of controversy. The topics he chooses are almost exactly those emphasised by Averroes in his *Paraphrase*. Cremonini's *De calido innato* (1626) and *De semine* (1634) deal with the status of the geniture and the role of male and female contributions to generation; and his *Apologia de origine et principatu membrorum* (1627) treats exhaustively the question of the primacy of the heart over the other members. Cremonini is more partisan in his opinions than Averroes or Nifo, who were doctors themselves. His works on generation seem to have arisen from a dispute between Cremonini and a doctor and anatomist, Pompeius Caimus (or Caimo) which may explain their monotonous Galen-bashing. And he seems to have taken Averroes as his guide to chinks in the medical armour. His printed works on natural philosophy contain three themes which are of importance to the narrower question of the *Generation of Animals*: the importance of celestial influence in the sublunar world; the divisions of natural knowledge and their order; and the relationship of medicine and natural philosophy.

In his forays into the disputed territory between medicina and physica the Italian natural philosopher Cesare Cremonini takes up the "many controversies between Aristotle and Galen in the books on animals", but in a spirit of controversy not conciliation. The topics he chooses are once again almost exactly the same as those emphasised by Averroes in his Paraphrase. In particular, Cremonini's two short treatises De calido innato (1626) and De semine (published with the former in 1634)[1] deal with the status of the geniture and the role of male and female contributions to generation; and his Apologia de origine et principatu membrorum (1627)[2] treats exhaustively the question of the primacy of the heart over the other members. Cremonini is more partisan in his opinions than Averroes or even Nifo, both of whom were after all doctors themselves. His printed works on generation topics seem to have arisen from a dispute between Cremonini and a doctor and anatomist, Pompeius Caimus (or Caimo) which may explain their monotonous Galen-bashing. And he seems to have taken Averroes as his guide to chinks in the medical armour. Some background is in order.

L. Deer Richardson, *Academic Theories of Generation in the Renaissance*,
History, Philosophy and Theory of the Life Sciences 22,
https://doi.org/10.1007/978-3-319-69336-1_11

Cesare Cremonini was born in 1550 at Cento in the Duchy of Ferrara and took his doctorate in law at the university there, following studies in humanities and philosophy. In 1591 he moved from the chair of philosophy at Ferrara to succeed Piccolomini in the second chair at Padua, graduating to the first chair in 1601. He held this post until his death in 1631, and was thus in a position to dominate the teaching of philosophy at the University for 30 years.[3] In 1623, Portinari described him as "the most famous philosopher in Europe, the most learned in the Platonic doctrine as well as in the Peripatetic…the ornament and the immortal glory of the University of Padua."[4] The French Aubrey, Gabriel Naudé, who met Cremonini at the height of his career (during or after his dispute with the anatomist Pompeio Caimo in the mid 1620's) reported that he was the most renowned professor that there had ever been in Italy, and that he lived in his palace in Padua like a Cardinal at Rome, surrounded by servants and supported, at the time of his death, by a salary of 2000 crowns a year plus the fees of 400 students.[5]

He was, clearly, an important figure. Despite this, relatively few of his treatises were ever printed. A large number, however, survive in manuscript, mostly in the form of copies or student notes. There are major collections in Padua and Venice, and a smaller collection in Paris.[6] They include works on a wide range of topics: physics, metaphysics, logic and psychology, with multiple commentaries on all the standard Aristotelian natural-philosophical works.[7]

There are also several manuscript texts which reveal Cremonini's interest in the border area between philosophy and medicine. They include a commentary on the Galenic psychological text, Utrum animi mores sequantur corporis temperamentum, an incomplete set of lecture notes on the topic "On controversies between famous doctors and philosophers" and expositiones of Parts of Animals I i and Generation of Animals IIiii.[8]

His printed works on natural philosophy are somewhat more accessible and contain three themes which are of importance to the narrower question of Cremonini and the Generation of Animals: the importance of celestial influence in the sublunar world; the divisions of natural knowledge and their order; and the relationship of medicine and natural philosophy.

It tends to be said of Aristotelian science that it disjoins completely the celestial and terrestrial worlds, "physics" and "astronomy". Cremonini's interest, however, is notably in areas where the two interact. This appears especially in his Disputatio de caelo (1613) and Apologia.. de quinta caeli substantia (1616) as well as in the manuscript Tractatus de coeli efficentia.[9] The efficient power of the heavens, according to Cremonini, is evident in the actions of light and motion and the celestial heat which results, and, in the case of generation, in the activity of celestial heat and spirit working in and through the seed.[10]

In the first section of his Disputatio de caelo, Cremonini argues, after consideration of the opinions of the Greek authorities and of Averroes, that the heavens are composed of matter and form, and not form alone. That being so, they cannot move themselves, and therefore require a soul. He takes Averroes' commentary on De caelo as an authority for the manner in which the heavens are animated. Later, in the third section of the work (Pars Tertia, De Motoribus Coeli Abstractis) he discusses the possibility of other non-material movers, which one may call intelligences, or gods. This is a theme which he pursues in several other works, printed and manuscript.[11]

His bold defence of the Platonic anima mundi and the opinions of Averroes in the Disputatio seem to have led to difficulties with the local branch of the Inquisition.[12] In his Apologia ..de quinta caeli substantia he is more cautious, while still sticking basically to his philosophical guns. In the work, he defends Aristotelian opinions on the quintessence against those of Plato in the Timaeus, while insisting that his arguments are not necessarily true, especially if they should happen to conflict with Christian doctrine, but only, "demonstrations, assuming Aristotelian principles"[13] In the body of the Apologia, Cremonini argues, on behalf of Aristotle, that the heavens are a simple body, of a different, incorruptible substance from the elements, and having their own, circular, motion. They are, furthermore, higher than the elements: none of which Plato is willing to allow.[14] He also suggests, as in the Disputatio, that the motion of the heavens implies the existence of an anima mundi, and that this celestial soul and body are united by an innate heat. The heavens act on the inferior world, essentially, through the power of motion; and at the end of de quinta caeli substantia he looks forward to a fuller discussion of the "efficient power" of the heavens.[15]

In de quinta caeli substantia, Cremonini emphasises the differences between the superlunar and sublunar worlds. But in other works, and particularly in the Tractatus de coeli efficentia (in which he fulfills his promise, though the work was never printed)[16] he emphasises their similarities. If the heavens are to act on the sublunar world, they must either touch it physically or be otherwise "proportionate" to it in some, though not necessarily all, respects. Cremonini argues that the stars are proportionate to elemental substances through their possession of matter and form, and through their participation in the four prime qualities.[17] The fullest treatment of this comes in the manuscript de coeli efficentia. Here, Cremonini distinguishes two types of celestial influence: efficient, but not formal (represented by the activity of light and heat); and formal and final, according to which the stars serve as types, similar to the Platonic ideas. Hence, the species or essential form of things is subject to celestial influences.[18] Cremonini also argues that the elements act, as it were, as the agents or instruments of celestial light and motion, by means of their active and passive qualities and especially the active quality of heat, which is 'directed' by the heavens.[19] It is through their action on heat as a qualitas formativa that the stars have an active role in generation.[20] This is a topic which Cremonini takes up again in his treatise on innate heat.

Cremonini seems to have taken Averroes as a guide to the structure of science in general, and not only as it applied to man. In his introduction to the Physics (1596)[21] Cremonini gives a breakdown of the divisions of natural knowledge and their correct order - that is, the order in which they should be learned, or taught - which he says is based on Averroes' preface to the work.[22] Except for the fact that he puts Theophrastus' books on plants before the animal books and not after, the order he gives is the same as that of Accorambonius, but there are differences in interpretation. The order he gives is not justified by reference to the four causes, but takes the form of substance preceding accidents or affectiones. Thus he identifies the group of minor works which follow De anima not with the formal cause, but as accidents of sensation, and similarly those which follow De generatione not with the efficient cause but with the vegetative faculty. His order, then, though it follows that of a progression through the causes as given by Gaza and Accorambonius, works downwards from intellect to the lower faculties of the soul such as nutrition and generation.[23]

He also points out the controversy concerning the correct order of De partibus animalium and De anima. Does discussion of the whole belong before or after its parts? This depends, says Cremonini, on your point of view. If you consider that the body comprises the essence of the animal, then a knowledge of its parts will more easily lead you to an understanding of its form or anima. But alternatively, if you consider the body as the instrument of the soul, then a knowledge of the soul is required to understand the body.[24] Cremonini himself follows Accorambonius and others in placing Parts first. But he mentions Avicenna, followed by Albertus Magnus and "the Latins" as fierce defenders of the alternative order.[25]

Also important to the consideration of Cremonini's thoughts on De generatione is his contention that man is the prime subject of natural philosophy, and not the heavens, or the beings of the inanimate world. As he explains in the Introduction to the natural philosophy of Aristotle, natural philosophy is the study of reality as it applies to being "conjoined with matter". Natural objects of knowledge are of three types: (1) disjoined from matter ie., supercelestial being; (2) conjoined with matter, ie., all sensible substances; (3) partially conjoined and partially separate. Man is in the third category, and this gives him a special importance as a subject for natural philosophy. It is reasonable to argue, as Aristotle does in the Physics, that the most important subject of natural philosophy is the heavens, "as divinest and most superior of all sensible beings". Cremonini concedes that the heavens are superior to man; nevertheless, "we, however, prefer to establish man as the most important subject of natural philosophy"[26] Man, and not the heavens, is the prime example of natural bodies, that is to say, bodies subject to generation and corruption. And so "the study of nature means in the first place the study of man, as it were the end of natural knowledge."[27]

Man, as Cremonini described him in his inaugural lecture of 1591, is the microcosm and epitome, the epilogue as it were, of the world. By understanding the composition of the human body, we can best come to understand the lesser beings of the natural world. And by following the Platonic injunction, "know thyself", the philosopher can come to see the rest of Nature, of which he is the image, as in a glass.[28]

This definition of man as the prime concern of natural philosophy sets a collision course with the area of study traditionally mapped out by medical writers. Cremonini seems well aware of this conflict, and pursues it directly, above all in his three published works defending the opinions of Aristotle on innate heat, on the seed, and on the origin and primacy of members. What, for Cremonini, is the proper relation of medicine and natural philosophy? If the knowledge of man is the prime object of natural philosophy, if "the study of nature means in the first place the study of man," then surely the doctor must give way to the superior understanding of the philosopher. Medicine begins, says Cremonini, where natural philosophy leaves off; but it is the latter which can claim to tell us of causes, including the causes of sickness and health.[29] Why is the natural philosopher interested in sickness and health? Because, says Cremonini, his interest is in causes, and the underlying balance of contraries:

> It is the business of a natural philosopher to examine the first fundamentals of sickness and of health, that is the beginnings from which sickness or health arise or in which they first appear, because it concerns a certain balance of the prime qualities, as dry, moist, cold.[30]

For "doctor" and "philosopher" in terms of Cremonini's argument, one can read "Galen" and "Aristotle" respectively. This identification appears in the introduction to his second series of lectures on controversies between philosophers and doctors: "We return to consideration of the opinions of Galen and of the doctors against Aristotle".[31]

It is tempting to associate these lectures, and the three printed treatises on generation topics, with the response by Cremonini to an attack on his theory of innate heat made by the Paduan professor of anatomy, Pompeio Caimo, in 1625–1626. The full titles of the treatises certainly suggest a fighting mood, and parallel the theme of the lectures: Tractatus de Semine pro Aristotele contra Galenum; Apologia dictorum Aristotelis de calido innato Adversus Galenum; Apologia dictorum Aristotelis de origine, et principatu membrorum Adversus Galenum.[32]

The dedication to the work on innate heat gives no doubt that Cremonini proposes to concern himself with a question fundamental to medicine, not alone to natural philosophy.

> The topic is not a meagre one. It concerns the fundamentals of the art of medicine, which leads to the body's sickness or health. A discussion of the heat which is called innate, and hence the life and death of man. [It is] an argument for Aristotle and against Galen; the former master of all knowledge, the latter one of the leaders of medicine.[33]

If Aristotle is Master of all learning and Galen merely a prominent doctor, there is no doubt in Cremonini's mind as to which authority he will support.

He is similarly direct in his introduction to the Apology on the principal parts. "We intend to study the structure of animals following the opinions of Aristotle" he says. Among the animals, he will concentrate on man, since his dispute is with doctors and medical opinions, and "Doctors concern themselves solely with man."[34] But he insists, though he may have harsh words for Galen, that his purpose is not glory, or even Galen-bashing, but the discovery of the truth.[35]

All three works have a similar, and somewhat boring, structure: thesis (Aristotle's opinion); antithesis (Galen's opinion); "objections of Galen against Aristotle, which are fundamental to Galenic opinions"; and finally, resolution of the argument in favour of Aristotle.[36]

To begin with the Tractatus de semine: its subject is not everything to do with generation, "but only that on which there is a notable controversy of major importance, namely as much as concerns the active and passive faculty which is found within it [the seed] for generation."[36] Moreover, Cremonini will confine himself to the action of these faculties, not in animals and plants, but only in man - unlike Aristotle himself.[37]

In chapter ii, he confronts directly the absurdities which he finds in Galen's account, in which both male and female produce seed having an active and passive, formal and material, part. There are two basic difficulties with the Galenic explanation, according to Cremonini.

The first concerns the duplication of efficient cause and the problem of a double soul. If female seed contained an active virtue, it would have a soul-making faculty; two souls would thus be produced, and the result would always be twins.[38] On the material side, the problem is similar; in Cremonini's phrase, "to a single form, a single matter is ordained". The form is thus the male contribution, the matter the female, as menstrual blood. In Aristotelian terms, the clinching argument against an equal and similar female contribution is that if the mother provided seed with both active and passive powers; the material from which the foetus was shaped; and a place for it to grow and develop, the male parent would be superfluous and parthenogenesis would in fact occur.[39]

In the following chapters, Cremonini deals with similar firmness with the other Galenic arguments for female semen: the evidence from anatomy, from the resemblance of children to their mothers and from the argument that certain parts, such as bone, originate from seed rather than blood. His chief addition to the arguments of Averroes is his suggestion (based, indeed, on Aristotle) that the basic difference between male and female is a difference in heat [eg. in G.A. IV i]. For the production of true semen, the degree of innate heat required is higher than for the production of menstrua; how, then, in one and the same individual can there co-exist two grades of heat? How can women produce both menstrua and seed?[40]

Cremonini's argument corresponds precisely with Averroes' paraphrase of Book I chapter 20; it is "an examination of the opinions of Aristotle and Galen on female seed, whether it has an active principle in generation." For Cremonini, as for Aristotle and Averroes, the answer is no; and his chief argument is a development of the Averroist one of the "double agent". Cremonini argues, not that if the female seed contained an active virtue, either efficient or material, men would be superfluous; but that two souls, or two sorts of matter, would be found in the uterus. The result in the first case would be twins; in the second, simple redundancy. In his conclusion, Cremonini tells us that he has put forward his arguments "in the manner of the digressio [or paraphrase] of Averroes."

In conclusion, Cremonini tells us that he has produced, not a general treatise, but a monograph: "only what Galen brings up against Aristotle on the subject of active and passive faculty." And he promises to return to a discussion of other topics in De generatione, in particular the question of resemblance which is dealt with in Generation of Animals IV 3. And this future work on generation (which does not appear to have been written), like the present one, will be presented "in the manner of the digressio of Averroes."[41]

Cremonini's Apologia...de calido innato is concerned with the question of the relationship between elemental or temperamental and innate or vital heat, and hence with the question of what distinguishes living from non-living. (See the chapter on innate heat for the opinion of medical writers on this same topic). He begins with a reminder of the Aristotelian doctrine of elemental mixture, as set out particularly in De generatione et corruptione and Meteorologica IV i, in which heat and cold play an active part in the constitution of bodies.[42] In this theory, the similar or homogeneous parts of animated bodies, such as bone, nerve and flesh, are distinguished by different degrees of elemental heat. How are these distinct temperaments reconciled and harmonised into one degree of heat, that of the temperament of the animal or plant as a whole?

> It follows.. that in the different parts containing different similarities there must be different heats.. how do these heats come together in one, since an ensouled compound is said to be one, and to have one heat for its preservation and nutrition, and others required for the business of living?[43]

One solution to this problem is to distinguish two types of heat, one, elemental or temperamental heat proper to the parts, and another, innate heat which is responsible for coordinating the actions of the body as a whole. This is the system presented by Jean Fernel in his Physiologia, for example.[44] Cremonini's solution is different: the same elemental qualities which go to form mixtures can also act as instruments of an external efficient cause:

> elemental qualities, by which elements come together in mixtures, are the instruments of an external efficient, which is seed in animates, and simply speaking, the heavens.[45]

Heat and cold thus act as instruments of the true formal cause, as is further explained in his Tractatus de coeli efficientia.[46] So innate or native heat is not different from elemental heat, but simply different in its mode of action in animate and inanimate bodies.[47]

This is the basis of Cremonini's interpretation of the passage, in Book II Chapter iii of Generation of Animals, in which Aristotle says that the substance in the seed which makes it fertile is different from and more divine than the elements and is analogous to the element of the stars. Whatever he may have said, Cremonini insists, "it is certain that Aristotle did not want any celestial heat, which was a quality of celestial substance, to be present in sublunar bodies, whether animate or inanimate."[48] For the properties of the quintessence are incompatible with those of sublunar substances, as indeed he had explained in his treatise de quinta caeli substantia. Aristotle meant to compare the action of the two agents, not their substances:

> It is certain, that if it be analogous it is not by its nature, because that would not be a proportion, but a conformity in substance and nature.. Moreover, the seminal heat is proportionate with the element of the stars; because just as the nature of the element of the stars is to form everything, so the heat's nature is to form certain species of animals, and it has an efficient power ordained under the heavens.[49]

Indeed,

> Elements are subordinate to the heavens, as instruments of all sublunar generation, because the heavens could not accomplish it without them; both generation and corruption require activity and passivity, which does not exist without contrary qualities in the same substance.[50]

The alternative to this position, as usual, is to be found in the works of Galen, in particular in On the Seed, On temperaments and in the short Hippocratic commentary Against Lycus, an exposition of Aphorisms 14 i. And, as usual, Galen's response leaves much to be desired, both in accuracy and taste; he goes for his opponents fangs out, says Cremonini, "like a viper attacking with poisonous bite."[51] Galen's argument, as Cremonini reports it, is that "innate heat is a kind of body, and composed of corporeal substance" and that "this corporeal substance is constituted from the first mixture of seed and menstrual blood, because it is necessary that this sort of heat, in which the foetus itself as a whole must consist, should be first produced, and first to take place."[52]

Cremonini produces a range of detailed arguments against this position, but his chief argument is that innate heat as Galen distinguishes it is superfluous:

> Assuredly, temperamental heat is in all those things, and has all those powers; as we said at the beginning; and these are just the same powers as innate heat has.[53]

If it is necessary for innate heat to inhere in some body, the most appropriate choice is the blood; otherwise, if it is considered to be a body 'in its own right', the two bodies, blood and innate heat, must occupy the same space, and both must be dispersed throughout the body to nourish the parts.[54]

Thus the seat of these faculties is the heart, and the blood, which in Aristotelian terms are the principles of generation and the first organs formed. This theme, which we also found in the Averroes commentary, is developed at length in the third of Cremonini's medical works, Apologia... de origine et principatu membrorum.[55] This somewhat lengthy treatise - over four hundred pages - argues exhaustively for two main points: that there must be a determined order in which the parts are generated; and that in this order, the heart must be first to exist. In the process, Cremonini opts generally for an Aristotelian system of physiology, based on Generation of Animals (especially Book II) and other of the "Animal Books". Among other writers, he cites John Philoponus and Averroes approvingly, especially the latter's Digressiones on Parts and Generation of Animals. For example, Chapter VI considers the arguments of Averroes for the primacy of the heart, based on its importance as a source of heat for the body. And in Chapter XXVIII he again draws on Averroes for an explanation of the relation of the heart and the vital faculty.[56]

According to Cremonini, Book II chapter 3 of Generation of Animals is one of the two chapters in the animal books "which should be frequently referred to, for an understanding of the soul and of substance"[57] The other is the first chapter of Parts of Animals, on which he also wrote a commentary which survives in manuscript. His commentary on Generation of Animals II iii, which exists in several copies,[58] repeats many of the themes which we have seen in the Averroes paraphrase, and in the printed works of Cremonini.

The first of these is the respective role of male and female as "principles of generation". In his introduction to II iii, Cremonini reviews the arguments of the last chapters of Book I and II i against pangenesis and against female seed. The female represents the material principle of generation, the male the formal; it is impossible that the contributions of the two sexes should be alike.[59] He follows Aristotle (G.A. II i 732a) in explaining the separation of the sexes in animals from the principle that "it is good to separate the superior from the inferior".[60]

The second, related question which he considers is the nature of the semen - that is, the male contribution - and its role in generation. The semen, for Cremonini as for Aristotle and Averroes, and in opposition to Galen, 'forms the foetus, but forms no part of the foetus': it acts only facultatively, not materially.[61] How does the seed act, how form the organic parts of the body 'by its faculty'?[62] These organic parts are the instruments of the soul, and the question which interests Cremonini, and occupies much of the first part of his Exposition, is the question of the generation of the soul.

> the parts are not organic except by means of the soul, therefore [we can] not explain how the seed makes them, unless it is explained how the seed makes the soul.[63]

This is clearly not an easy question. Is the geniture itself ensouled, and if so, in what sense? It is clear to Cremonini as to Aristotle that fertile seed, plant or animal, has life, a facultas vitalis, in comparison with infertile seed. Aristotle had offered three possible solutions to this question, which Cremonini lists. Either soul preexists in the geniture, or it enters from outside, or, in Aristotle's own solution, the vegetative and sensitive parts of soul pre-exist, the rational part is infused.[64] In Christian Aristotelianism, this rational part had become identified with the immortal soul of Christian doctrine, and the precise moment of its entry debated. In discussing this question, Cremonini considers both the Aristotelian model and the alternative presented by neo-Platonic philosophy. The Platonist solution of a soul entering from the outside presents the problem of transmigration of souls, and Cremonini seems in the end to adopt a broadly Aristotelian or Averroist position. As the geniture develops, the three souls appear successively, so that the foetus lives first with the life of a plant, then of an animal, before becoming fully human.[65] Development of the soul keeps pace with the 'organisation' of the body, with which it is intimately involved:

> the conceptus is as it were as organised as is necessary for receiving form from them [the souls].[66]

The vegetative and sensitive souls are closely associated with the male seed; the first is within it, 'not yet separated', the second arises from it. Cremonini here repeats the point made in De semine, that seed - and here again he confines his meaning to the male contribution in animals, ignoring the geniture of plants, in which the two sexes are conjoined - has a vis animastica, a soul-making faculty. "Of the sensitive part [of the soul] there is no doubt that it arises solely from this same masculine power."[67] that is from the soul-making faculty of the male seed.

The question of the origin of mind, or of the rational soul, is more difficult. If the mind or intellect is not present in the geniture but enters from outside, when, and whence, does it enter? Cremonini seems to attempt to 'tie down' even the intellect to physical substance, making it conform more closely in its origins with the two lower souls. He uses the Platonic idea of the vehicle by means of which the soul or intelligence enters into inanimate matter to suggest that the male seed may function as a kind of vehicle, a medium which 'draws in' the immaterial soul and without which it cannot be received in the material part of the conceptus. And he also seems to suggest that even the intellectual soul cannot act without the body, for two reasons. First, the mind is corporis informativus, the power which gives the body form.

And secondly, he suggests that in so far as the use of intellect involves the imagination, it may require a location in the body, just as walking requires feet.[68] If one leaves aside the mind, the physical dependence of soul or anima is even clearer:

> In so far as an action is corporeal it cannot exist without a body, or before body, and it cannot enter from outside...with the exception of mind action is corporeal...below mind no soul is before body, and no soul can enter from outside.[69]

The problem of the successive development of the faculties of the soul, so that the foetus lives first 'in the manner of a plant' (ad modum plantarum), then animal-like, before attaining the full exercise of its humanity, leads Cremonini into a digression on the question of species. A man is distinguished from a horse by his essentia, his essence or being, but it is not necessary to suggest that different species have different 'grades of being'. Rather, the special way in which the horse, for example, is constituted is what distinguishes it essentially from all other animals.[70] But Cremonini has just suggested that a foetus is animal (ie. possessed of the sensitive soul) before it attains its specific identity as horse, or man. And he also seems to suggest that the presence of the rational soul in a human infant, the 'special and peculiar constitution' which proves it truly human, may only be indicated by signs of intelligence, such as the first smile. But this presents a problem. From the beginning of its development the foetus is organised in a characteristically human way, with the actualisation of humanity as its end. But there can be no organisation as human without a human soul. It seems inescapable in logic that the rational soul, which acts as forma corporis informans, the 'forming form' of the body, must be present to constitute the foetus in a particularly human way, from the earliest stages of its development.[71] If the foetus acquires the general nature of an animal before developing a specific human nature together with such human attributes as reason, how is it that it develops organically in a specifically human way, long before such mental attributes can be demonstrated? The animation of the foetus is a question which, as we shall see, was also dealt with by medical writers. Cremonini seems to seek to solve this dilemma by identifying the anima as closely as possible with the "informing form" of the body, in a manner which again recalls Averroes. This "Arabist" solution is one which, as we have seen, Accorambonius tried to avoid - not entirely successfully.

For Cremonini, as for Averroes, the major questions raised by the Generation of Animals are: the respective roles of male and female in generation; whether the semen is materially part of the conceptus or not; the order of formation of the parts; and the question of when, how and whence the soul enters the body. These are specific, detailed, questions, which, however, ignore much of the 'practical' content of Generation of Animals - the sections on the variations in structure of the sexual organs in various animals and on sex determination and resemblance, for example. The treatment which Cremonini gives them seems characteristic of one important, and relatively distinct, strand of Renaissance generation theory, that which derives from the "Averroist" tradition of natural philosophy in North Italian universities of the later Renaissance - a tradition which has often been identified particularly with Padua.[72]

Notes

1. Apologia dictorum Aristotelis de calido innato. Adversus Galenum. Venice, 1626. (Hereafter cited as Cremonini, De calido innato). Tractatus de Semine.. pro Aristotele adversus Galenum, [Leiden, 1634] pp. 131 e.s. (Cited as Cremonini, De semine)
2. Apologia dictorum Aristotelis De origine, et principatu membrorum. Adversus Galenum. Venice, 1627. (Cited as Cremonini, De origine membrorum)
3. L. Mabilleau, Étude historique sur la philosophie de la Renaissance en Italie: Cesare Cremonini. Paris, 1881, pp. 1–8; M.A. Del Torre, Studi su Cesare Cremonini; cosmologia e logica nel tardo Aristotelismo padovano, Padua, 1968, pp. 15–17; Lohr, 1975, 28(4):728.
4. Mabilleau, op. cit. p. 7; quoting Portinari: "C'est aujourd'hui le plus fameux philosophe d'Europe, le plus savant dans la doctrine platonicienne aussi bien que dans la peripateticienne, grand orateur et poete illustre, l'ornement et la gloire de l'Université de Padoue."
5. Gabriel Naudé, in Naudeana et Patiniana, ou singularitez remarquables, prises des conversations de Mess. Naude et Patin. Paris, 1701, p. 45.
6. Further discussion and lists of Cremonini's productions can be found in Mabilleau, op. cit., pp. 62–85 (chapter 3, L'oeuvre de Cremonini) and 367–386 (Catalogue General des Oeuvres Manuscrites de Cremonini); Del Torre, op. cit. pp. 20–31 (Gli scritti del Cremonini); and Lohr, op. cit. (1975) pp. 728–739. Mabilleau catalogues three collections, those of the Marciana Library in Venice, the Biblioteca Universitaria in Padua and the Bibliothèque Nationale, Paris, and arranges the collections under general subject headings such as natural philosophy. Del Torre follows him in listing titles by library, and adds a manuscript from the library of St. Giovanni di Verdara in Padua, as well as one or two other titles. Lohr's list is given by title and is accordingly difficult to compare, but includes manuscripts in other libraries and is accordingly the most complete.
7. See Lohr, (1975) and note 11, above, introduction to Part II.
8. Op. cit. note 6, especially Mabilleau, p. 381. I have been able to examine microfilms of two versions of the Expositio of Generation of Animals II iii, those in the Museo Civico, Padua (CM. 494 (XVII) item 2) and the Biblioteca Marciana, (Lat. VI 179 (XVII) fs. 75–142). Both are copies, in poor hands and with many mistakes.
9. Caesaris Cremonini Centensis Disputatio de coelo:... Venice, 1613. Apologia dictorum Aristotelis de quinta coeli substantia adversus Xenarcum, Joannem Grammaticum et alios. Venice, 1616. Manuscript, Tractatus de caeli efficentia, Biblioteca Marciana, lat. VI cod. 176 (dated 1628) and Biblioteca Universitaria, Padua, cod. 200 (xvii) 1; My remarks are based on the transcriptions using both these sources, given by Maria del Torre and her discussion of the text, op. cit. pp. 63–88, "L'inedito 'De Coeli Efficentia'". The Tractatus de caeli efficentia is one of several surviving manuscripts on topics dealing with the nature of the

heavens (cf. Lohr, pp. 733–4 for a list) and was probably written later than the two published works; it is promised by Cremonini as a successor to his Apologia… de quinta coeli substantia (1616) at the end of that work (p.202) and again in 1619 in a letter to the Inquisitor at Padua (Mabilleau, p. 71).

10. Cremonini, Disputatio de coelo (1613), Prima Pars, II, i-iii, pp. 33–40; V, 87 ff, especially ch. iii, 91 ff., on the animation of the heavens; and ch. iv, p. 95 ff.
11. Ibid. Pars Tertia, esp. 287, 302.
12. Mabilleau, op. cit. ch. ii, esp. 39–52. His errors were tabulated in three areas: on the nature of the heavens, of the human soul, and of God. (p. 41).
13. Apologia de quinta caeli substantia (1616) p. 4: "Dicuntur nempe demonstrationes, suppositis eiusdem Aristotelis principiis." The same caution appears in his definition of philosophic truth as fallible, based only on sense perception: "si alicubi diximus ipsum [Aristotle] versari in rei veritate, id intelliximus secundum eum modum veritas diminutum, et fallacem, qui potest haberi ex sensaris.." (Ibid., p. 6).
14. Ibid. pp. 8–9
15. See note 9, above, and discussion by Mabilleau, op. cit., pp. 227–241, esp. 232 ff, 235. Mabilleau traces this discussion of the celestial action back to Averroes' commentary on Metaphysics XII in Destructio destructionem disp. III, and to Conciliator, differentia VI, on the influence of the stars, and thence back to Ptolemy's Centiloquium. (Op. cit., 243, 253–4).
16. See also a vernacular version of this argument, "Cesare Cremonini nel Primo delle sue Meteori a diffessa per Aristotile Contro li Astrologhi Giudiciarii," p. 131 e.s. of Raccolta medica et astrologica, divisa in dio discorsi. Venice, 1645
17. Ibid. pp. 131–7.
18. Mabilleau, op. cit. 254–5, refs.; Del Torre, op. cit., pp. 63–88, especially 68, 83.
19. Del Torre, pp. 83, 84.
20. Passage transcribed, Ibid, p. 85, note 1.
21. Explanatio Prooemii Librorum Aristotelis De Physico Audito. Padua, 1596.
22. Ibid. 21r. "Sunt, quae veniunt explicanda, ea, quae Averr. in suo prologo enumeravit.."
23. Ibid. 26v-33r, "Partitio Philosophiae Naturalis et ordo partium". The general principle governing his order is given on 31v: "Substantia nempe animalium quaerenda est, deinde affectiones eorundem."
24. Ibid. 35r. "dispositiones corporales Animalis duobus modis possunt considerari, vel pro ut corpus essentiam animalis constituit, et sic corporis cognitio ob rationes supra positas magis confert ad cognitionem animae, quam cognitio Animae ad cognitionem corporis, vel pro ut corpus est Animae instrumentum.. et sic magis deservit Anima ad cognoscendum corpus."
25. Ibid. 34v for Cremonini's own position; 34v-35r for the opinion of "the Latins" and their opposition to change: "qui mutare eum ordine voluerit, Latini mordicus hoc defendunt" (35r).
26. Ibid. 21v. "Nos autem..libentius hominem pro subjecto praecipuo naturalis philosophiae constituimus.."

27. Ibid. "Naturalis igitur contemplatio primam intendit hominis cognitionem, quasi terminus naturalis cognitionis.."
28. Excellentissimi Caesaris Cremonini Centensis Lecturae Exordium Habitum Patavii. VI. Kalend. Februar. MDXCI. Quo is primum tempore philosophiae interpres ordinarius eo est Profectus. Ferrara, 1591, pp. 15–16: "corporata membrorum mole compositorum vultum imitatur; qualitatum temperamento exprimit primorum corporum proprietates; sicque illa omnia, quae Mundus habet, quasi epilogo ut loquuntur, in homine conclusa fuisse inimadvertimus, eaque de causa eundem parvum mundum non ab re nominamus; quapropter dum ipse se ipsum videt, postquam intus alia continet, fieri prorsus nequit, quin ad res illas de vera earum facie internoscendas, quarum simulachra in se ipso, veluti in quodam speculo adumbrata perspexerit.."
29. Cremonini, De Physico Auditu (1596) 33v. "..completa est naturalis contemplatio, et incipit medicina, ab iis, quae de causis sanitatis et morbi philosophia est contemplata."
30. Ibid. 36r: "Physici interest prima morbi, valetudinisque fundamenta perscrutari, idest a quibus initiis oriatur, vel in quibus primum versetur, vel valetudo, vel invaletudo quia in quadam primarum qualitatum utpote siccitatis, humoris, frigoris mediocritate versatur."
31. "Revertimur ad consideranda dicta Galeni et medicorum contra Aristotelem"; quoted by Mabilleau, op. cit. in his Catalogue, p. 381
32. For full references, see note 1, above. On the list of works by Cremonini made by his friend and contemporary Tomasini, De semine appears, revealingly, as "Expositio in digressionem Averrhois de semine contra Galenum pro Aristotele" and De origine membrorum as "Tractatus de animalis structura, seu de principatu cordis." Ibid. pp. 73 and 76, n. 3 and 5.
33. Cremonini, de calido innato, p. 3. "Res ipsa non adeo exigua. Agitur hic fundamentum artis Medicae, quae aut in perniciem, aut in corporum salutem. Sermo de calido, quod vocat innatum; hinc vita hominum, et interitus. Disputatio pro Aristotele contra Galenum; ille totius sapientiae, hic inter Medicinae Principes."
34. Cremonini, De origine membrorum, p. 1: "Medici in solo homine versantur".
35. Ibid. p. 2, similar sentiments can be found in de calido innato, p. 3, De semine p. 134.
36. As set out at the beginning of De semine, p. 136.
37. Ibid. 134: "sed illud solum quo est controversia praecipua et majoris momenti, quantum scilicet spectat ad facultatem activam et passivam quae pro generatione in eo reperitur." 135; repeated in conclusion, p. 219.
38. Ibid. p. 146: "si semen [that is, female seed] haberet facultatem activam, haberet facultatem animasticam, quia haec facultas tendet ad productionem animae; quare si sunt duo semina distincta, sunt etiam ibi duae facultates animasticae,atque ideo duo semper deberent generari."
39. Ibid. 148, 216.
40. Ibid. pp. 208–9.
41. Ibid. p. 219: "solum ista quae de virtute activa et passiva Galenus objicit Aristoteli." Cremonini describes his plans for a future work in these terms: "Est

modo Averrois digressio explicanda, quod faciemus, Deo dante, in principio studii futuri; ubi etiam, si considerabitis, pleniorem explicationem illorum quae attinent ad similitudinem natorum, de qua agit Aristoteles, de generat. animal. cap. 3. [G.A. IV 3 deals with resemblance] eiusdem capitis explicationem habebimus."

42. See the chapter on elements and temperaments, Part III. Chap. 14, below.
43. Cremonini, de calido innato, p. 10. "Sequitur..quod in diversis partibus sint diversi calores continentes diversas similaritates.. quomodo is <$$$> ti calores in unum veniunt; ita ut dicatur mistura animatum esse unum, et habere unum calorem ad se ipsum conservandum nutritione, et aliis requisitis ad opera vitae?"
44. See below, Part III.
45. Ibid. (1634 edition) p. 18. "qualitates elementares, per quas elementa veniunt ad mistionem, sunt instrumenta externi efficientis, quod est semen in animatis, et simpliciter loquendi Coelum."
46. Ibid. pp. 12–14 (1626)
47. Ibid. p. 19. "Calidum modo temperamenti in animatis non solum habet ista facere, sed habet usum pro animae operibus, et potest agere instrinsice in suum subjectum pro conservatione et sui ipsius, et totius animati."
48. Ibid. p. 58 (p. 87, 1634). "certum est, Aristotelum nolle ullum caelestem calorem, qui sit qualitas de caeli substantia, esse in corporibus sublunaris sive animatis, sive inanimatis."
49. Ibid. "Certum est, quod si respondet proportione non est de illius natura, quia non esset proportio, sed esset convenientia in substantia, et natura. Et quod proportionatur alicui; est diversum ab illo, cui proportionatur, nec potest dici esse de illius substantia. Habet autem calor ille seminalis proportionem cum elemento stellarum; quia sicut elementum stellarum se habet ad omnia facienda, sic iste calor se habet ad faciendum animatum in taligenere, et habet vim effectivam sub coelo ordinatam."
50. Ibid. p. 59 (89–90, 1634) "elementa sunt coelo subordinata, tanquam instrumenta ad omnem generationem sublunarem, quia coelum sine istis non poterat hoc praestare; et generatio, et corruptio acquirit actionem, et passionem, quae non est nisi per qualitates contrarias."
51. Ibid. p. 69. "Galenus autem invehitur contra illum [i.e. Lycus] more solito absque ulla modestia. Adoritur illum quasi vipera impetens morsu venenato."
52. Ibid. p. 70 "Calidum innatum esse corpus quoddam, et substantiam corpoream compositum. Istam corpoream substantiam constitui ex prima concretione seminis, et sanguinis menstrui, quia oporteat huiscemodi calidum, in quo totum ipsius foetus debet consistere, esse primo productum, et primo consistere."
53. Ibid. p. 75. "Calor certe temperamenti est in omnibus istis, et habit omnes illas facultates, quas diximus in principio, et sunt praecisae facultates calidi innati."
54. Ibid. pp. 75–76.
55. See note 1 for full title.
56. Cremonini, De origine membrorum, Averroes: chapters VI, pp. 22–25, XXVIII p. 110, XLII p. 165; Philoponus, ch. XLI, p. 161.

57. "Duo sunt capite in libris de animalibus quae frequenter adduci soient pro animis nam et substantia cognoscenda.." Cesare Cremonini, Expositio super caput tertium libri secundi de generatione animalium, Museo Civico di Padova, M.S. CM 494 (XVII) Item 2. fol. 54r. [Wellcome Microfilm 210a] I have had access to a microfolm of another version of this manuscript M.S. Lat. VI 179 (XVII) [= 2825] fols. 75r-142v, Biblioteca Nazionale Marciana, Venice. [Wellcome Microfilm 212a]. On the whole the Venice manuscript is even more illegible, and citations will be taken from the Padua copy unless otherwise noted.
58. In addition to the two copies cited above (notes 8 and 57) Lohr, 1975 p. 739, lists another version in the Biblioteca Universitaria, Padua, M.S. BU 200 (XVII)
59. Wellcome Microfilm 210a, 54r–v. This part of the Expositio corresponds to G.A. I ch. 17 e.s, II ch. 1.
60. Ibid. fol. 55r. On this point, see also Cesare Cremonini, De semine, chapter III, pp. 159–160, which is an attack on Galen, De semine, I ix (K. IV 543–6, esp. p. 544). Here Galen considers the parallels between the development of plants (in which the sexes are generally not separated) and animals, and points out that Aristotle is forced by his separation of material and efficient causes into the two sexes to explain generation differently in plants and animals. Elsewhere in book II of G.A., however - in particular, II iii 736b 12–15, and II iv 739b 34 - 740a 4 - Aristotle uses the analogy of plant generation to explain animal generation, another respect in which Cremonini follows him.
61. Ibid. 56r - 58r. "semen non sit pars constituendi conceptus sed agat solus active non corpulentia (56r)".
62. Ibid. 58r: "quo modo semen sua facultate possit facere partes organicas."
63. Ibid. 58r. "membra non sunt organica nisi per animam, ideo non [possumus] declarari quo modo semen faciat illa, nisi declaretur, quo modo faciat animam."
64. Ibid. 64v. (based on G.A. II iii 736a 24-736b 29.)
65. Ibid. 58v - 63r.
66. Ibid. 63r: "quidam conceptus est sic organisatus ut sufficiat pro recipienda illarum informatione"; a somewhat circular statement.
67. Ibid. 58v. "semina sive genitura, habent vim animasticam..de parte vero sensuali non est dubium, qd provenit a vi sola ipsius masculi.."
68. Ibid. 64v, 65v-66r.
69. Ibid. 65v. "proprium actio est corporalis non [possit] esse sine corpore..non [possit] esse ante corpus, et non [possit] extrinsecus venire…mente excepta actio est corporalis..primo infra mentem nulla anima est ante corpus, et nulla anima pot; extrinsecus venire."
70. Ibid. 60v.
71. Ibid. 60r (signs of rational soul), 61r, "non possit esse humana organisatio sine anima humana", 67v. rational soul as forma informans of body.
72. See in particular the works cited in note 8, above, by Gilbert, Randall, Jardine, Wightman, Nardi; this traditional association, as Dr. Charles Schmitt points out

in "Paduan Averroism", goes back at least to Renan, (Paris, 1852, chapter 3 in particular). Modern scholars have challenged this association while pointing out the continued importance of the Great Commentator in the Renaissance. Among the wealth of studies one might mention, in addition to the above, C.B. Schmitt, Critical Survey and P.O. Kristeller, "Renaissance Aristotelianism" and "Paduan Averroism and Alexandrism."

Chapter 12
The 'Anti-commentary' of Bernardino Telesio (1509–1588)

Abstract In a class by itself is the *De Rerum Natura juxta propria principia* of Bernardino Telesio. This is a commentary on the natural philosophical works of Aristotle, and a rejection of the Philosopher Telesio argues that the world is constructed of two substances, heaven and earth, shaped by paired opposing qualities of heat and cold. Generation involves *spiritus* with its qualities of heat and motion. The two seeds contribute to the foetus by fusion, not as agent and patient. Telesio rejects Aristotle's claim for the primacy of the heart in generation. He tackles a much trickier question: the operation of the soul in generation. The role of *spiritus* in Telesio's writings seems designed to counter a major problem of unity. Galen's solution, the multiplication of 'faculties' to control the body's diverse operations, compounds the problem. *Spiritus* is one, and performs the actions of the soul throughout the body. Telesio's monistic system, with its emphasis on unity, has considerable claims to originality. He is included because his work follows the standard format of Aristotle commentaries, and treats many of the same problems.

The list of Renaissance treatises which touch on the generation debate is much longer than the list of commentaries on De generatione animalium proper. As we have seen, commentaries on Generation of Animals formed a relatively small group in comparison with commentaries on other Aristotelian works, such as the Physics, which were required texts for the arts course. The above chapters provide what I hope is an adequate sample of natural philosophical opinion on certain 'classic questions' concerning generation.[1] Generation was also a favourite topic in Renaissance debate literature of the quaestiones or disputationes type. Some of these, such as Valles' Controversarium medicarum ac philosophorum,[2] Caesalpino's Quaestionum Peripateticarum and Book VI of Scaliger's Exotericarum (as well as Cardan's De Subtilitate, to which it was a reply) touched on generation questions in the course of a much more widely-ranging debate on varied natural-philosophical topics. Others – for example Franzosius's Tractatus de Semine, Alberti's De Nutritione and the polemics of Mundinius versus Hofmann,[3] are disputationes in which the major 'dispute' concerns the respective roles of male and female in generation. Lengthy as they often are, they focus on a very small area.

L. Deer Richardson, *Academic Theories of Generation in the Renaissance*, History, Philosophy and Theory of the Life Sciences 22,
https://doi.org/10.1007/978-3-319-69336-1_12

More comprehensive than any of these, and in a class by itself, is the De Rerum Natura juxta propria principia of Bernardino Telesio: in one sense a commentary on the natural philosophical works of Aristotle, in another sense a rejection of the Philosopher – an anti-commentary, as it were. Contemporaries, such as Caspar Hofmann, sometimes divided the combatants in the generation debate into Aristotelians, Galenists and Telesians; Temkin's Galenism uses a similar three-part division. It was just this sort of alternative system which Telesio hoped, and claimed, to provide.[4]

In De Rerum Natura, Telesio presents the thesis that the world is constructed of two substances, heaven and earth, shaped by paired opposing qualities of heat and cold. The first four books deal with the inanimate world, Book V with animals and plants, and Book VI with man's physical nature, including the question of generation. In these last two books, Telesio takes up the major points of debate found in contemporary commentaries on De generatione.[5] At the beginning of Book V, Telesio reviews his argument so far:

> The primary bodies of the world, and those of which other beings are constituted, are heaven and earth; the former constituted by the highest form of pure heat, the latter by cold, pure and most intense.[6]

But living things are different from, and nobler than, the inanimate objects of the natural world, generated by the opposing qualities of heat and cold. For the living generate of themselves, from their own, particular seeds.[7] The substance of the new individual is drawn out from the seed by the operation of spiritus, which for Telesio has many of the formative powers which he attributes to heat in the natural world.[8] Telesio calls it 'the substance of the soul' and makes it responsible for sensation and motion. Indeed, he assigns to spiritus many of the activities which Aristotelians gave to the soul or anima: a point to which I shall return.[9] Telesio locates spiritus in various places in the body, without distinguishing it into animal, vital and natural spirits as medical writers usually did. Both here and in the posthumously published Quod animal universum ab unica substantia gubernatur he places it in the cerebral ventricles and the nervous tissue of the body. But he also associates it with the 'fibres' which Galen used to explain the actions of the natural faculty; and with the blood. It is the last, elaborated to form true semen, which is most closely associated with the process of generation.[10]

Generation for Telesio involves spiritus by virtue of its qualities of heat and motion, especially the former. Heat is present in the seed, formed from blood in the body of the parent, and increased by the process of elaboration. The former blood becomes hotter and more full of spiritus: part becomes more rarefied and spirit-like, while the rest becomes thicker and more viscous, bitter instead of sweet in nature. For proof of this, he refers us to the ancients, and in particular Aristotle (G.A. II iii 736 b), who spoke of the heat in the seed and associated it with pneuma or spiritus. The heat for generation comes also from the uterus, which itself has a supply of spiritus-rich blood.[11]

Generation is only one of the operations of spiritus, according to Telesio. It is also responsible for the other actions of the natural faculty as described by Galen, working through the Galenic fibres. By analogy, it is also responsible for the activities of plants. Telesio associates the denser plant-spiritus with the fibres of the plant cortex: which explains why if you ring a tree it dies. But spiritus controls the higher animal functions as well: sensation, motion and even some of the lower mental functions, including emotions. In Book VI, Telesio gives us a detailed account of the formation of the parts of the human body and the action of spiritus within each. Books VII to IX consider spiritus in relation to man's sensitive, intellectual and moral qualities.

The last 10 chapters of Book VI are devoted to summarising the main points of the Aristotelian argument on the 'classic questions' of generation (chapters 31–33) and refuting them (chapters 34–40). As Telesio summarises them, the Peripatetic arguments are:

1. semen is the male contribution, menstrua the female one;
2. in viviparous animals, seed is formed at the time of coition, as a result of friction;
3. testes contribute nothing to the production of seed;
4. the male contribution is agent only, dissipates after coitus and forms no part of the foetus; the female contribution is material only;
5. "thus the soul is rightly the work of the male, the body of the female."[12]

Telesio's refutation of these points relies heavily on Galen for its authority and detail – as does, for example, his description of the 'use' of the genital parts in chapter 19 of Book VI. Semen is elaborated in the testes of those animals which have them, and prepared in advance, not at the time of coition; furthermore, the impetus to the sexual act comes not from a potentia generativa residing in the heart, but from the amount and qualities of the seed produced in the testis, particularly its 'biting', acrid quality and its high spiritus content – both of which promote sexual desire.[13]

Females as well as males emit a 'moisture' which is a true fertile semen, although it is less concocted and concentrated than male seed, because of the lesser heat which women possess. The existence of female seed is proved by the presence of female 'testes' and by the pleasure which both sexes feel at the time of 'emission'. As with the male, however, this pleasure does not mean that female seed is necessarily fertile; the question of what determines the fertility of seed will be considered further below.[14]

The male seed provides, not the soul or agent power of generation, but a part of the foetus; the white, bloodless, seminal parts of Galen. And the difference between male and female contributions to generation, like the difference in appearance of male and female genital parts, is not due to a difference in kind, but, "as Galen tells us", simply due to the different degrees of heat in the two sexes. According to Telesio, the two seeds contribute to the foetus by fusion, not as agent and patient – as can be observed in the eggs of fishes and other ovipara, including birds. Thus, "the foetus cannot be constituted from a soul dwelling in the male seed", as Aristotle's theory implies.[15]

Telesio also rejects Aristotle's claim for the primacy of the heart in generation in favour of the testis; 'nervous tissue' (genus nervosum); and the uterus itself, heated by the blood. In Quod animal, he similarly attacks the Aristotelian notion that the heart is the primary organ of the body and source of the nerves.[16]

As well as refuting these specific points in Aristotle's account, Telesio tackles a much trickier general question: the operation of the soul in generation. What do we mean when we say that seed is fertile? How, for example, can we distinguish a fertile hen's egg from an infertile one, since both are produced in the same way and seem identical in composition as well as appearance? If the difference is in the presence or absence of soul or anima – in what sense can we say that undeveloped geniture is ensouled? This problem is related to that raised by the phenomenon of spontaneous generation: what is the common factor in successful generation of both types?

At the beginning of Book V of De Rerum Natura, Telesio distinguishes between an 'animal soul' (which he identifies with spiritus, and so with the action of heat) and a human or immortal soul. The latter, as he defines it, is 'drawn out' from the seed by spiritus; created by God and implanted or infused in the individual human body by Him; incorporeal and immortal; and the 'universal form' of the body. This is a somewhat confusing definition, especially as when Telesio speaks of the activities of the soul within the body, he is most likely to be speaking of the anima animalium, the anima which man shares with the other animals and which Telesio and his followers described as neither the form of the body nor its temperament, but as substantially spirit, dwelling within the body and directing its actions.[17] In Quod animal he describes this 'animal soul' more fully as the 'spirit contained in the ventricles of the brain' and gives a lengthy refutation of the Galenic three-faculty system as well as of the Aristotelian notion of the heart as centre of mental activities. There is one centre, not three, and it is the brain, not the heart; the stomach, bladder and other organs are governed not by a natural faculty as Galen claimed but by soul. And it is spiritus which is the 'substance of the soul' (substantia animae) and which performs the functions and operations of the soul in the body, by virtue of its powers or qualities: heat, tenuousness, motion and the 'power of sensation' (vis sentiendi).[18]

The role of spiritus in Telesio's writings seems also designed to counter a major problem caused by the traditional definition of the soul as the 'form' of the body. The problem is that of unity. It is brought about because of the diverse structure and function of the body's parts and the further changes which occur in growth and development. As the anima is form of the whole, is it necessary that parts of the soul, acting in diverse ways, be forms of the different parts of the body? Galen's solution, the multiplication of 'faculties' to control the body's diverse operations, compounds the problem rather than solving it. Spiritus, however, is one, and performs the actions of the soul throughout the body by virtue of a uniform set of qualities.[19]

If the relationship between anima and spiritus is a somewhat blurred one – at best, as D. P. Walker has pointed out, the conventional anima is left with a much reduced range of functions[20] – the relationship between spiritus and heat as Telesio defines it also raises problems which relate directly to the question of generation. For Telesio argues that fire, animal heat and celestial or solar heat are of the same nature, differing only in intensity. There is thus, he argues, an exact parallel between spontaneous and seminal generation – not, as Aristotle claimed, because both require the participation of a 'special kind' of heat, nobler than that of the elements, and associated with the quintessence; but because all heats are fundamentally the same. Fire can generate as effectively as the heat within animals or heat of the sun.[21] But why, if no special form of heat is required for generation, are creatures not created out of every bonfire? Telesio attempts to get round this difficulty by a careful delimitation of fertile conditions, which apply to both sorts of generation. In the first place, the heat, of whatever kind, must be perfectly balanced and moderated. This is particularly critical in the case of the higher animals and man, and explains why they cannot generate spontaneously. But elemental heat can act in exactly the same way as the sun if it is not too 'vehement' but warms softly and gently, as in the incubation of hens' eggs by artificial heat.[22]

Fertility has other predetermining conditions, which explain why neither women nor dunghills produce new life at every opportunity. In the case of human generation, Telesio follows the formulae of the Hippocratic tradition, much as it appeared in contemporary women's books. If the seed from either sex or the uterus which receives it is too hot or too cold, too moist or too dry, if the uterus fails to retain the seed or to receive it in the first place, if the seed is too thick or too thin or lacking in spiritus, conception will not take place.[23] In spontaneous generation similar rules apply. Because the medium, earth, is not uniform, but differs in its degree of lightness or heaviness, it has different predispositions – which help to determine the type of animal produced, or whether anything is produced at all. Just as in sexual generation, with which it forms an exact parallel, the media must be perfectly balanced, neither too moist nor too dry, and possessed of the proper amount of heat and spiritus.[24] As we have seen, degrees of heat also constitute the essential difference between the sexes and their respective contributions to generation.

Telesio's monistic system, with its emphasis on unity, has considerable claims to originality. And, according to Neal Gilbert, he differs from the other commentators considered in this section in not finding the Averroist tradition, with its basis in Arabic sources, of much appeal. An accomplished Greek scholar, he apparently preferred to go back to classical sources, and retired from the academic world to a monastery to write De Rerum Natura. Gilbert calls him an 'Aristotelian revisionist',[25] and indeed he is included in this section because his work follows the standard format of Aristotle commentaries, and treats many of the same problems in the chapters devoted to generation: the roles of male and female, the action of the soul, and the importance of heat and spirit.

Many of these same questions were the concern of the medical writers who are the subject of Part III. But the works which I have chosen to consider in this last section are, for the most part, of a different type than those in Part II. Instead of commentaries, I shall look at a group of medical textbooks which approach the questions of generation from a somewhat different perspective: that of the system of theoretical medicine, or physiologia.

Notes

1. One commentary by Odone, mentioned by Lipenius, I have not been able to trace.
2. Franciscus Valles, Controversiarum medicarum ac philosophorum, Alcala, 1556; Andreas Caesalpinus, Quaestionum Peripateticarum Libri V..., Venice, 1593; Julius Caesarus Scaliger, Exotericarum Exercitationum..., Paris, 1557; Hieronymus Cardanus, De subtilitate, Nuremberg, 1550. These 'quaestiones' treatises seem to have been produced at least partially with the needs of arts teaching in mind (cf. Lohr, 1973, 21: 228–32). Scaliger deals with the generation of man in Book V, chapters 5–6 of his Exercitationes, and the British Museum has a group of published theses on generation topics, defended by students at Jena and Wittenberg between 1645 and 1652, which take Scaliger's comments as their starting point (B.M. 1173.k. 8 (1–20)). Caesalpino treats questions of generation in Book V of his Quaestiones; this book deals with Aristotle's animal works and follows earlier books dealing in order with questions arising from the Physica, Metaphysica, De Caelo and De generatione et corruptione. Question 1 (fol. 104v-109v) looks at spontaneous generation, question 2 (109v-115r) at the roles of male and female in generation. In both Caesalpino provides a generalised commentary on the standard Aristotelian texts which relate to these points of debate. Valles looks at questions such as the role of male and female and the nature of the seed in the context of a discussion on 'parts'.
3. Hieronymus Franzosius, Tractatus Apologeticus De semine Pro Aristotele Adversus Galenum, Verona, 1645; F. Aloysius Albertus, De Nutritione, Augmento, et Generatione, Disputationes.., Venice, 1627; Mundinus Mundinius, Disputatio...de semine, Treviso, 1609; De genitura pro Galenicis adversus Peripateticos..., Venice, 1622; Ad Disputationem de Genitura Additamentum Apologeticum, Venice, 1622; Caspar Hofmann, De Generatione Hominis..contra Mundinium, Frankfurt, 1579. Their late date and somewhat sterile combative nature are two reasons why I have not treated them in more detail, but all have interesting points to make. Hofmann clearly refers to an earlier work of Mundinius than those I have seen, or, more probably, to an earlier edition.
4. Bernardino Telesio, De Rerum Natura juxta propria principia Libri IX, Naples, 1587. Hofmann complains in the introductory letter to De Generatione Hominis that the topic is divided into warring sects, "quibus pueris suis persuadeant, se Aristotelicos esse, Galenicos, Telesianos, & nescio quos alios.." (Fol.A4vo.) For modern recognition of Telesians as a distinct group, cf. Temkin, Galenism, pp. 145–8; Walker, Magic, Part III ch. 6. Telesio's choice of title may reflect the popularity of Lucretius's work in the Renaissance (the British Library has 19 editions before 1600, many in multiple copies) but the systems of the two natural philosophers have very little in common, and Telesio is certainly not an

atomist. Indeed, the arrangement and logical structure of the book owe more to Aristotle and his commentators than to Lucretius.

5. Telesio's system as a whole is too large and complex to be considered in detail here; all I propose to do is draw out certain aspects which relate to the commentary on generation as presented in early chapters. The Telesian literature includes Francesco Fiorentino, Bernardino Telesio, Florence 1872–4; Giovanni Gentile, Bernardino Telesio, Bari, 1911; N.C. Van Deusen, Telesio, the first of the moderns, New York, 1932. According to the useful article by Neal Gilbert in the D.S.B. (1978, 13: 277–280) this is the most complete work in English, but there are useful insights in Temkin and in Walker, op. cit., especially pp. 189–192.
6. De Rerum Natura, p. 177: "Prima mundi corpora, et quae reliqua constituunt entia, coelum esse, & Terram; & a summo illud, integroque calore, a summo hanc, integroque frigore constitutam.."
7. Ibid. "Non siquidem, ut entia reliqua, e terra a sole, sed e propriis seminibus, & a seipsis generantur."
8. For spiritus in general, see the works by D. P. Walker cited in the Introduction, note 48; for Telesio, his Magic, pp. 189–192.
9. De Rerum Natura, V 186–204; 210.
10. Ibid. V pp. 182–3, VI 240–244. In plants, the fibres are again associated with spiritus: VI pp. 258–9. Quod animal universum ab unica Animae substantia gubernatur, adversus Galenum (cited as Quod animal..) was published posthumously by Telesio's disciple Persio, in Varii de naturalibus rebus libelli ... 1590. See especially chapter 6 (6v-8r) where Telesio first sets out his own position in opposition to Galen's three-organ one: "spiritum in cerebri ventriculus contentum animalium animam esse."
11. Ibid. V 183–4; VI 229–30.
12. Ibid. chap. 33, pp. 264–5.
13. Ibid. chapter 19, p. 252 for Telesio on the 'use of the parts'; ch. 34, pp. 266–7, on spiritus and desire.
14. Ibid. ch. 35, 267–8; for heat, VI 1229.
15. Ibid. ch. 36, p. 268; the difference in heat which distinguishes male from female appears earlier, in chapter 18, pp. 250–1; cf. chapter 38 p. 271 for the argument against Aristotle: "foetum ab anima masculino semini inexistente constitui non posse."
16. Ibid. ch. 40, 273–4. Elsewhere, in Quod animal.. he attacks the Aristotelian notion of the heart as the primary organ and source of the nerves.
17. Ibid. V 1–3, pp. 177–180; Temkin, Galenism, pp. 145–8; Walker, Magic, pp. 189–192.
18. Ibid; also in Quod animal.. especially chapters 6 and 7 (6v-8v) and the argument quoted in note 10, above.
19. This problem of unity is a major theme of Quod animal.., in De Rerum Natura, Telesio takes it up in most detail in V 4–7 and 33–36, pp. 180–185 and 216–222.
20. Walker, op. cit.

21. De Rerum Natura, VI 20253–4: "..Quoniam igitur igneus etiam, ubi moderatus est., animalia & ipse, veluti solis, animaliumque calor, generat, … & immoderatus animalium, solisque calor ad animalium generationem nihilo minus, quam igneus, ineptus apparet.."
22. Ibid.; also, for parallel between heat of incubation and uterus, VI 1229. Telesio gives other reasons why man cannot generate spontaneously: the lack of suitable material for complex organs, insufficient quantity of spiritus and the belief that man was made in God's image. (ch. 23, p. 256).
23. Ibid. VI 20 pp. 253–4.
24. Ibid. VI 21 pp. 254–5.
25. D.S.B. 1978, 13: 277–80 (op. cit.)

Part III
Medici et Philosophi: Generation in Textbooks of Theoretical Medicine

Abstract Many of these same questions were the concern of the medical writers who are the subject of Part III. Instead of commentaries, I look at medical textbooks which approach the questions of generation from the perspective of theoretical medicine, or *physiologia*. I approach this relationship as it appeared in discussions of a natural-philosophical question – the physical constitution of the natural world – which was at the same time basic to contemporary theoretical medicine and to medical theories of generation. The question of generation was one of the most important of the problems which they shared.

Chapter 13
The 'Compendia Tradition' and Jean Fernel (1497–1559)

Abstract I consider how theoretical questions were treated in a textbook form which I call the compendium. The compendium had topics: on the materials of which living things were built, elements, temperaments, humours and the simple and composite parts; and the processes according to which they operated: heat, spirits and the soul.

Fernel's *Physiologia* contains a section on each of these topics and one on generation. I believe that the *Physiologia* became the model for later writers into the seventeenth century, but he did not invent the form. The First Fen of the *Canon* of Avicenna (980–1037). covers the topics used by Fernel. Fernel's earliest important Western predecessor appears to have been the fourteenth century Paduan medical and philosophical writer, Peter of Abano, known as Conciliator.

Jean Fernel wrote two major works of theoretical medicine. The first was published as *On the natural part of medicine* (1542) and in 1554 as *Physiology*, the first section of a general medical work, *On medicine*. The other (1548) was a dialogue, *De abditis rerum causis* (*On the hidden causes of things*). I examine the questions Fernel and his contemporaries ask on one particular topic, generation, within one particular context, the 'compendium'.

In each section of this thesis, I have looked at the question of generation from the viewpoint of a particular group of writers, and a particular type of text. Part III looks at the second of the two groups of theoretical writers which I identified in the Introduction: doctors who wrote about generation in treatises of theoretical medicine.

We saw in Part II how Renaissance natural philosophers defined and tackled certain problems raised by the phenomena of generation – the role of male and female, the nature of the seed, the activities of heat, spirits, and the soul – within the context of commentaries on Aristotle's De generatione animalium. One would expect that these questions would be important to the doctors who wrote treatises of theoretical medicine, too. Such men, who usually taught theoria in the universities, were themselves, like their colleagues in natural philosophy, graduates of the arts course. They might well describe themselves as philosophi et medici on their title pages. Their interests went beyond the compounding of remedies for the cure of disease, which was the main concern of the practising doctor.

L. Deer Richardson, *Academic Theories of Generation in the Renaissance*,
History, Philosophy and Theory of the Life Sciences 22,
https://doi.org/10.1007/978-3-319-69336-1_13

Philosophical doctors might approach a question such as the generation of man, on which medical and philosophical interest overlapped, in a commentary. As in natural philosophy, this was most likely to be based on a popular work of theoria, used as a set text. One of the best examples is Galen's Ars medica, which was very popular as a teaching text and in turn inspired a flood of commentaries. In an earlier version of this chapter I considered the contributions to the theoretical debate on generation made by eight Renaissance commentators on this work.[1] In Part III, however, I want to look chiefly at the way in which theoretical questions concerning generation were treated in what became a standard textbook form for theoretical medicine (or physiology, as it came to be called); a form which I call the compendium.[2]

Treatises on the theoretical or natural part of medicine went by different names, but often they had a common structure, as the list in Appendix II suggests. The compendium as I define it was built around a common series of topics, which were central concerns of theoretical medicine. Like their predecessor, Galen, philosophical doctors were interested in broader questions than those of therapy; and in how the body worked, not just the fact that it did work in a certain manner. They therefore considered the materials of which living things were assumed to be built: elements, temperaments, humours and the simple and composite parts; and what we might call the processes according to which they operated, heat, spirits and the soul with its faculties.

Fernel's Physiologia contains a section on each of these topics and one on generation. I believe that the Physiologia became the model for later writers well into the seventeenth century, but he did not invent the form. For a start, this set of topics corresponds roughly with one of the three traditional divisions of theoretical medicine in the Middle Ages, the so-called 'naturals'.[3] But where did these traditional categories come from? Arabic systematisers of Greek medical theory seem a likely answer. Georgius Valla's De Inventa Medicina, a relatively early Renaissance compendium, suggests the Isagoge of Joannitius as one source. The Isagoge was a Latin translation of an Arabic ninth century text, and was printed frequently in anthologies early in the Renaissance.[4] Another, somewhat more substantial text, which, like the Isagoge codified and simplified Galenic medical theory, was the Canon of Avicenna (980–1037). Nancy Siraisi reports that this was available (and important) for medical teaching in fourteenth century Padua, and the large number of editions and commentaries printed in the Renaissance suggest that its importance as a set text for theoria continued into the early seventeenth century. K. E. Rothschuh has pointed to the Canon as the particular model for Fernel's Physiologia, though without comparing other examples as I have done.

The first six theses of the First Fen of the Canon deal in order with the definition of medicine; elements; temperaments; humours, normal and pathological; members or parts; and finally with the natural, vital and animal faculties. Generation is not included as a separate topic in Avicenna's scheme; it is considered much later, in the chapters of Book III which describe the male and female genital organs.[5]

Avicenna's system was repeated in commentaries on the first Fen, such as those of Jacopo da Forli or Jacobus Foroliviensis (d. 1413) and Jacobus Montanus (1498–

1552). Both follow Avicenna's order and include the same set of topics. And both were printed after Fernel's work, although the former at least was certainly composed earlier. Avicenna commentaries appeared both earlier and later than this, however. The commentary on Book I Fen i of Hugo Senensis (d. 1448) appeared in print in 1498 and 1517, and the commentaries of Dino del Garbo and Gentile da Foligno, discussed below, also saw publication around the turn of the century. At the other end of the time scale, commentaries on the same section of the Canon were published by Ponce de Santa Cruz in 1622 and Sanctorius in 1626.[6]

But Fernel's earliest important Western predecessor appears to have been the early fourteenth century Paduan medical and philosophical writer, Peter of Abano, known as Conciliator from the title of his most famous work. In Conciliator, Peter adds the topic of generation and spirits to Avicenna's list; otherwise, he keeps to the Canon's order.[7] Peter has most often been considered by historians as a natural philosopher, but he was a doctor as well. In addition to Conciliator – the intention of which is clear from its full title, The book called Conciliator of philosophical and especially of medical differences – he wrote other books on medical topics. Almost all of them survived and were printed in the Renaissance, often in multiple editions. They include additions to the Practica of Mesue the Younger (In librum Johannis Mesue), a Physionomia, a treatise on poisons, a commentary on the materia medica of Dioscorides, plus translations of a number of Galenic works and of a pseudo-Hippocratic work on medical astrology.[8] His commentary on the pseudo-Aristotelian Problemata also includes substantial sections on human physiology and disease, some of them relevant to generation.[9] He seems also to have taught or at least recommended the 'animal works' of Aristotle to his pupils, in particular Historia Animalium and De sensu. His successors at Padua included famous doctors who themselves became commentators in this same tradition, men like Dino del Garbo and Gentile da Foligno.[10]

But it is not only as a doctor, as a self-proclaimed conciliator of philosophical and medical differences and an important link in formulating the standard topics of the 'natural part of medicine' that Peter is of interest. He has been claimed as the founder of the 'Averroist' school at Padua, notably by Renan and Peter's biographer, Sante Ferrari. Randall followed this tradition in his discussion of Peter's contribution to the development of scientific medicine at Padua in the fourteenth century.[11] More recent scholars have challenged not only Randall's thesis, but the claim for Peter's Averroism on which it was partially based; Siraisi provides a useful summary of the debate. Nardi and Thorndike, for example, have doubted both Peter's heterodoxy and his difficulties with the Inquisition – two important claims of the Averroist party – and pointed out that his views can be associated with Arab writers other than Averroes. Thorndike, in his History of Magic, argued that Renan himself had admitted that Peter knew neither Colliget nor the other medical works of Averroes, and that in any case Renan's definition of Averroism was so loose (medicine = Arabism = Averroism = astrology = incredulity, in Thorndike's summary of it) as to be meaningless. Nardi argues from an examination of Peter's opinions, expressed in works such as the Expositio Problematum Aristotelis, that his views on the soul and its generation were relatively orthodox, and that some of the extreme opinions attributed to him were only cited by him as part of his encyclopaedic coverage of the range of available views on any given topic.[12] One effect of the scholarly debate on Peter's 'Averroism' has been to focus attention on the metaphysical elements in his writings, including such questions as the origin of the human soul and the nature of species which touch on the generation problem.

Peter's orthodoxy or lack of it, and his brushes with the Inquisition, genuine or alleged, are not of particular concern to me here. My purpose in raising the question of his 'Averroism' is to place him within the tradition of commentary and 'conciliation' which I identified among the natural philosophers in Part II. What role did he play in establishing the categories and classic questions of theoretical medicine in the Renaissance? Is it similar in any ways to the role which, I have argued, Averroes' Paraphrase and Colliget played for natural philosophical commentators?

The available evidence suggests that it was, and that the influence may have been quite an important one. Like Colliget, with which, according to Nancy Siraisi, Peter was indeed familiar, his Conciliator attempts to reconcile the opinions of Aristotle, Galen and their followers on a range of topics, including those which concern the material and facultative makeup of the body and the problem of generation. But because of the focus of attention on Peter's metaphysical opinions, scholarly interest in his medical works has been largely bibliographical.[13] These studies do at least reveal that, like the medical and philosophical works of Averroes, Peter's writings were surprisingly popular in the Renaissance. Manuscripts survive of most of the Galen translations mentioned in Conciliator and listed in later compilations; a number of these were printed in the Galen Opera of 1490 and 1515.[14] Both Conciliator and the Expositio Problematum Aristotelis also survive, in whole and in extract, in multiple manuscripts. The Expositio was printed in 1475, 1482, 1501 and 1520. Conciliator had an even more impressive run. It was printed eight times before 1500 and had at least three sixteenth century editions, in 1521, 1526 and 1565.[15] In 1516 Symphorien Champier produced a (critical) commentary on Conciliator, which was included in the 1521 and 1565 editions.[16] As late as 1615, Gregorius Horstius published Conciliator enucleatus, a paraphrase of the text with the original questions, which went into a second edition in 1621.[17] Less well-known medical works also were unexpectedly popular; the British Library holds 12 editions of In librum Johannis Mesue, for example.[18]

The obvious focus for a study of Peter's influence on the questions asked within theoretical medicine in the Renaissance is his major work, Conciliator, written probably between 1303 and 1310. In the prologue, Peter explains that he has grouped his 200 plus differentia or quesita into two groups, theoria and practica, the two standard teaching divisions in the sixteenth as in the fourteenth century. The first ten differentia, however, cover questions common to both theory and practice. They are largely concerned with the definition of medicine, and with establishing its place in the scheme of knowledge: the theme of the first thesis of the Canon.[19] Is medicine a scientia, that is, a body of knowledge based on general principles and having a method? Or is it an art or craft, based primarily on experience? Is it necessary for the doctor to study and to know the investigations of the other scientiae? Another of the questions which deals with the relationship of medicine and natural philosophy is differentia 6, 'An corpus humanum sit medicinae subjectum'. As we have seen, later Paduan natural philosophers, such as Cremonini, were to deny medicine an exclusive claim to the human body as subject matter.[20]

Peter's technique in Conciliator is that of the commentary, or rather of the quaestiones tradition as described by Lohr.[21] Rather than attempting to cover a topic such as the definition of medicine comprehensively, he draws out the 'classic questions' which it raises, and puts the authorities for and against before giving, in most cases, his own judgment. It was a standard technique, but applied in this systematic way, it is easy to see how it would have been useful and popular as a guide to medical teaching. Similarly, one must assume that the questions he chooses were not new even in 1300, but had already been debated in medical and natural philosophical circles for long enough to give them a status as 'classics'.

In Book II of Conciliator, 'Quesita ceterum de theorica cum eorum appenditiis', Peter turns to the topics covered by Theses 2–6 of the Canon, Book I Fen 1. Quesita 11–16 consider the elements, with such questions – debated by natural philosophers as well as by doctors in the sixteenth century, as we shall see from Chapter 2 – as whether fire is hot; whether earth or water are colder, air or water wetter, air hot or cold; whether the elements act within mixtures in potency or in act. Quesita 17–28 are on complexiones or temperaments, including the question of whether men have a hotter temperament than women; 29–33 deal with humours.[22] These are followed by four differentia on sexual generation, 'Quesita circa generationem ex spermate dependentem.' The questions included here are all familiar from our study of the natural philosophical tradition: does seed come from all parts of the body? Are testes necessary to generation? Is the male geniture a constitutive, i.e. material, part of the embryo? Does female seed enter into the generation of the embryo? The authorities he cites – Aristotle, the Problemata, Hippocrates and Galen, Averroes's commentaries and Colliget, Avicenna's De animalibus and Canon – are equally familiar. Once again, his approach is based on a careful balancing of conflicting opinions. And his answer often seems to try to find a middle road. For instance, on the question of panspermatism he concludes that the entire body does contribute to the production of seed, "not indeed in bulk, but in force"; similarly, he suggests that the male contribution to generation, heat and spirit, is quantitative and not just qualitative.[23]

He continues with questions on the remaining topics of the compendium: members or parts (38–47); virtues or faculties (48–58)[24]; a single question on spirit, "whether heat and spirit are the same"; and finally questions on actions or operationes. Here he considers, in general terms, the rules governing the processes of bodily action, and compares the levels of operation which he described in earlier sections: is action within the body from quality, or from substance? Do the activities of the principal or official parts – heart, brain, liver and testes, as Galen described them – proceed from their elemental and humoral composition or from their temperament?[25]

In the sections on temperaments, parts and faculties, as well as in the separate questions on generation, Peter considers relevant aspects of the generation debate.

For example, the first of the questions on faculty is, "whether the seed is animated/possessed of soul".[26]

Having dealt with the 'naturals', he continues with the 'non-naturals' – factors outside the body, such as food and drink, sleep and waking, which affect health and disease; the 'praeter-naturals', or causes of disease; and the means of cure.[27]

I have already said that more straighforward Avicenna commentaries were produced by Peter's successors at Padua, including two, Dino del Garbo and Gentile da Foligno, who were pupils of Taddeo Alderotti, a Florentine doctor who taught medicine at Bologna between c. 1260–1295. Those of interest to us include Dino's commentary on the so-called 'golden chapter' on the generation of the embryo (Canon Book III Fen 21, I chapter 2); and Gentile's on Book I (as well as II, III, V and Fen I of Book IV). All of these were published in the late fifteenth and early sixteenth centuries, often in multiple editions.[28] As already mentioned, a successor of Dino and Gentile, Jacopo da Forli (Jacobus Foroliviensis) also produced commentaries on the First Fen of Book I and on the 'golden chapter'. Both were again published in the Renaissance, the latter in a collection which included the commentaries on the same chapter by Dino del Garbo and his son Tommaso.[29]

Renaissance writers thus had a number of earlier models to draw on, most of them in commentary form and in print relatively early in the Renaissance. These commentaries in compendium form (and Conciliator in particular) seem to have been as important in providing a framework for discussion in theoretical medicine as somewhat similar commentaries were in natural philosophy.[30] What is perhaps most surprising about this is the continued popularity of medieval commentaries based on Arabic ones, despite the well-publicised contempt of Renaissance medical scholars for both; this is a question which Durling takes up in his Census and which I dealt with at greater length in the Introduction. The second point to note is the way in which these earlier models were used, and here one can detect differences from medieval examples. These commentaries provided a structure and categories which a Renaissance writer could use to organise his material, while adapting them to quite different purposes. Neither Fernel's Medicina, nor, for example, Sanctorius's Commentaria could be called "commentaries on the Canon of Avicenna" in the sense in which the commentaries of Dino, Gentile, Hugo Senensis or Jacobus Foroliviensis could, and the same might be said of Peter of Abano, who again uses the categories of the Canon, and adds to them, for his own purposes. And as well as adding spirits and generation to the compendium categories, additions which Fernel took up, it might be argued that Conciliator contributed to the continued use of the quaestiones format and the aim of reconciling authorities in Renaissance texts; not only in examples such as Valles' Controversiarum..libri decem, Montanus's Medicina universa and Bertinus's Medicina..methodice but in other works with similar title-page aims: Cardanus's Contradicentium medicorum libri duo (1564) for example.

If we look now at a selection of textbooks of the compendium type written or printed in the Renaissance, it may make this point somewhat clearer. Appendix II lists 26 works by 21 authors (17 of the Renaissance) published between 1471 and 1668, which follow the format I have described, with some variations. They are listed in order of their publication (and usually, though not always, of their composition) beginning with the Canon and Conciliator. The format column in the table shows only the portion of the work which corresponds to the Canon I 1 Theses 1–6 (or to Fernel's Physiologia). Many of the texts go on to consider disease, hygiene and therapy as well.

This list was compiled from a number of sources, but the most important single bibliographical source was the Bibliotheca Realis Medica of M. Martinus Lipenius (1679), using the subject headings "Medicina" and "Physiologia" as a guide. Not all texts found under these headings were in compendium form, of course, but the list includes those which were and which I have been able to trace and examine.[31] One characteristic of this group, which is obvious from the list in Appendix II, is the very small number of Renaissance compendia which appear before the publication of Fernel's De naturali parte medicinae in 1542, and the cluster published immediately after. The two earlier examples, Cornarius's Epigraphe (which is not a compendium proper, but contains a table on the naturals in compendia form) and Valla's De inventa medicina, both clearly derive from Arabic and medieval models, in particular the Isagoge of Joannitius. But the 'bulge' of compendia published in 1547, 1555, 1556 (two examples) and 1558 (two examples), suggest that Fernel's two treatises – particularly, it appears, the Physiologia – had an influence on the form in which questions of theoretical medicine were presented; an influence which parallels that of Conciliator.[32]

Even if we suggest that the texts in Appendix II seem to take Fernel as a model, there is one point worth noting. Few of these texts include generation as a separate topic, as both Conciliator and Fernel do. Instead, most consider generation in the context of one of the other sections, often, like Avicenna, members or parts.[33] Sometimes, too, the virtus generativa is grouped with the other faculties or virtutes.[34] There are some compendia which follow the Fernelian model more closely, however. Cornarius' Medicina sive Medicus (1556) divides medicine into five parts. The first part, which considers the nature of man, he subdivides into three further parts: (1) the elements of which man is constituted; (2) the theory of generation and the formation of the foetus; (3) internal and external parts, as revealed by anatomy. The first subdivision is the "res naturales sive secundum naturam, quae hominis corpus constituunt": i.e. elements, temperaments, humours, simple, composite and principal parts, faculties and spirits. Taken together with the other two subdivisions, these cover the ground of the Physiologia quite thoroughly.[35]

The Opera medicinalia of Petrus de Peramato published in 1576, contains short treatises on elements, humours, temperaments and faculties, followed by a Liber de Semine. Two other examples, published a few years apart, are Wecker's Medicinae Utriusque Syntaxes (1583) and Bertinus's Medicina…Methodice (1587). Both follow Fernel closely. Wecker's is a curious book, composed entirely of tables, which illustrates the scholastic passion for division and definition at its extreme. Bertinus quotes Fernel frequently and respectfully and is clearly much influenced by him.[36]

Jean Riolan the Elder, who published commentaries and praelectiones on each book of the Physiologia (except Book I) is the clearest example of influence. It may be that the further 'bulge' of compendia which appear in the 1580s is due to the publication of his Commentaries in the 1570s, although that may be stretching the case. But the evidence of Riolan, Wecker and Bertinus particularly, of this group, suggests that Fernel had a second generation of followers; indeed, his influence between 1575 and 1610 or thereabouts seems, on this evidence, even stronger than in his lifetime.[37]

A "second generation" of commentaries on the First Fen also belongs to this period: those of Oddis (1575), Costaeus (1589), Ponce de Santa Cruz (1622) and Sanctorius (1626). It seems likely that all were produced with the needs of the theoria syllabus in mind. The examples on this list published in the 1630s and later: Campanella's Medicinalium…libri septem (1635); the Medicina universa (1665) and Institutionum medicarum compendium (1666/7) of Blasius; and Schenckius's Synopsis Institutionum Medicinae (1668) follow the standard compendium format only in part. As Blasius and Schenckius illustrate most clearly, they seem to be moving towards a three-part division of the 'naturals' into spirits (air-like), humours (liquid) and parts (solid) – a division by physical state, as it were. To these Blasius adds the soul as form of the human body, and Campanella the traditional elements and the mind. By the second third of the seventeenth century, then, the compendium form is, if not dead, at least disintegrating and undergoing metamorphosis, to produce a new set of categories.[38]

The focus of attention in Part III, then, is the topic of generation as it was presented in a series of treatises and monographs of theoretical medicine, particularly those published between 1542 and 1600. Into this period, and particularly between 1550 and 1580, fall a surprising number of compendia and also of monographs on single topics taken from the compendium: elements, temperaments, spirits and innate heat, the soul. I would argue that the model of Jean Fernel's Physiologia is at least partially responsible.

Jean Fernel appears in the title and on the first page of this thesis as, in one sense, the inspiration for the questions I have chosen to ask throughout. The time has come to introduce him. Fernel was born near Clermont in about 1497 and died in Paris in about 1558 – both dates are disputed. After graduating from the arts course in Paris in 1519, he seems to have become disillusioned with the teaching of the schools and for some years devoted himself to the private study of philosophy, mathematics and the classical writers favoured by the humanists, Cicero in particular. He eventually returned to the university, however, and graduated in medicine in 1530. He became famous both as a teacher in the Paris faculty and as a practising physician, but for our purposes he is chiefly notable as the author of two major works of theoretical medicine, both of which were classics in their day and continued in importance well into the following century.[39] The first was originally published in 1542 as Fernel's earliest medical treatise, On the natural part of medicine. It reappeared in 1554 with the title Physiology, as the first section of a general medical work, On medicine or Medicina, which also included sections on pathology and therapeutics.[40] The other treatise, first published in 1548, was a speculative work in dialogue form, De abditis rerum causis (On the hidden causes of things) or Dialogi, as Fernel called it. In it, two friends, Brutus and Philiatros, go to consult a third, the doctor Eudoxus. The question they put to him concerns the 'hidden causes' of certain diseases, and in particular Hippocrates' question, "Is there not something in disease which is divine?"[41] But in fact the work ranges widely over the philosophical and theoretical questions underlying medicine, and provides a different perspective on questions which Fernel treats in a more orthodox 'textbook' context in the Medicina. Fernel's modern biographers, in particular the English physiologist Charles Sherrington, writing in 1946, have tried to argue away the less 'scientific' side of Fernel's output, his interest in topics such as astrology and his appeal to occult or 'hidden' causes. He thus ignores De abditis rerum causis almost entirely, claiming that it was an early work (composed c. 1538, although published later than De naturali parte medicinae) and that Fernel later rejected the opinions contained in it.[42] Whatever the relative dates of composition of the two works may have been, there seems very little evidence for the view that the Dialogi was an aberrant piece of work. Certainly, despite its speculative and sometimes unorthodox content, it became nearly as popular as the Medicina, which rapidly attained the stature of a classic and was used as a textbook for over a century. It went through 19 sixteenth century and 16 seventeenth century printings, while the Dialogi, which was often reprinted in the same or a companion volume, went through 27 editions and reprintings before 1700.[43]

When I began this thesis, a former teacher, Theodore M. Brown, wrote to me from the Institute of Advanced Study at Princeton to cheer me on, suggesting that Fernel's "impact on medical study and physiological thought in the mainstream seems to have been enormous, but much in need of study."[44] It is difficult to explain this neglect, given Fernel's immense prestige in his own day and his influence on succeeding generations. Fernel's complex system of physiology, involving recourse to elements, temperaments, anatomical parts, humours, faculties and spirits, with several levels of explanation, is in many ways a summing up of classical, medieval and Renaissance medical thought. An attempt to examine Fernel's impact on his contemporaries and successors thus leads one directly into the intellectual traditions, the institutional and teaching priorities of sixteenth and early seventeenth century medicine. But very few serious attempts have been made to work out the details and the strategy of Fernel's system in a way which makes clear what he was trying to do; how (and how well or badly) the components fit together and how close he came to producing a plausible and useful explanation of phenomena. And in considering Fernel's output and his methodology, one needs to take into account, I believe, not only the 'textbook medicine' of the Medicina, but also the complementary, and equally popular, De abditis rerum causis.

Most earlier accounts of Fernel belong in the 'life and works' category, and more recent ones concentrate for the most part on his pathology and therapeutics, but there are exceptions. K. E. Rothschuh and Magnus Schmid have written recent articles on aspects of Fernel's physiology, and James Bono's thesis should do much to redress the balance of Sherrington's outdated biography and to place Fernel's work in its proper intellectual context, much more comprehensively than I can hope to do here.[45] What I should like to do, in the remainder of this section, is to examine the questions Fernel and his contemporaries ask on one particular topic, that of generation, and within one particular context, that of the 'compendium'. I have chosen this approach as a means of narrowing and focussing an otherwise enormous field, but also, as I argued in the Introduction, as a means of structuring it and comparing like with like. As in Part II, I shall be looking for the kinds of problems which writers chose to tackle and the authorities to which they appealed. We shall then be in a better position to compare the similarities and differences between this group and the natural philosophers in Part II, and to make some attempt to account for them.

The question of generation was one which Renaissance medical philosophers could approach in a variety of ways, and, one might say, on a variety of levels. It could be considered as a separate topic, or within one of the other topics concerning "the elements of which man is constituted". Most practically, it could be considered descriptively or anatomically, as in Fernel's Book I, on Parts, the corresponding sections in other compendia, or a straightforward treatise of anatomy. But the composition of the human body could be considered theoretically as well as practically or anatomically. Generation therefore appears in discussions of the elements and their temperament in the body. It belongs, too, in any discussion of the processes by which the body acted, and for Fernel this meant at least two levels of explanation, those discussed in Book IV of his Physiologia, on spirits and innate heat, and Book V, on the faculties of the soul.

The remainder of this study considers these topics in turn, comparing Jean Fernel with his contemporaries and successors writing on generation within the context of theoretical medicine and in particular of the compendium as I have defined it. Parts is considered briefly and bibliographically in Appendix III, since anatomy is one aspect of Renaissance medicine which has received considerable attention from modern scholars, and also because anatomies are among the 'practical' works on generation which I excluded in the Introduction. The following chapter therefore begins this comparison with the most basic theoretical level at which medical writers – and natural philosophers – spoke of generation: the level of elements and temperaments.

Notes

1. Unpublished paper, "Eight Renaissance commentators on Galen's Art of Medicine". Galen's Ars medica or Art of medicine was also known to Renaissance readers as the Ars parva or (particularly in Arabist versions) as the Tegni or Microtechne, with various alternate spellings. For details of the numerous editions and commentaries, see Kuhn, I cxv et. seq.; Durling, Census, pp. 230–305. The Ars medica, Hippocrates' Aphorisms and First Fen of the Canon of Avicenna were set books in theoria at Padua from the thirteenth century until 1767. Since the examination was to expound a passage from a set text, printed commentaries on all three works, usually based on lectures, were likely to be popular. These points and others relevant to the question of the relationship between medicine and natural philosophy are made by Jerome J. Bylebyl, "The School of Padua: humanistic medicine in the sixteenth century", pp. 335–370 in Charles Webster (ed.) Health, medicine and mortality in

the sixteenth century, Cambridge, 1979. I am grateful to Dr. Vivian Nutton for recommending this article to my attention. Bylebyl also mentions that, at least at Padua, "philosophus et medicus" was a recognised joint degree: a detail that Dr. Gwenneth Whitteridge was also kind enough to remind me of. For a more general account of the relation of medicine and natural philosophy in the Renaissance, see also the Introduction, above. The Canon is treated in more detail below.

2. Like Lewis Carroll's Humpty Dumpty, I use compendium to mean what I say it does, rather than in its usual sense of an abridgement or short cut, but I hope my meaning will be clear. By compendia I mean works of theoretical medicine having a certain structure: a structure which is common to Book I Fen I of the Canon of Avicenna; the section "Quesita..de theorica" of Peter of Abano's Conciliator; and the De naturali parte medicinae and Physiologia of Jean Fernel, all of which seem to have had an influence in the Renaissance. The order of topics in the three works is not identical, but together they can be considered as a standard listing of what was meant by "the natural part of medicine". In Fernel's order they are: parts; elements; temperaments; spirits (and innate heat); faculties; humours; and the procreation of man. I do not use 'compendia' in a disparaging sense, unlike Thorndike (History V pp. 5–6 and 151–2) who speaks of certain natural philosophical compendia as "writing down to dull students" and comments, "Their titles are trite, their contents brief and unoriginal." And unlike Andrew Cunningham (op. cit., 1974) I have not made my classification by title, as he does in the case of 'Institutes'. In fact, very few of my 'compendia' have the word 'compendium' or 'physiologia' in the title; and those which do, for example Cornarius's Epigraphe..compendio tractata and Blasius's Institutionum medicarum compendium, are not particularly good examples. "Physiologia" on the title page can be even more misleading. It can be used as a synonym for physica or natural philosophy, as it is by Hieronymus Brisianus, Physiologiae libri ii.., Venice, 1596 and Georgius Valla in the section "De Physiologia" of his De expetendis, et fugendis rebus opus.. [Venice, 1501; unpaginated].

3. The 'naturals' were traditionally the components of which the body was made up, while the non-naturals were other factors which contributed to the body's health or lack of it but were not part of the body. They were a somewhat mixed lot, including 'inputs' such as food and drink, mental and physical states such as sleep, waking and 'passions of the soul' and actions such as rest, exercise and coitus. The third group were the praeternaturals or causes of disease; by the mid-sixteenth century they had usually acquired more modern titles, i.e. physiology, hygiene or therapeutics and pathology respectively. The division appears to have been based on Galen's Ars medica, chapter 23, as codified by Arabic commentary. (Andrew Cunningham, op. cit. 1974 and seminar paper, University College, London, 13 February, 1974, "Development of the 'Institutes of Medicine'".)

4. Georgius Valla, De Inventa Medicina, Strasbourg [1529?] Fol B 4 vo. Joannitius or Johannitius was the Latinised name of Hunain ibn Ishak, Al Ibadi (809–879).

The British Library has a 1487 Venice anthology containing his Isagoge and several translations together with works by Hippocrates and Galen. (IB. 21,384). The Wellcome Institute Library has editions, published in Articella, of c. 1476 (Padua); 1483 and 1502 (Venice); 1506 and 1510 (Pavia); 1519 and 1534 (Lyons) which give some indication of the work's popularity in the early Renaissance and of the spread northwards from the early Italian centres of publication. By Fernel's day, Articella were beginning to be superseded as 'critical editions' of Greek medical authors became more widely available in direct translations from the Greek; Durling's Census illustrates this well.

5. Avicenna, Liber Canonis, Hildesheim, 1964 (facsimile of Venice, 1507 edition) Book I, Fen I, Theses 1–6 cover the 'compendia topics'; Book III, Fen 20 and 21 concern the male and female members and their function in generation. Siraisi, Arts, 142, 155, compares the popularity and availability of works by Avicenna and Averroes; she gives Avicenna the edge. The Canon appeared in the Renaissance, in the translation of Gerard of Cremona, from 1472; the Wellcome Institute has the editions of 1479, 1486, 1489–90, 1490, 1491, 1507, 1523, 1527, 1528, 1556, 1564, 1595 and 1608 (not all complete) plus additional part-editions (W.I.H.M. Printed Books, I, cat. 569–581, 588); I have used the facsimile of the 1507 Venice edition cited above. K. E. Rothschuh, "Das System der Physiologie von Jean Fernel (1542) und seine Wurzeln", Verh. XIX int. Kongr. Geschichte der Medizin, Basel, 1964, pp. 529–536. I am grateful to Dr. Walter Pagel for bringing this paper to my attention.
6. Jacobi Foroliviensis...Expositio et Quaestiones in Primum Canonem Avicennae..Venice, 1547; Jo. Bapt. Montani..In Primi Lib. Canonis Avicennae Primam Fen..Venice, 1558. I have not found earlier editions. Montanus' Methodus medicinae universalis (1558, in Opuscula; 1565, in an edition cited by Lipenius which I have not seen; 1587, as Medicina universa..) also follow Avicenna's order. Hugo Senensis, Expositio in primam et secundam Fen primi Canonis Avicennae..Venice 1498; Expositio in primam Fen...cum quaestionibus ejusdem, Venice, 1517; Antonio Ponce de Santa Cruz, In Avicennae primam primi..., Madrid, 1622; S. Sanctorius, Commentaria in primam Fen primi libri Canonis Avicenna, Venice, 1626. Jacobus Foroliviensis, Montanus and Sanctorius also wrote commentaries on Galen's Art of Medicine, published in 1508 (and 1547), 1554 and 1632 respectively. Bylebyl, op. cit. pp. 156–8, mentions two other commentaries produced by lecturers at Padua (in addition to Sanctorius's). Oddo de Oddi (or Oddis) (1478–1558) wrote commentaries on all three of the theoria set texts, which were published posthumously by his son Marcus: Oddo de Oddis, ..in primam totam Fen primi libris Canonis Avicene.. expositio, Venice, 1575. Morgagni lectured on theoria until 1715 but his lectures were not published until this century.
7. Petri Aponensis..Liber Conciliator differentiarum philosophorum praecipueque medicorum appellatus... I have consulted three editions of Conciliator: those of Venice, 1483 and 1521 (W.I.H.M. 4940, 4941) and Venice, 1504 (Cambridge University Library, Sel. 3.63). References below are from the 1521 edition.

8. Lohr 1972, 28: 329–30, and references, provides a brief but useful introduction to Peter and his output. Also important are the bibliographical essays by Lynn Thorndike, in History, II pp. 917–926; "Manuscripts of the Writings of Peter of Abano", Bull. Hist. Med. 1944, 15: 201–219; "Translations of Works of Galen from the Greek by Peter of Abano", Isis, 1942, 33: 649–653, plus those cited below and in Thorndike and Kibre, Catalogue of Incipits (op. cit., II 1 note 4, above).
9. Nancy G. Siraisi, "The Expositio Problematum Aristotelis of Peter of Abano", Isis, 1970, 61: 321–339, especially 324–9, 335–6. See also the Thorndike references above and also his "Peter of Abano and another commentary on the Problems of Aristotle" Bull. Hist. Med., 1955, 29: 517–23.
10. Siraisi, Arts, pp. 116, 127–8, 140 e.s.; George Sarton, Introduction to the History of Science (cited as Sarton, Introduction) III (1) 837–838, 848–851.
11. Renan, Averroisme, pp. 326–7, Randall, School of Padua (both op. cit.); Sante Ferrari produced two accounts of Peter: I tempi, la vita, le dottrine di Pietro d'Abano, Genoa, 1900; Per la biographia e per gli scritti di Pietro d'Abano, Rome, 1918. A similar argument is put forward by Erminio Troilo, Averroismo e Aristotelismo Padovano, Padua, 1939.
12. Siraisi, Arts, pp. 136–9, note 143; "Expositio", p. 322, note 3; Thorndike, History II 887–8; Nardi, Saggi, especially chapter 1, pp. 1–17 and 2, pp. 19–74, which deals especially (pp. 38–69) with Sante Ferrari's claims and the question of Peter's alleged heresy.
13. Siraisi, Arts, p. 116 and references in notes 8 and 9, above.
14. Thorndike, "Translations", op. cit. note 8 above; also other references.
15. Thorndike, especially History, II, 919, 921.
16. Symphorien Champier, Cribratio..et annotamenta in Galeni Avicennae et Consiliatoris opera .. (1516), as cited in Brian P. Copenhaver, Symphorien Champier and the Reception of the Occultist Tradition in Renaissance France, The Hague, 1978, p. 18. Thorndike, History V 119–121, gives an earlier date (1514) and different title. The thesis by Holmes which Copenhaver cited looks at this question in some detail. Cribratio was appended to at least two later editions of Conciliator, those of 1521 and 1565. Champier also wrote on Peter's exposition of the Problemata. He seems to have objected mostly to Peter's favourable attitude to magic and astrology as well as to certain errors of fact. Thorndike suggests that he got the idea from Giovanni Pico della Mirandola, who complained that Peter's errors would fill a book. (Thorndike, Ibid.; Copenhaver, esp. 148, 150–2, 213) but the trouble he takes to refute particular errors, his respectful references to Peter as an authority in other works (for example the Dyalogus) and the company in which he places him suggest that for him Peter was not just one of "the more notorious proponents of magical medicine" (Copenhaver, 213; for Dyalogus, pp. 243–330).
17. Conciliator Enucleatus..Compendium Opera..Gregori Horsti .. Editio nova. Giessen, 1621. The 1615 edition is also in the B.L.
18. B.L. Printed Books, vol. 188, p. 213; their dates range from 1471 to 1623.

19. And, of course, of a number of Galenic works. This became a usual beginning, even for those compendia which were not direct commentaries on the Canon: Fernel considers the definition of medicine in the preface to his Physiologia. For other examples, see the compendia by Valla, Fuchs, Cornarius (1556), Montanus (1558 and 1587) and Campanella (references above, note 6 and below, notes 32 and 38). Theoretical works such as Julius Alexandrinus, De Medicina..Dialogus, Zurich 1557, focus on these 'quesita communia' (what is medicine? what are its limits? how should it be taught?) in much more detail.
20. Ed. cit. (1521) fol 1r: (Prologus) "Quesita communia decem utriusque partis medicinae theorice et practice., An necessarium sit medico ceteras scire speculationis scientias..(etc.)". The text does not have question headings in this form, but begins with one side of the argument, e.g. "Quod necessarium non sit medico ceteras scire speculationis scientias ostendit.." (fol. 3r-17v). For Cremonini, see Part II, Chap. 11, above.
21. Lohr 1967, 23: 313. He also includes the abbreviatio or compendium in his list of types of exegesis, but in the standard sense of a work which notes briefly the contents of various source works. For medieval compendia, i.e. "flores" of practical medicine, see Luke Demaitre, "Scholasticism in Compendia of Practical Medicine, 1250–1450", in Science, Medicine and the University: 1200–1550. Essays in Honour of Pearl Kibre, eds. Nancy G. Siraisi and Luke Demaitre, Manuscripta, 1976, 20(2): 81–95.
22. Ed. cit. fol. 17v-25r (elements); 25r-40v (temperaments); 40v-50r (humours).
23. Ibid. 50r-55v; "sperma a toto corpore demandari non quidem mole sed vigore" (50v); 52r-53r for question 36, "An genitura sit pars constitutiva embrionis".
24. Ibid. 55v-67v (parts); 67v-83v (virtues or faculties).
25. Ibid. 83v-85r, "An calor et spiritus sit idem.."; 85r-93r (operationes or actions, a topic which Renaissance writers usually grouped with faculties).
26. Ibid. 67v-69r, "An sperma sit animatum." Another relevant question, "An masculis sit femina calidior", comes in the section on temperaments: 39v-40v.
27. Ibid., 94v-247v.
28. Sarton, Introduction; Siraisi, Arts, 116, 127–8, 140 e.s.
29. Op. cit. (1547), note 6, above. Jacobi de Forlivio/Scriptium/Super Avicenna Canonis Lib. III Fen XXI.. Venice, 1489 Jacobus forliviensis, expositio cum questionibus supra aureum capitulum.., Venice, 1502 quaestiones eiusdem doctoris super idem capitulum.. Expositio Dini de Florentia supra capitulum de generatione embrionis cum questionibus.. expositio super capitulum de generatione embrionis.. Thomam del Garbo.. (with text and commentaries by Dino on de natura puerorum of Hippocrates).
30. Franciscus Valles, Controversiarum medicarum et philosophicarum libri decem, Alcala, 1556; Georgius Bertinus, Medicina.. Methodice absoluta.., Basle, 1587. I wrote this before reading what Dr. Bylebyl (op. cit. 1979, pp. 368–370) has to say about the uses of Avicenna's Canon in the Renaissance, but I was glad to see that our opinions reinforce one another. He suggests that the Canon retained its place at Padua as a set book, despite the humanist movement, because no other single work covered the same topics so neatly; but

that by the mid sixteenth century commentaries had altered in purpose: their intent had become "not to expound Avicenna's views, but to show what the ancients had taught on these subjects, and how Avicenna had distorted their doctrines." He also points out, as I have above, that in later commentaries the Canon served as "a fairly flexible format for discussing whatever the lecturer wished".

31. M. Martinus Lipenius, Bibliotheca Realis Medica.., Frankfurt, 1679. I am also grateful to Professors D. P. Walker and M. Screech, both of whom have kindly offered much valuable bibliographical assistance and loaned me copies of their own indices; and I have of course benefitted from the catalogues, and the collections, of the British Library and Wellcome Institute Library.
32. Janus Cornarius, Universa rei medica EPIGRAPHE.. compendio tractata.., Basle, 1529; Valla, op. cit. Leonhard Fuchs, Institutionum medicinae.., Lyons, 1555, revised version Basle, 1618; Janus Cornarius, Medicina, sive medicus.., Basle, 1556; Valles, op. cit. The same could be argued for the publication at this time of the posthumous works of Foroliviensis and Montanus.
33. Montanus considers the 'classic questions' of generation at some length in the section "De membris" of his Commentaria (op. cit.). Lectio xliiii, pp. 544–553, compares the arguments of Aristotle and Galen on the question of female seed, concluding, "Rationes igitur Aristoteles sunt meliores rationibus Galeni." (p. 553). Later, in the second chapter, pp. 586–605, he takes up the question of generation again, and identifies four questions which must be answered before his original question - on the spermatic or sanguineous origin of the parts - can be dealt with. This, he says, has led to a difference of opinion between doctors and philosophers on a matter which concerns them both, and the four fundamental questions which he identifies (whether the members called spermatic are generated from seed or blood; whether the female emits seed; if so, whether this mixes with male seed; if it mixes, whether the female's contribution is passive and the male's active, or not) are all familiar from Part II. Once again, Montanus determines in favour of Aristotle. There is a similar discussion of generation at the end of the section on 'Parts' in his Universa medicina, pp. 204–210 (1587). Valles, op. cit., II chs. 5–8 and 12, similarly deals with generation in chapters on parts, and the same is true of Joannes Costaeus, Disquisitionum Physiologicarum.., Bologna, 1589, pp. 365–394. Costaeus, who was first professor of medicine at Bologna, also wrote a separate treatise on generation: De Humani Conceptus Formatione.., Pavia, 1604 (first edition, Bologna, 1596). In the Disquisitionum, his treatment is very like that of Montanus: again he begins from the question of the origin of the spermatic and sanguineous parts, and asks almost the same questions (are the testes principal parts? Do women contribute menstrua or seed? Is the male's contribution facultative or material?). Again he agrees with Aristotle, and, he says, with medical opinion; except that he considers the male contributes both form and matter to the conceptus, while the female seed serves only to nourish the male: a position which could also be derived from Galen without much difficulty.

34. For example, Montanus (1587) pp. 214–216, "De virtute generativa", where his discussion again arises out of an area of controversy between Aristotle and Galen concerning the origin of the parts; also Valles, op. cit. II ch. 15.
35. Janus Cornarius, op. cit. 1556, p. 12: "Dividitur autem Medicina in partes quinque. Primam, quae naturam hominis considerat.. Prima quidem [dividitur] in tres: quarum prima elementa hominem constitutionem tractat. altera generationis et foetus formationis rationem tertia & internas & externas corporis partes respicit, in resectione, aut ossium disparatione."
36. Petrus de Peramato, Opera Medicinalia, San Lucar, 1576; Joannes Wecker, Medicinae Utriusque Syntaxes, Lyons, 1583; Bertinus, op. cit. (note 30).
37. Jean Riolan (the Elder) Ad librum Fernelii de Spiritu et Calido innato..commentarius, Paris 1570 and 1576 (I have used the latter); ..de elementis..commentarius, Paris, 1575; de temperamentis..commentarius, Paris 1576; de facultatibus animae..commentarius, Paris, 1577; de procreatione hominis.. commentarius, Paris, 1578. Also, ..de abditis rerum causis..commentarius, Paris, 1589. His Opera omnia, Paris, 1610 pp. 1–116, contains these in the order of the Physiologia, and adds a commentary de Functionibus et Humoribus (p. 45–75). It also includes Riolan's Universae medicinae compendium, in which the 'compendium topics' are treated in Fernel order, though with parts in fifth place, after humours, and functions last, omitting generation. Another edition of these commentaries is Praelectiones in libros Fernelii physiologicas, et de abditis rerum causis.., Paris, 1601. The impression of Fernel's popularity and importance in this period is confirmed by other evidence, much of it provided by Sir Charles Sherrington, The Endeavour of Jean Fernel, Cambridge, 1946. The Vita Fernelii of Guillaume Plancy was published in the sixth edition of the Universa Medicina (1607) and in most later editions; both author and subject were already dead. Other briefer biographies appeared in Sambucus (1574), Thevet (1584), Gallus (1590) and de Thou (1604). (Sherrington, Appendix, Note 1 and 2, pp. 147–171.) Sherrington also lists additional versions and editions of Riolan's works on Fernel, published between 1570 and 1626 (Ibid. p. 59 and Note 9, pp. 179–180) and his "List of editions of the writings of Jean Fernel" (pp. 187–207) not only shows numerous editions and printings of the Medicina (reissued in 1567 as Universa Medicina, with Plancy as editor) and Dialogus, in this period, but also, between 1571 and 1664, separate issues and translations of individual parts of the Pathologia and Therapeutica, translations of these into French and English and separate publication of two short treatises on fevers and lues venerea discovered by Plancy among Fernel's papers after his death (Ibid. esp. 200 ff for details). There are also consilia and letters among the publications, and even an attributed work on ancient medicine; the whole process of popularisation of an Authority seems to have changed little since the invention of printing.
38. Oddis, Sanctorius, Ponce de Santa Cruz, op. cit. note 6; Thomas Campanella, Libri VII. Medicinalium.., Lyons, 1635; Gerardus Blasius, Medicina universa.., Amsterdam, 1665; Johann Theodor Schenck (or Schenckius), Synopsis Institutionum Medicinae.., Jena, 1668. As both Rothschuh (op. cit., 1964,

pp. 531–2) and Schmid (note 45, below) point out, Fernel himself divided the partes similares into three types: substantia solida, substantia carnosa and substantia spirituosa; this may be the origin of the three-part division found in Campanella, Blasius and Schenckius. Boerhaave's Institutes provide one example of the later development of the 'compendium' loosely defined.

39. My chief sources for Fernel's career have been the two standard biographies in English and French: Sherrington's and Louis Figard, Un Médecin Philosophe au XVIe siècle, Paris, 1903. All later biographers depend largely on Plancy (whose Vita is translated in Sherrington, pp. 150–170) for the details of Fernel's life.
40. Joannis Fernelii.. De naturali parte medicinae.., Paris, 1542; Medicina, Paris, 1554. I have used the latter; there are very minor changes between the two versions.
41. Jean Fernel, De abditis rerum causis.., Paris, 1548. I have used the edition of Venice, 1550; hereafter referred to as D.A.R.C. The pagination in the copy used is faulty, and references will be given, where necessary, with the true page number in brackets after the printed one. Fernel tells us in the Preface (D.A.R.C. 3r–v (5–6) that this question is one on which he has long pondered, and that the book in a sense grew out of it; but he does not give a reference, either here or when the question is first introduced (4v, (8)) and I am not certain of the passage to which he refers. In Hippocrates' most obvious consideration of the 'divinity' of disease, in de morbo sacro, he concludes in favour of a naturalistic explanation of the cause of all diseases, but adds "These things are divine so that there is no need to regard this disease [epilepsy] as more divine than any other; all are alike divine and all human."
42. Sherrington, op. cit., especially Preface, ix-x; 19–41. The aspect of the occult from which Sherrington is most concerned to dissociate Fernel (whom he presents as "a reformer, and…a factor in the creation of our Medicine of today") is astrology. Fernel was a keen mathematician in his youth and wrote three treatises on the subject: Monalosphaerium (1527); Cosmotheoria and De Proportionibus Liber (both 1528, according to Sherrington). Plancy reported that in his later life Fernel repudiated his early interest in mathematics and in astrology, and Sherrington uses this claim as part of his evidence for the early date of D.A.R.C. and its supersession by the more orthodox Medicina. See also Walker, "Astral Body", for the importance of 'occult causes' for Fernel.
43. For editions, see Sherrington, Ibid. pp. 189–200, plus the part-editions mentioned above. D.A.R.C. also inspired two commentaries, by Riolan (op. cit.) and Jacques Aubert, Progymnasmata in Joan. Fernelii ..de abditis rerum causis.., Basle, 1579; as Sherrington (p. 22, note 5) mentions.
44. Personal communication.
45. In addition to the biographies of Sherrington and Figard and the references in the former, pp. 147–171, the 'life and works' category includes Pierre Alberic Capitaine, Un grand medecin du XVIe siecle, Jean Fernel, Paris, 1925; A. Herpin, Jean Fernel, medecin et philosophe, Paris [1949]. More recent articles on specific aspects of Fernel's pathology and therapeutics include

Antoinette Stettler, "Jean Fernel, Felix Platter und die Begrunnung der modernen pathologischen Theorie", Gesnerus 1977, 34: 331–5 and references; Elio de Caprariis, "Vecchio e nuovo nella fisiologia e nella patologia nervosa di Jean Fernel", Episteme, 1972 6: 294–304 and references; Owsei Temkin, "Fernel, Joubert and Erastus on the specificity of cathartic drugs" in Science, Medicine & Society, 1:61–8.

See also Rothschuh, op. cit. (1964) and "Technomorphes Lebensmodell contra Virtus-Modell (Descartes gegen Fernel)" Sudhoffs Archiv 1970 54(4): 337–354. Rothschuh's account of Fernel's physiology, like Thomas Hall's, op. cit. vol. I, ch. 13, is a useful summary but somewhat mechanical in its description. And see Magnus Schmid, "Die Lehre von den Homoiomerien in der Physiologie Fernels", Sudhoffs Archiv, 1957, 41: 317–335, for an account of one aspect of Fernel's physiology. James Bono's thesis I have not seen.

Chapter 14
Elements and Temperaments

Abstract This chapter deals with the first of the compendia topics important to generation.

The Elements

'Elements' were usually defined as the simplest of material constituents. The four writers I have chosen are Fernel (1542); Gaspar Contarenus or Contarini (1483-1542: published 1548); Jacobus Sylvius (1550) and Joannes Baptista Montanus (1498-1531: published 1554). For these writers, the physical world was made up not of one element, but four. On the whole they lie within a common tradition, and at some points are strikingly similar. I think there are three major points. The first is that all of our examples assume some form (or forms) of a four-element theory, based ultimately on that of Aristotle. Secondly, there are areas of confusion among three possible forms of the four-element theory itself, confusions which these writers reflect in various ways. This confusion began with Aristotle. The third point: the suggestion, made most strongly by Fernel, that the elements in their classical form are *not enough* to explain the material composition and structure of the world, much less the life-activities of plants and animals. A second (or even third or fourth) layer of explanation is necessary: temperament, innate heat and spirits, the soul; or the divine actions of the quintessence and of 'aethereal heat'.

Temperaments

Temperament explains the body's function. The commentaries of Sylvius and Montanus are very short. Contarini studies temperament types describes the relationship of the four primary qualities with greater complexity and precision than Galen. Function requires a soul, according to Fernel. This makes it difficult to explain the function of the body by appeal to temperament. Whatever the level of explanation – temperament, faculty, soul – it is ultimately related to the one bit of evidence the doctor has at his disposal, the actions of the living body in health and disease. Galen created confusions of terminology (such as the multiple meanings of the verb 'to heat'). These were not of great moment to those (such as Montanus and Sylvius) who were chiefly concerned to interpret the Galenic texts for medical students. But they presented greater difficulties for those (such as Contarini and Fernel) who approached the problem on a more sophisticated philosophical level. Unlike temperament, which can be explained by touch as well as by its actions, and related to the assumed structure of the rest of the physical world at the level of the elements, soul is an unknown substance. Its use advances physiological understanding very little.

L. Deer Richardson, *Academic Theories of Generation in the Renaissance*, History, Philosophy and Theory of the Life Sciences 22, https://doi.org/10.1007/978-3-319-69336-1_14

This chapter deals with the first of the compendia topics important to generation, taken from Books II and III of Jean Fernel's Physiologia. What light, then, can theories of the elements and temperaments held by medical writers in the middle years of the sixteenth century shed on their theories of generation, and those of their contemporaries?

The Elements

'Elements' – unlike 'temperaments' – is a term which was shared by medicine and natural philosophy, and could be used to describe the material constitution of bodies in the natural or the medical world. Elements were usually defined as the simplest of material constituents; the theory in its most extreme form reduced the material world to one element only. Variations on this theory were held by certain Pre-Socratic philosophers, by Plato and by the atomists[1] Any of these systems present us with the problem of explaining the variety and complexity of the physical world we see around us by changes in a single substance; the greater the contrast between the simplicity and uniformity claimed for the element (or elements) and the diversity and complexity observed in the world, the more difficult it becomes to bridge the gap.

This may help to explain why, for the sixteenth century medical writers I shall be discussing in this chapter, in particular Jean Fernel, the theory of elements and temperaments inherited from the Greeks was not considered to be sufficient to explain the structure and function of the human body. For one thing it did not take into account the sequential nature of medical explanation: in particular, the importance of order and sequence in animal generation, growth and decay. But there were other points of difficulty as we shall see.

The four writers I have chosen all wrote commentaries or treatises on the elements, published around 1550. In order of publication, they were Fernel (1542); Gaspar Contarenus or Contarini (1483–1542: published 1548); Jacobus Sylvius (1550) and Joannes Baptista Montanus (1498–1531: published 1554).[2] For these writers, and for the vast majority of their contemporaries, the physical world was made up not of one element, but four. They inherited the Empedoclean system as adapted by Aristotle in his works of natural philosophy. On the medical side, the major sources were the Hippocratic treatise De natura humana[3] and Galen's De elementis and other works.

The problem of the elements comes, I have suggested, in the attempt to explain various and complex phenomena in terms of a few simple, primordial bodies. In addition to the theories of the atomists and of Plato, there were at least three basic models available in the sixteenth century. All three drew their explanation not from the shape, size or (physical) movement of the elements but from their qualities. And all three assumed four elements: earth, air, fire and water, which in various combinations made up the physical world. But despite these similarities, logically the three systems were completely incompatible.

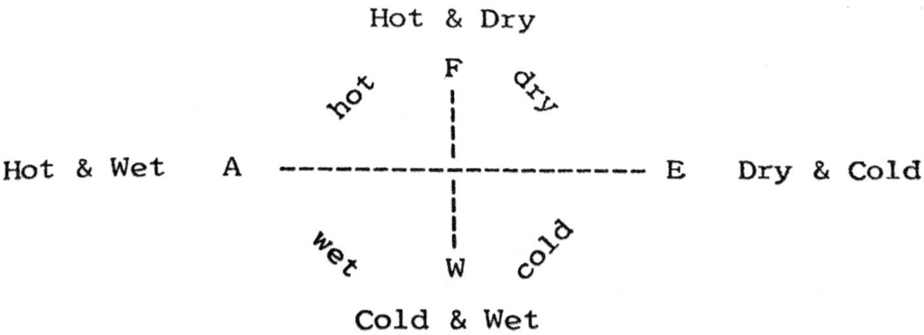

Fig. 14.1 "Aristotelian" (change between pairs of qualities)

The Aristotelian system has four elements, each possessing two complementary sensible qualities, out of a list of four. There are two pairs of opposites, hot/cold and wet/dry, and so four possible complementary pairs: hot/wet, hot/dry, cold/wet and cold/dry. Aristotle defines one kind of motion, alteration, as a change in these qualities and transmutation of one element into another as what takes place when one quality of a pair alters into its opposite. This alteration of qualities – which is also described as a motion from active to passive, or from actual to potential – explains not only the complete interchangeability of the elements themselves, as diagrammed in Fig. 14.1, but also the various physical qualities of the compounds which they make up.[4]

The second system is associated by Galen with the Pre-Socratics and Plato, and by Renaissance writers with the Stoics. It is really a single-element theory in disguise and is criticised by Aristotle on these grounds. The four elements, earth, air, fire and water, differ not because of their opposing qualities but because of their differing densities. Fire is the most rare element, and earth the most dense. Transmutation from one element to another is more limited (Fig. 14.2) and takes place when a certain element, for instance water, is rarefied or made more dense and becomes air or earth respectively. Heating and cooling has an obvious part to play in this process.

The third system of elements is that in De natura humana. Galen in his commentary on it (De Elementis Book II) notes that Hippocrates' use of the term 'element' is confusing, since he uses it to mean not only earth, air, fire and water, but also the 'elements' of the human body, i.e. either the humours or the simple parts. What is also important is that neither the natural nor the human elements can transmute from one to another, since each element represents a single, distinct quality: hot, cold, moist or dry. In an apparently 'simple' substance, such as blood, all four qualities are present indiscriminately. More complex bodies – for instance, a developing embryo – arise when these qualities are separated out according to mechanical affinities: hot to hot, moist to moist and so on. An excess of one quality or humour can be treated by administering a purge which will attract the troublesome humour by affinity and cause its expulsion.[5] (Fig. 14.3)

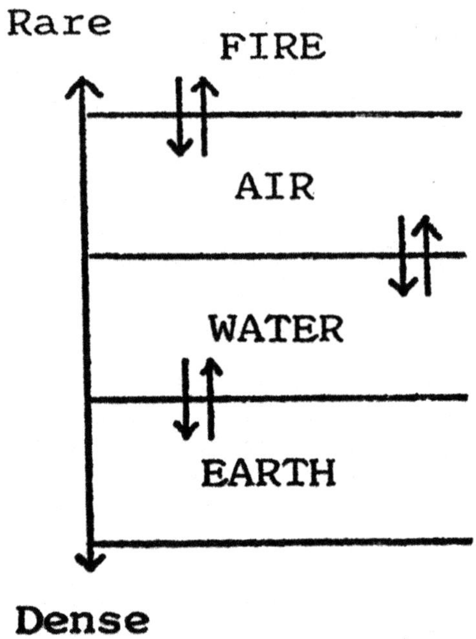

Fig. 14.2 "Stoic" (changes in density)

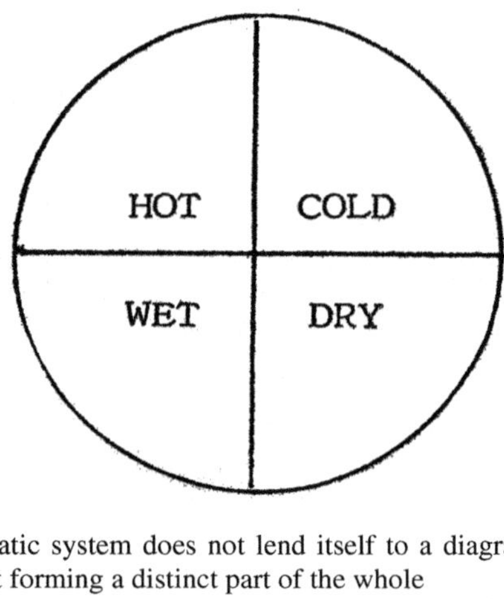

Fig. 14.3 The Hippocratic system does not lend itself to a diagram, but could be shown as follows, with each element forming a distinct part of the whole

The first stage in telescoping these three distinct systems is achieved by Galen himself, for in the commentary De elementis he juxtaposes the two-quality transmutable elements of Aristotle and the single-quality non-transmutable ones of Hippocrates. In the Paris edition of 1550 which I have used, a translation of De natura humana is appended to that of De elementis. This further emphasises the confusion between the two books, a confusion which reappears in later commentaries.[6]

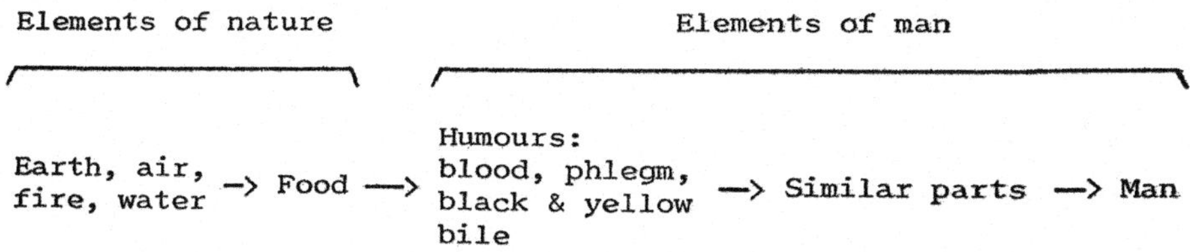

Fig. 14.4 Relation between nature and man in Galen

Galen begins De elementis with a definition, based on Aristotle: an element is the smallest possible portion of any substance of which it is the element. He goes on to eliminate several rival theories. Galen names Plato, Thales, Anaximander, Anaximenes and Heracleitus among those who imagine that there is one element only. He rejects such systems on the arguments of both Aristotle and Hippocrates. The former is that qualitative change requires more than a single element; the latter, similarly, that the presence of pain and disease proves that the body is made up of several, opposing, elements.[7]

Again following Aristotle, Galen also rejects the Democritean 'elements' or atoms. Having cleared away the opposition, Galen goes on to derive the familiar four elements from four primary sensible qualities: hot, cold, wet, dry.

If the universe is formed of these four elements, it follows that the animal body is also. However, the elements appear in animals not in pure form, but as mixtures: qualities, such as heat and moisture, which we can perceive are 'reminders' of the elements contained within. The concept of mixtio is an important medical borrowing from natural philosophy and the distinction between mixtio and temperament presents many difficulties. The compounds which make up animals and plants must contain some proportion of all four elements: just as, according to Aristotle, all compounds must. For example, plant seeds, which appear to grow from earth and water alone, must also contain air and fire in their mixtio. For, says Galen, earth + water = mud, and not bean sprouts.[8]

Man's generation from these physical elements is not direct, but can be traced back through several stages:

> [Man] is composed of those first and most simple sensible elements, which we call similar parts: namely fiber, membrane, flesh, fat, bone, cartilage, ligament, nerve, marrow and also from all the others, each and all of which likewise retain their form completely. And these also are generated from certain other proximate elements, that is blood, pituita, yellow and black bile: which also have their origin in those things which we eat or drink; moreover these are generated from air, fire, water and earth..[9]

We thus have no less than three sets of 'elements', the relationships of which can be diagrammed as follows (Fig. 14.4):

This explanation comes not from Aristotle but from De natura humana, as we shall see.

What is important in the basically Aristotelian argument of Book I, is the primacy of the qualities. The elements are identified by their two primary qualities: fire, for example, is hot and dry in the highest degree. Transmutation and mixture both take place by an interchange of qualities. Galen also leans towards the Aristotelian definition of mixtio, in which qualities, not substances, 'mutually interpenetrate'.[10]

At the same time, body must generate from body, and not from quality, and qualities are inseparable from the bodies in which they inhere. And it is in this context that, in Book I, he criticises the Hippocratic system of single-quality elements. Hippocrates, says Galen, is the source of verbal confusion, since he "calls his elements by the names" of hot, cold, wet and dry. But this is either mere multiplication of terms – calling one thing by two names – or else it represents a false attempt to apply substantive status to a mere quality.[11] But he goes on to discuss this same Hippocratic system with evident approval in Book II.

De natura humana and Galen's commentary on it are so closely parallel in their arguments that it is convenient to treat them together. Hippocrates, notes Galen, demonstrates hot, cold, wet and dry – the qualities which join in pairs to make the Aristotelian elements – to be "the common elements of all things." Related to them in the way described above, are the elements of man or humours: blood, black and yellow bile, and phlegm. All parts of the blooded animal derive from blood: but within this blood are all four of the elemental 'juices', or humours. And the process of generation is simply the separating out of the quality-particles in the mass of the blood according to their affinities:

> Whether therefore all things are produced from the same substance or whether instead nature, the wise maker of things, produced and formed the foetus first of blood, which flowed into the uterus from the mother: it drew together what was thick to form the solider parts of the body, what was thin to make the moister parts, what hot, the hotter and likewise what was cold to the colder. And indeed it seems much more in keeping with nature that the foetus should be formed right from the start, and subsequently that each part should grow and be nourished from its own particular material, as appropriate. For blood, like milk, seems to be a simple substance, but reason tells us that neither blood nor milk are simple.[12]

Hippocrates emphasised the impossibility of deriving one element or humour from another by insisting that if all of a particular humour could be removed from the body, the animal would immediately die: it could not be replaced by any of the other three.

The set of correspondences which Hippocrates then draws between the natural elements, the elements of man, the seasons, the ages of man and classifications of disease do not concern us directly.[13] The unity which the theory brought to a variety of disparate systems was one of its great attractions; but for the sixteenth century writers, to whom we can now turn, these correspondences seem no longer valid, or only in part: too many discrepancies have crept in.

Jacobus Sylvius' commentary on the elements is appended to this 1550 edition of Galen and Hippocrates. It is more than a mere paraphrase, but like his authorities Sylvius is interested in the elements as they apply to medicine. And his work can be seen as an attempt to answer the question, "what is the role of the elements in generation?" In pursuit of this, he deals first with the elements as a whole, and then with each in turn.

He begins – as does Fernel – from the simple and organic parts which make up the human body, and in particular from their spermatic and sanguineous 'principles'. These in turn derive from the four elements, but, as we have seen, indirectly:

> Now seed, both of man and woman, and maternal blood, are the principles of our generation, the former as form and the latter as matter. Both are made up of the four elements, but with the four humours composing the blood as an intervening step.[14]

According to the natural philosophers, elements are "the first and most simple principles of nature", principles which cannot be divided further, except 'scientifically' (physice) into form and matter. But this definition, says Sylvius, is of limited use to doctors, for whom the order and sequence of the building up of living things, compositio, and their dissolution or resolutio, are of primary importance.

He notes that "element" means something different to a natural philosopher, and to a doctor. For the former, it is earth, air, fire and water; for the latter, using the term 'loosely' (abusive loquentibus), it is the elements of man, that is, the four humours and the simple or similar parts, parts which are 'simple to sense'. Sylvius adds another distinction: elements can be considered as substantive or qualitative. If one is speaking of them in terms of their substance and nature (as a natural philosopher would) they are earth, air, fire, water: if in terms of their activities (medically, as Sylvius himself will treat them) they are heat (fire) moisture (air) cold (water) and dryness (earth).[15]

He goes on to explain this. When objects differ in their essence (as do air and fire) the different form which each takes is a consequence of its properties (proprietatibus consequentibus) and just as the properties of a thing determine its form, so these properties arise from its 'most perfect action' – e.g. heating in the case of fire. The most perfect action of a body thus comes close to defining its essence. Action in turn is related to quality. Just as for each organ or instrument of the body, one primary part is the efficient cause of its particular action, so for each element one primary quality is the efficient cause of the action of the element; the other is supportive only. For the element, as for the organ, there are other, secondary qualities (or, for the organ, parts) which contribute to its action in some way. Sylvius names these secondary qualities as "tenuitas, raritas, levitas, splendor, perspicuitas ... crassities, densitas, gravitas, obscuritas, opacitas.[16]

Clearly for Sylvius it is the qualities which are of prime importance in determining the action of an element, and it is at these which one must look to decide the role of a particular element in generation. And, by including rarity and density among the active and supportive qualities, he is moving, as we shall see, towards the "Stoic" system.

Sylvius goes on to consider qualities as the efficient cause of natural actions in the case of each of the four natural elements in turn, beginning with fire. He speaks of the action of fire, air, water and earth in generation, and not of blood, bile and phlegm, despite his objection that the categories of the natural philosopher are inappropriate to the doctor.

Fire is the first and most important of the elements active in generation, and as its prime action is to heat, so its prime quality is calor. Sylvius defines calor or heat in terms which identify it very closely with all forms of activity: it is the subject of all interchange, all change of state, all alteration. In a more narrow sense, it is the agent within the semen which separates out the different portions to form the brain, heart and other organs. In this activity he distinguishes two kinds, or two aspects of heat:

> The divine heat itself, indwelling in the seed of animals and plants, labours to build these things; but I added "divine" because heat, insofar as it is heat, separates and disperses what is different, and draws together what is like; as far as it is truly divine, it distributes the cerebral portion in the upper part, and separating the cranial portion from this same semen, it puts it into place around the brain; it separates the heart from the central portion of semen, distinguishing it from the lungs and thorax, and surrounds the heart with this same pulmonary and thoracic portion of the semen; and distinguishes itself in the rest of the body by the same perfectly equitable distribution… This same divine heat within the semen, or that hot spiritus which moulds us, hollows out veins, arteries, and the other channels and cavities, both those which are clearly visible and those which are hidden…attaches them and opens them up, so that they are best fitted for action and use…. What does it not do? being considered by some either as a kind of soul, or as its first and most important instrument, from its outstanding powers in generating, forming, living. These effects indeed, although they are inseperable from that diviner, unobstructed heat, are not exclusive to it, since, as shall be mentioned presently, the other qualities of heat and those of all the other elements take part in them, but that fiery energy, always and everywhere active, whose origin is the heavens, sustains the primary part in this formation.[17]

This description recalls, not the four lower elements as presented by Aristotle, but the passage in the Generation of Animals (736b, 35ff) in which the 'spirit which is in the semen' is compared with the element of the stars. It recalls also Galen's description, in De semine i, 8, of the process of differentiation; in both cases the spiritus within the semen is active in the process. The innate heat, in its relation to celestial and elemental heat, and to the spiritus, will be treated in more detail in the next chapter.[18] But we can just ask: what, for Sylvius, is the relation between divine heat and the heat which is a quality of elemental fire? They are clearly not opposites: in fact, they act in similar ways, but the divine heat is more active, and more intelligently active, than its elemental junior partner. Qua heat, its action is mechanical separation and combination; qua divine, it is an instrument of the soul; but the former powers seem to derive from the latter. Sylvius suggests that all 'heat', all activity, participates in the divine to a greater or lesser extent. Fire is after all the element nearest the celestial sphere in the Aristotelian universe. And we shall find similar arguments, more fully spelled out in natural philosophical terms, in Contarini.

The second quality of fire is siccitas, or dryness. This too has a role to play in generation. In the first place, the dryness of fire increases the action or effect, of its heat; secondly, it counters the natural moistness of the principles of generation, semen and blood, and helps prevent the portions of the semen which calor has separated out from flowing back into the indiscriminate mass.[19] This solidifying action is even more important when dryness is a prime quality, in the case of earth.

What of the accessory qualities, which are, for fire, levitas, tenuitas, raritas, splendor, perspicuitas? Sylvius divides them into active qualities and inactive ones, and it is the active – levitas, tenuitas and raritas – which really concern us, since they are both the active agents of transmutation in the "Stoic" system of elements, and the basis on which Sylvius distinguishes the two active and efficient elements (air and fire) from the two passive and material ones (water and earth). For Sylvius, these active accessory qualities are also the source of the 'hidden powers' of ethereal fire, which, though it is far less dense and thus would seem to have far less 'push' than earth, yet is far more active and effective in altering other bodies. Moreover, these are the qualities of spiritus, which as we have seen is related to divine, ethereal heat in some way. The existence of spiritus is evident both from its active qualities – for instance, from the speed of nerve transmission – and from its inactive ones – for example, the shining of the eyes visible in a dark room.[20]

The action of fire requires something to be acted upon, and this is, broadly speaking, the role of the other three elements, especially of the two 'lower' ones.[21] Air, the second of the elements, is defined by Aristotle as moist and warm, and since fire is hot in the highest degree, the prime quality of air must be moisture. This quality is more obviously appropriate to water, and Sylvius reports a certain amount of controversy on the matter.[22] Sylvius himself follows Aristotle; the role of air in generation is to provide active and penetrating moisture to foster growth, and in this it is comparable with spring, the season of generation:

> Moreover it [spring] is much moister than winter, and from this you will observe that it is most apt for germination and the generation of animals, neither of which takes place without abundant moisture, more abundant than the native heat of the seed or the sun can consume. For if all of it were consumed by one heat or the other, neither generation, nor growth by nutrition, would take place. Air then moistens everything most penetratingly and rapidly and powerfully, aided by its heat, and by its subtlety, lightness and rarity; water, having the opposite qualities, cannot moisten penetratingly or powerfully.

Air possesses the same accessory qualities of rarity, subtlety and lightness as fire, but in weaker form.[23] Earth and water, however, have the "opposite qualities" of heaviness, density and solidity. And it is this distinction between the two pairs which is the real basis of their different actions, not the differences in their Aristotelian qualities of heat, cold and so on. Though Sylvius ostensibly follows the Aristotelian (and Galenic) system of transmutation by qualities, it seems fairly clear that for him air, for example, is not only moist where fire is dry, but also less rare and less active than fire. The ratio of active to passive, as of rare to dense, decreases as we go down the scale, with earth nearest to pure material (and densest) just as fire is nearest to pure activity (and rarest).

The real distinction between the activity of the two sets of elements becomes clear when Sylvius turns to the role of water in generation. For the action of water is also to moisten: its quality of humiditas serves to aid the moisture of air and to counter the dryness of earth and fire. However, it is humidissima in a different way from air: air moistens most rapidly, water most thoroughly over a longer period. More importantly, the function of water, as one of the cold and thereby passive elements, is to form a partnership with earth in the material composition of the body. And Sylvius suggests that air and fire, having formed the mixture, are dissipated in the same way as the male semen in the Aristotelian theory of generation:

> And this [water], together with earth, yields most of the material for bodies whose constituents are generally like earth and water together. And there are also, apart from motion, active substances or spirits, which are more air-like and fiery: then just as earth prepared by water makes mud, suitable for taking on any form, so the first principles of our generation,

> male and female semen and menstrual blood, are both composed of the four elements: but blood [is] more watery and earthy, semen indeed, although at first more airy and fiery, the great power and substance of these elements having been soon dissipated, the parts made from it are watery and earthy as a result.[24]

From this analogy, the role of earth in generation is relatively clear. It, like water, forms part of the 'mud' on which the higher elements can act. And its particular virtue in forming bodies comes from its extreme durability – the result of its dryness – which however makes it difficult to 'work' unless combined with other, more ductile elements, water in particular.

In mixtures, as in generation, the active and the passive elements play distinct, even opposing roles. Air and fire act as a kind of proximate form, earth and water as a kind of prime matter. The lighter elements, Sylvius thinks, are either consumed in the effort of preparation or return to their proper spheres; but they do not remain to form part of the mixture. The most important elements in the work of generation are fire (as a kind of efficient cause) and earth (as the material subject par excellence) but the others are important too, and all four must always be present. All four are necessary in part because, for Sylvius as for Aristotle, the opposition of qualities is essential for successful mixtio. And the more perfect the balance which can be achieved of substances and of qualities, the more durable is the compound formed. Man, for instance, is a moderately durable compound. And his health or disease is similarly related to the balance or imbalance in the 'human elements' or humours of which he is composed.[25]

In discussing the role of the Aristotelian and Galenic elements in generation, Sylvius strays rather a long way from his models. Like Aristotle, he associates the actions of the elements with their qualities, and like Aristotle he distinguishes between facultative and material aspects. But he chooses in the end to attribute the action of the elements in generation to other qualities, the "Stoic" ones of density and rarity, levity and gravity. In so doing, he has committed himself to at least two 'layers' of explanation. If the lower elements, especially earth, are a kind of prime matter and the higher elements, especially fire, are a kind of efficient cause, still the latter do no more than give 'a kind of form' to the former, preparing it to receive further elaboration. The higher efficient cause of generation, as Sylvius hints in his discussion of fire, is the soul, with innate heat and spiritus as its almost indistinguishable instruments.

The Methodus..de Elementis of J. B. Montanus, written some time before his death in 1551 and published in Vienna in 1554, is much simpler in its argument and purpose. It is designed as a summary of the argument of Galen's De elementis and of the De natura humana, and was almost certainly intended for the use of students. According to Montanus' editor, a knowledge of the elements was essential to all serious students of medicine:

> For indeed, if they do not understand the elements, temperaments, humours, spirits and faculties of the human body, together with their actions, natures and functions, plus the construction of all the parts (which anatomy teaches) how (I submit) ignorant of the naturals, will they have certain knowledge of the operations and influence of the praeternaturals?[26]

From this we can expect that Montanus will treat the elements in their medical context, as part of the 'naturals' and as a necessary preliminary to the understanding of the contra-naturals, and so of the causes of disease.

In the Methodus itself, Montanus confirms this. Medicine is an art and must therefore seek some end: in its case, health. Health in turn implies that the actions of the body must maintain a certain proportion. But as these proportions are not immediately perceptible to sense, one must have a method for recognising them; and this leads one inevitably to a knowledge of the elements, since these are the principles of which nature is composed. For Montanus, as for Galen, the elements are the basis for the medical theory of temperaments, and, like Sylvius, he also wrote a commentary on Galen's De temperamentis.

Montanus follows this introduction with a brisk summary of the two standard medical authorities on the subject, De elementis and De natura humana. The questions he raises are already familiar to us: are there one or several elements? If several, how many and of what kind? Do the elements mix as prime qualities or as bodies? In Book II he discusses the humours, concluding that all four are necessarily present in healthy blood, because blood nourishes (and therefore re-creates) all parts of the body. Related to the primary elements in a certain proportion, the humours are the four elementa propinqua of man. The universal proportion existing between humours and elements is shown, according to Montanus, by what one might call the co-ordination of their actions, and particularly by the relation between the seasons of the year and the seasons of human life, as argued by Hippocrates.[27]

Towards the end of the companion Methodus de Facultatibus naturalibus, Montanus returns to the question of the elements, and to the way in which they transmute and join to form mixtures. He is clearly aware that there is a major difference of opinion on this point between Aristotle and his followers on the one hand and Zeno and the Stoics on the other. The conflict, as he describes it, covers two separate but related questions. The first is the question of whether qualities or substances are altered when elements combine: Hippocrates and Aristotle voting for alterations in the secondary qualities of the elements, Zeno for alterations in their substance. The second concerns the generation of one element from another, or transmutation. According to Aristotle, elements are mutually generated from the four qualities (as in Fig. 14.1). According to the Stoics, says Montanus, the elements are generated by the "diffusion and contraction of matter": i.e. by changes not in their qualities, but in a universal substance (Fig. 14.2).[28] Montanus does not himself take sides, but it is interesting to note that he sees the opposition between these two systems more clearly than his three later contemporaries, all of whom produce what one might call 'mixed models'.

Montanus' nearest contemporary, Gaspar Contarini, is the only one of the four to write more as a natural philosopher than as a doctor, although he is careful to point out that his treatise will be of special use to physicians.[29] As a survey of natural philosophical arguments on the subject, his treatise, together with Sylvius', provides a useful background to Jean Fernel, who combines medical and natural philosophical themes.

The difference between Contarini and my other three examples is suggested by his qualifications, and by his sponsors. He is described on the title page as a Cardinal, and as one of the most distinguished philosophers of his age. His book, published posthumously, was dedicated to a fellow Cardinal, Marcellus Cervinus, by Joannes Gaignaeus, doctor of theology and Chancellor of the University of Paris.[30] Contarini's work is divided into five books, of which the first two deal with the elements, book three with their mixtures, and books four and five with the temperaments. It seems convenient to consider the first three here, and the fourth book later on, in the section on temperaments. Book five deals with the 'sensible consequences' of temperament – colours, odours and so on – and is largely irrelevant for our purposes. In the first book, Contarini sketches out a general theory of the elements, based largely on the authority of Aristotle; in the second, he takes the elements one by one, like Sylvius, but with more emphasis on their meteorological and other natural-philosophical operations and less on their medical ones.

As we have seen, both Sylvius and Montanus, being primarily interested in a medical theory of the elements, based their commentaries on Galen, and on Galen's own sources, in particular the Aristotle of De generatione et corruptione, De caelo and the Physica, and the Hippocrates of De natura humana. But the key work for understanding Contarini's idea of the elements is not De elementis or De natura humana, but the Aristotelian corpus and in particular the Meteorologica.[31]

Contarini's aim, he says, is to bring to light the ideas of the early philosophers on the subject of the elements and their mixtures, and to draw together arguments from a wide range of sources. The sources he cites include a high proportion of Aristotelian works and commentators; so his definition naturally follows Aristotle's. Again following Aristotle and Galen, he goes on to reject the notion of an element which is itself built up from more primitive bodies: the atoms of Democritus and Epicurus, the triangles of Plato in the Timaeus, or the 'seeds' of Anaxagoras.[32]

The most striking feature of Contarini's argument in Book I is his suggestion that two sorts of elements are involved in the composition of the material world: the four familiar ones, which serve as the material cause of bodies, and a celestial element (or elements) which represents the efficient cause. This argument comes in a discussion of necessary cause: which, he says, can either lead to the emptiness of divinatory astrology, or to the more fruitful recognition that

> these elements and the inferior world exist as a kind of matter, but indeed celestial bodies and superior intelligences are the efficient cause, which certainly induce motion by their light, and by that warmth, different somehow from elemental heat and fire, which is a follower of light, and which we may rightly call aethereal warmth. The Platonists and more ancient philosophers thought this to be a certain aethereal and most tenuous body, penetrating all, which some of them according to Aristotle were also accustomed to say was the great seed-bed of nature, whereas no body, even a mathematical one, however exceedingly rarefied, can penetrate some other body, at any rate there is no such body permeating these inferior elements, but the celestial body is aethereal and distinct from the elements and has a natural motion which is circular: nevertheless light and motion flow back into the elements from this body, and also what we call aethereal heat: not a body but an aethereal quality, which penetrates and permeates all elements. This heat is recalled in the book of the generation of animals, in which it is said not to be elemental heat, but analogous to the ele-

> ment of the stars. Since elements are matter which is formed by celestial bodies and their intelligences, it is quite clear that many things take place in the elemental world, the cause of which we cannot refer to the natures and forms of the elements themselves: but one must refer the cause of all this to the celestial bodies, whose power and action the elements receive.[33]

The suggestion that there are two sets of elements, one celestial and one material, and in particular two fires, is a feature which Walter Pagel and Marianne Winder have studied in the work of Paracelsus, and of certain Neo-Platonic Renaissance natural philosophers, beginning with Ficino. None of these writers, however, seem to suggest that their 'higher elements' have a role to play in the generation of ordinary corruptible bodies, including the body of man and animals. Rather, they are the elements which, in a strictly analogous sense, go to make up the 'body' of the heavens, the angels and so on. Moreover, their theories appear to derive largely from Neo-Platonic and Gnostic sources.[34] But Contarini, though he mentions Plato and the "Platonici", Pythagoras in particular, bases his own argument squarely upon the authority of Aristotle: Meteorologica I ii 339a 20–33 in particular, where Aristotle declares that

> This world necessarily has a continuity with the upper motions: consequently all its power and order is derived from them. So we must treat fire and earth and the elements like them as the material causes of the events in this world…but must assign causality in the sense of the originating principle of motion to the influence of the eternally moving bodies.

The motion of the elements is, according to Aristotle, derived from the circular motion of the quintessence (I iii 340b 12–19). And from this motion of the elements, moving to, and resting in, their proper place, the parts of animals, and indeed of the universe, are formed, argues Contarini, following Aristotle. Their motion is in turn the consequence of their intrinsic form, and this, according to Contarini, is to be identified not with their qualities of heat, cold and the like, but with their quality of heaviness or lightness which corresponds to their form.[35]

At the beginning of Book II, Contarini turns to the nature of fire. Fire is the element which in every respect approaches nearest to the nature of the celestial element, the aether or quintessence. The motion of fire, especially of the upper layers of the fiery sphere, is influenced by its nearness to the quintessence and partakes of its circular movement. In any case, the proper natural motion of fire, i.e. upwards, tends towards the celestial sphere, "as to its proper place". The heat of fire is of all qualities the most active, and altogether in its activity, its rarity of substance and paucity of matter and its possession of the qualities of heat and light it comes closest, of all the elements, to resembling the celestial body and to the nature of a form. According to Contarini, each element is equal in total matter, but fire is much rarer, much more dispersed than the others. He supports this with two arguments: one is the observation that a tiny piece of wood (itself much rarer than pure earth) produces an amazingly large flame; the other derives from the attempt to determine the extent of the sphere of fire ex astrologorum rationibus, i.e. from measurements of the distance between the earth and the moon. We have already seen the idea that rarity of matter approximates to form, and density to 'true' matter, in Sylvius.

The special, quasi-formal nature of fire thus sets it apart from the other elements: a relationship which Contarini explains by means of analogy. As form: matter, and the celestial body: the elemental sphere, so fire: the other elements.[36]

Contarini's treatment of the relation of fire to the other three elements, like his notion of the nature of fire itself, is similar to Sylvius', published two years later. This resemblance extends to Contarini's discussion of the other elements; for instance, his definition of air as "a body being somewhat hot, and exceptionally moist, denser than fire, but much rarer than water and likewise water vapour; which is not so effective in action as fire, for indeed heat is weakened by moisture."[37] Here, as in Sylvius, we have the statement that qualitatively, as well as in its relative density and degree of activity, air is intermediate between fire (hot) and water (wet).

Water, the element beneath air, is according to Contarini absolutely heavy, while air is absolutely light. Water is the primarily cold and wet element; but its density, rather than its qualities, seems to be important in transmutation. Water vapour is aqua rarefacta, and differs from water or air by its degree of rarefaction or condensation.[38]

Earth, like water, has an important role to play in the nutrition and generation of animals, though Contarini devotes much of his attention to meteorological, rather than medical, examples. But he produces an analogy between the macrocosm and microcosm which suggests the wider context within which Renaissance ideas of animal generation need to be considered:

> And in the same way as the animal body is warmed throughout by spirit, so this earth contains within itself many aethereal vapours and fiery exhalations, because of all of which it can generate without mixture and serve as nutriment.[39]

In the third book, which deals with the composition and mixture of elements, Contarini considers the answers of Aristotelian writers – Avicenna, Averroes, and Thomas Aquinas for example – to questions such as, does the elemental form remain in act in a mixture, or not? Do sensible qualities represent the true form, substance or nature of the elements themselves? Contarini considers that mixture takes place when the forms of the elements are 'broken up' or checked (refracto), neither existing in perfect act within the mixture (as Avicenna and Averroes argued) nor 'corrupted' (as the Latin Peripatetics said). This weakening of the elemental form can take place in many ways: one is by changes in the qualities of the elements, another by rarefaction and condensation. Whatever the process, the result is that the elements are no longer able to retain their full elemental identity and are able to be 'conflated' into a new, compound form by the action of celestial heat.

Do the prime qualities of the elements remain in these mixed bodies? Contarini thinks they do, because certain mixtures 'have a complexion' to heat, cool, dry and so on, which they would not be able to do, "unless there was in them the power of heat and of the other prime qualities." Indeed, he adds, the prime qualities are as it were the elements of the temperaments or complexions, which are made up of qualities in the same way as mixtures are made up of elements. This point will become clearer when we turn to the temperaments and their part in generation.

Of these prime qualities, heat is the most important: because, as we have seen, it "aemulates the excellence of celestial heat." Heat is the most active quality in generating mixtures, and in this it acts as "the first and nearest instrument of celestial qualities and powers", i.e. of the aether or quintessence and its motion.[40] The phrase is very similar to one which we shall see used of innate heat and spirit in relation to the soul in animal generation. As Contarini points out, the two processes are much alike:

> The generation of mixed bodies is not totally unlike the generation of animals and plants, although they differ in not a few respects. For just as in the generation of animals there is a certain material in which the soul (which is first to exist) is induced, material which is shared by the natural heat proper to the soul, by which it forms the members; so in the generation of a mixed body there is a form which is induced in its matter, and is first extant in the natural heat. Likewise, just as in the generation of animals, apart from the heat in which there is the soul, there is also the semen, drawn from the parent, in which there is a power of generating the animal; so in the generation of all mixtures there is the celestial quality, in which as in some great seed-bed the power of generating all mixtures is drawn out from the celestial bodies and those intelligences that move the heavens, like a kind of art derived from that supreme art, and [by it] the elements in the generation of mixtures are arranged, and the form of the mixture induced.[41]

Mention of the celestial intelligences and of a magnum seminarium of all forms suggest an admixture of Platonic philosophy in Contarini's work. But a few pages later he develops this analogy in terms which bring it more closely into line with the description of animal generation given by Aristotle, suggesting that just as the male semen does not remain to form part of the foetus, so these celestial qualities 'disappear' once their part in forming the mixture is over.[42]

What Contarini, like Sylvius, seems to be describing is a two-tier system of generation. At the lower level, the active and formative role is taken by fire (among the elements) or natural heat (among the qualities). But just as in the generation of animals we must look beyond the animal being generated to find the cause of its generation, i.e. the parent, so in the case of mixtures we must look beyond the world of elemental bodies to find the ultimate cause of their being the way they are. The process can also be seen as forming two stages. In the first stage the elements are prepared for the action of the celestial qualities. In this activity fire plays much the most important role, acting, says Contarini, as a kind of form in relation to the matter of the other elements. The properties of fire derive from its qualities, and though the most important of these is heat, they also include rapid motion and extreme rarefaction: active qualities which it shares with the spiritus of Renaissance physiology. In fact, the most obvious way of producing fire, or rather extreme heat, is by producing the rarefaction with which it is associated.[43] This explains the 'generation' of fire; but Contarini suggests that a similar motion, rarefaction and change of qualities is the mechanism of refractio, the first stage in the production of a mixture, in which the elements are prepared for amalgamation into new compounds by the celestial qualities and their instrument, heat.

A similar two-tier system is to be found in the writings of the French medical writer, Jean Fernel, on the elements. Like Contarini, Fernel was a sophisticated commentator on the problems of natural philosophy, although, like Sylvius, he often used his knowledge of natural philosophy to tackle problems of medicine and physiology. His account of the Aristotelian elements is in Book II of the Physiologia, but also forms an important theme in De Abditis Rerum Causis. Fernel's emphasis in both works differs somewhat from that of the other writers we have been considering. In the first place, both are treatises rather than commentaries. And instead of asking what the elements do in terms of medicine and natural philosophy, he is much more concerned to emphasise what they cannot do. This theme, that "multa esse in philosophia extra elementorum ordinem" is most fully developed in De Abditis Rerum Causis, but I think readers of the shorter and more coventional treatment of the Physiologia would have been expected to keep the other work in mind.

De Abditis Rerum Causis is a dialogue among three natural (and medical) philosophers, two of whom, Philiatros and Brutus, consult the authority of a third, Eudoxus (who comes closest to representing Fernel). Broadly speaking, the first book is devoted to natural philosophy, the second to medicine, including the question of generation. Elements, as we have already seen, are a subject which belongs in both categories, and Fernel discusses the theory of the elements in both parts of the work. In the preface to Book I, he explains his reasons for choosing the dialogue form. The subject of "hidden causes", bringing as it does questions of divinity into natural philosophy and medicine, is "full of contentions" and for that reason should be presented as a debate, with all shades of opinion represented, rather than as a straight narration.[44] In fact, Brutus, who is cast as Eudoxus' opponent, is no more convincing as an independent figure than most devices of this sort. But as well as being fashionable – probably because of the influence of Ficino's Plato editions – dialogue has the same advantage as the medieval guaestiones form: it allows a range of opinion to be included.

Before turning to the elements themselves, it is important to take brief note of Fernel's method, and of the presuppositions on which his philosophy, both natural and medical, is based. Philosophy, says Fernel, seeks after causes: even though the reasons it arrives at are probable ones only. In pursuing this search, the natural philosopher must go beyond the evidence of the senses, if his conclusions are to be more than superficial and incomplete. For the true beginnings of things are "inward, hidden, and far removed from the senses": inaccessible to the eyes and ears, and attainable only by the mind.[45] Those who see the anatomist as the most successful exponent of the life sciences in the Renaissance, and his naive reliance on observation as the most commendable scientific method, may be inclined to discount this. But Fernel's approach is less exceptional among successful physiologists than one might think.[46] In any case, his largely rationalistic approach to how one attains truth has implications for his discussion of the elements. "There is much in philosophy outside the order of the elements": much, that is, outside the range of the senses. For the elements, although abstractions which exist in the natural world only in impure form, are sensible by definition: that is, they are defined by the possession of certain sensible qualities, those of hot, cold, dry and moist. The heart of Eudoxus-Fernel's argument in the Hidden Causes is that just as a true scientific method must move beyond observation to reasoning based upon it, so one must move beyond the attempt to determine the properties of living bodies by the study of their elements and temperaments – their sensible qualities of heat and cold – and examine instead their form, the true source of their powers, which is divine in origin.

The total form of a living thing does not derive from its elemental composition, or from its temperament, argues Eudoxus against Brutus. In so arguing, he opposes any attempt to deduce the powers of a thing from its qualities of cold and hot, which he attributes to certain Aristotelians. And he substitutes for this a multilayered explanation: living things take on a succession of subsidary forms, which serve as preparations for the reception of a total form, or soul. Only the lowest of these preparations involves the elements: that of permixtione, in which all four elements combine to form the simple parts and to determine their individual temperaments. At the second level of preparation, these simple parts are "composed" into organs of definite temperament: and at the third, the structure of the entire animal is established, and the parts of the organism are integrated with the help of innate heat and spiritus. With the final reception of a total form or soul, divinely implanted, this complex and integrated structure becomes at last an active living organism.[47]

In this blueprint of the process of generation, Fernel, like Sylvius and Contarini, is describing the hierarchy of forms, which we can imagine as existing along a continuum. At one end, the lower and more imperfect forms, i.e. the elements, are so dis-organised as to be very close to pure matter; at the other end, that of "total form" or soul, the degree of organisation is so high that the result is immateriality. But there is no attempt to associate 'degree of form' with rarity of substance among the elements themselves. The distinction between Fernel's account and those of Sylvius and Contarini is that the "higher" or rarer elements, fire and air, are not elevated into a kind of form, while the lower and denser are denigrated to a kind of matter. All four elements are of equally low status among the material causes of living things, though innate heat, which Fernel distinguishes from elemental heat, does have a more formal and active integrative part to play.

If we look at Fernel's system in more detail, with particular reference to the elements, several points can be made. One concerns the grounds for distinguishing between the various levels, or preparations, as Fernel calls them. In particular, Eudoxus is challenged to explain the difference between the levels of element and temperament. He does so largely on practical grounds: it is convenient, he says, to distinguish the material consistency of a thing (the result of elemental mixture) from its temperamental quality. The latter is in fact more easy to determine than the former, but they clearly vary independently:

> PHILIATROS: Indeed, to my way of thinking, the method (modus) of temperaments is inexplicable.
>
> EUDOXUS: You should say that much more in [reference to] conformation: for that this or that temperament is hot, when the hot and fiery element dominates the rest of the elements in a mixture, is well enough known. And likewise if you feel it cold to the touch, that the cold element dominates: but by what reason it happens that this or that are [cause of] the shape of this or that thing, indeed, you can neither explain nor follow mentally. For we distinguish many things, which although they have the same temperament have dissimilar shapes, and others which are similar in shape but differ in temperament; which is furthermore proved more fully and abundantly from medical opinions.[48]

The distinction between the structure of a living thing and its form is a more important point, which relates to Eudoxus's, and Fernel's, insistence that the form does not derive from the elements. The successive preparations by which we can build up in our minds the plan followed in generation represent, not a material hierarchy, but a hierarchy of forms: as Fernel explains in the heading to Book I, chapter iv, "The forms of individual parts in a composite are like kinds of preparations leading to the total form, and these are multiple."[49] Furthermore, the total form is associated not with the structure of the body or its individual parts, but with its life-functions. It differs, therefore, from the form of the parts, which are structural and instrumental with respect to the whole, and which do not (like the total form) disappear with the death of the individual. The total form or soul is "that by which it is made one, and by which it lives (agat) its life".[50]

In Book II, Fernel strengthens his argument that form does not derive from the elements with quotations and arguments from the major medical authorities: Galen, especially the Galen of De placitiis, De usu partium, and the generation works; Hippocrates, whose concept of "nature" Fernel identifies with "form"; Plato; and Aristotle.[51] Fernel's own position is best summarised by the passage from De generatione animalium, which he quotes, in which Aristotle says that while hard and soft, dense and rare and other conditions of the parts can be derived from the principles of hot and cold, "the reason why flesh is flesh and bone is bone" cannot be so determined. Instead, "formam rei coelitus demitti prodit." Even the form of the simple parts cannot be derived from the elements; and form is divine in origin.

In developing this argument, Fernel argues against all those who would attempt to derive all causes from the elements, whether the earth, air, fire and water of the Hippocratic 'dogmatists' or the atoms of the Democritean 'methodists'. Equally, he opposes any attempt by medical writers to refer all aspects of medical explanation to the temperaments. Both the material composition of the human body and its qualitative or temperamental makeup are relatively insignificant causes, in relation to the question, 'by what cause is the animal integrated, made one, and by what cause does it live?' These powers do not arise from its material composition, but from its form, which is "wholly divine". Instead, the material or elemental composition, like temperament, acts as "a kind of instrument" of form – an instrument being defined as something which allows the 'workman' to work more efficiently in a given medium.[52]

In the Second Book of the Physiologia Fernel approaches the problem of the elements from a somewhat different angle, and one which is more like that of the three commentators we have already discussed. But De Elementis is a treatise, rather than a commentary; it has its own concise and logical form, and its own purpose, despite its basis on the authority of Galen and Aristotle. Moreover, I think the limitation put on the powers of the elements in De abditis rerum causis still has to be kept in mind as a background to this briefer treatment of the subject. After a preface, in which Fernel again emphasises the importance of seeking to understand events by looking for their causes, moving from what is perceptible to sense to what is perceptible to reason (resolutive method) and back again (compositive method), he turns to the "classic questions" of natural philosophy and medicine: of what elements are the parts made? how does mixture take place? and so on.

Fernel's elements are the same as everyone else's: the familiar earth, air, fire and water. They are the 'material' of which the simple parts are composed, and all four of them together go to make up the parts and all compound substances. His definition of "element" is both resolutive (the most simple body, i.e. the most minute division of a species) and compositive (that out of which all other bodies are composed). Both definition and discussion derive from Aristotelian philosophy: thus, the elements themselves can only be 'broken down' into form and matter, and they are defined by opposing pairs of tangible qualities. Equally, they can be defined within a Hippocratic context: elements are the material of which the simple parts are made, or, to follow the process in more detail, elements → food → humours → blood → semen → simple parts; just as in Galen's summary of De natura humana.[53]

In chapter vi, Fernel goes on to consider a more controversial point, which is also discussed in De abditis rerum causis. Do the forms of the elements themselves remain within the compound, or are they destroyed to make way for the new form of the mixture itself: the form of flesh, or bone, for example? In both works Fernel, following Aristotle, concludes that the elemental forms remain, but exist in potentia only. Philiatros, in the Hidden Causes, provides a practical reason for this decision: if the substance of the elements were destroyed every time they came together to generate a compound or mixtio, the body generated would be unable to revert to its elemental parts, and very soon there would be no elements left to make new compounds.[54]

This brings us to a further consideration of the way in which elements combine to form mixtures. Mixture, as Fernel describes it in chapter vii, is more likely to occur if bodies are 'susceptible', i.e. if their substance is easily divisible. Moreover, there is no possibility of mixture without the conflict of mutually opposing qualities. The elements themselves fulfil both criteria by definition, but Fernel seems to be using this to explain why certain substances found in the natural world combine more easily than others. Both prerequisites for mixtio are familiar from Galen and Aristotle. Mixtio itself is defined as the concretion of several substances into one. The elemental qualities are not destroyed, nor do they remain as distinct entities within the mixture. Rather, much as in Contarini, the elemental powers are weakened or 'repressed' to allow a new form to emerge – that of the 'mixed body' or simple part, e.g. bone or flesh.[55]

Finally, Fernel asks, is it the substance of the elements, or their qualities, which are "implanted all in all" in this concretion or mixtio? He concludes, as he had in the Hidden Causes, that the qualities alone combine all in all; the forms or substances remain within the mixture, confused, conquered, in a state of potentiality, ready to return to their original strength on the dissolution of the compound.[56] Like Contarini, he is following Aristotle – rather than the Arabic or Latin Aristotelians – in this opinion.

Fernel's theory of the elements attempts to further the understanding of physiological phenomena by looking at their causes, and in particular to lay the foundation for the, more narrowly medical, theory of temperaments. His discussion of the elements in both treatises is still based on the theories of Aristotle and Galen. But he makes, much more clearly than these authorities or his contemporaries, the point that the elements are of minimal importance in determining the total form and activity of a living organism. Like the celestial qualities and aethereal heat of Sylvius and Contarini, the most important determinant of how and why an animal lives – the soul – is outside the elemental world. Does this beg the question, by pointing for answer to questions in natural philosophy and medicine to something outside them? Galen, and Aristotle, may well have thought so. But for Fernel, the elements as they have been defined are simply inadequate to answer the questions he wishes to ask.

If we compare the theories of the four writers studied, we can see that on the whole they lie within a common tradition, and at some points are strikingly similar. Montanus rarely ventures into philosophical speculation, and Fernel, having argued that the true motive force of generation lies elsewhere, is less interested in developing a notion of the manner in which the elements act and interact. Thus, the most detailed accounts are those of Contarini and Sylvius, one approaching the problem as a natural philosopher, the other as a doctor. Despite this, they come together on a number of points: the gradation of elements into 'higher' and 'lower' according to their rarity or density and their assumed degree of activity; the association of the higher elements, especially fire, with formal activity and with the celestial qualities and celestial heat; the suggestion that these higher, formal elements do not make part of the finished mixture or geniture. Fernel's understanding of the relationship between elemental heat and innate or generative heat is similar to theirs, although this does not appear clearly from his Physiologia alone.[57] He shares with Contarini the idea that the elemental forms are weakened or "repressed" but not destroyed when mixture takes place, and with both Contarini and Sylvius the concept that elemental mixture – like generation – takes place at several levels, or in several stages, and cannot be thought of as a single operation. There is a further common feature, which forms the justification for this chapter: the stated or implied analogy between the process of mixture and the process of generation, which appears in all four writers.

What then can we conclude from this survey of four major commentaries of the mid-sixteenth century, all intended for a medical audience? I think there are three major points to be made. The first is that all of our examples assume some form (or forms) of a four-element theory, based ultimately on that of Aristotle. All four are familiar with other forms, from classical authorities: the single-element theories of the Pre-Socratics and the many-element theories attributed to the atomists. None mention, even in passing, the three-element system of the alchemists (or of Paracelsus) or the hierarchy of elements set out by the Neo-Platonic and Gnostic writers studied by Pagel and Winder. Thus they are working within a definite, and decidedly orthodox, system, which was, at the same time, the system accepted by the vast majority of their educated contemporaries, medical and otherwise.

Secondly, I think we can see that there are areas of confusion among three possible forms of the four-element theory itself, confusions which these writers reflect in various ways. This confusion began with Aristotle, who uses both qualitative change and change of density to explain transmutation in various works, and who also suggests that the higher or rarer elements, such as fire, are closer to form, the lower and denser, such as earth, closer to matter. It continued with Galen, who discusses the Aristotelian model in Book I of De elementis, and the Hippocratic one in Book II. Montanus, as we have seen, reproduces Galen's doublethink but distinguishes clearly between Aristotelian and Stoic arguments. Contarini and Sylvius both find it equally difficult to avoid confusing the "elements" of natural philosophy and those of medicine. In addition, they both adopt features of the Stoic explanation of transmutation by changes of density, while all their definitions are based on the more orthodox Aristotelian one of qualitative change. The first error arises from the common use of terms – elements, earth, air, fire and water – which mean entirely different things in the writings of Hippocrates and of Aristotle. The second is perhaps more complex. On the one hand, it can be argued that both Contarini and Sylvius are good Aristotelians in their natural philosophy, following the Aristotle of the Meteorologica, which describes a hierarchy of elements and their transmutation by rarefaction and condensation. Certainly Aristotle is used by both as a clinching authority. But the strength and something of the flavour of their argument derives also, I suspect, from the Platonists. Contarini in particular devotes considerable space to a discussion of the Timaeus, and he cites Pythagoras and other 'Platonici'.

Fernel is something of an odd man out, presenting his solution to the problem of the elements in considerably more negative terms. But like the others, he combines the definitions of Hippocrates and Aristotle, discussing the elements in both their medical and their natural-philosophical context. And like Sylvius and Contarini, he attempts to solve the problems of the facultative or 'ordering' aspect of the composition of material bodies in general and living bodies in particular. Contarini and Sylvius attempt to answer this question in part by reference to powers above the elements, the aethereal heat or quintessence, or celestial qualities and intelligences; and in part by 'promoting' the higher elements, fire and air or fire alone, into quasi-formal roles. But for Sylvius at least the higher elements do no more than prepare the mixture to receive "a kind of form", and it is this point – that elemental mixture is the lowest step in a series of preparatory forms, culminating in the acquisition of a 'total form' or soul – that Fernel emphasises. His terms are Aristotelian, but to explain the higher steps in this process he turns not to the quintessence of Aristotelian natural philosophy but to the innate heat and spirit of Galenic physiology, and ultimately to the Christian-Platonic soul.

This then is the third point: the suggestion, made most strongly by Fernel, that the elements in their classical form are not enough to explain the material composition and structure of the world, much less the life-activities of plants and animals. A second (or even third or fourth) layer of explanation is necessary: temperament, innate heat and spirits, the soul; or the divine actions of the quintessence and of 'aethereal heat'. For the major Renaissance writers we have been discussing, "there is much in philosophy outside the order of the elements."

Temperament is the medical descendent of the four Aristotelian qualities, which, as we have seen, combine in pairs to form the four Aristotelian elements. These qualities can also be distinguished as active or passive, and this gives us the pairs hot-cold (active) and wet-dry (passive). When living things are generated, hot and cold, acting on moist and dry, form the homogeneous bodies (which Galen calls the similar or simple parts). This is so because the physical qualities of hot and cold and "the mixture of their motions" – in other words, temperament – are precisely what differentiate the homogeneous parts from one another, and so define them. That is, the parts are distinguished from one another simply by being hotter, colder, wetter, dryer, softer, harder; and similarly by the other sensible qualities which Aristotle reduced to the four primary ones in De caelo II.[58] Galen's theory of temperaments compares the simple parts in exactly these same terms, defining them by their place on a scale from hot to cold and from wet to dry.

Aristotle does not clearly distinguish in practice between his elements and the pairs of qualities which compose them; Galen, defining 'temperament' as distinct from 'element' in Book I of De temperamentis, is similarly imprecise:

> Bodies are said to be put together from the just proportion [ex..temperatura] of hot, cold, dry and wet; of these, those which are so in the highest degree are taken to be the elements themselves, air, fire, water, earth.[59]

All bodies are formed of a tempering or 'balancing' of contrary qualities; elements are the name we give to bodies which possess certain of these qualities "in the highest degree".

In the fragment known as de facultatum naturalium substantia, Galen sets forth his confident opinion that "all the bodies which this earthly globe contains are composed from the four elements, blended with one another in proper proportion [contemperatis] as a whole and not, as Empedocles thought, broken down into the most minute particles."[60] Temperament deals with the relation and balancing of contraries in the whole as a whole; it is thus peculiarly adapted to the study of living things as they are, not broken down into their parts, and hence to medicine. In fact, as we shall see, Galen takes it to be the basis of certain and practical medical knowledge. A mixture is quite clearly a substance; temperament is a way of describing the state or balance of qualities in that compound substance and of comparing it with others. The distinction between qualities used in a temperamental sense and in other senses is confusing, as Galen points out. For temperamental descriptions are a special, highly technical use of words – such as "hot" or "dry" – which can also be used in other ways. In the first place, it is restricted as to subject: temperament is used, medically, to describe the state of man and of other animals and plants in their relation to man, as terms of comparison and as medicines. So "warm and moist" is a temperamental description when applied to the qualities of a man, but not when applied to air, for example. But calidum can be used to describe a man's elemental or temperamental heat, his natural or innate heat (when this is distinguished from the other) or his acquired or "alien" heat: heat produced by (temperamentally) hot foods, by exercise, or by fever. Only the first of these is a temperament properly speaking. Again, when we say that a body is hot, we may mean simply that it contains a high proportion of the elements fire and air, i.e. that it is hot "absolutely". Or we may mean that in the mixture of which it is composed, the quality of heat is dominant over the quality of cold, i.e. that it is relatively or dominantly hot. So "hot" can be used to describe the structure of a body in several ways and at several levels.[61]

Moreover, temperament is not a system based on absolute levels of hot, cold and so on, but primarily a system of comparisons. So the same body can be temperamentally "hot" or "cold" depending on the terms of the comparison. Temperament can be comparative within a given genus or species: this bear is hotter than the average bear; or it can be used to make comparisons between species: men are hotter than bears. Both comparisons and absolute statements, such as "bone is cold and dry" (a statement which is true of all bones with respect to the other parts of the species to which they belong) are made with reference to what Galen calls the symmetron, i.e. the point of balance or medium between extremes of temperament in an individual, a genus, or a species. If a subject is related to different symmetra, the resultant temperament may be different. For example, a man may be cold and dry, compared with the mean or symmetron for the species man. But he may at the same time be hot and moist, compared with the symmetron of some other species. A man's temperament is not an absolute state: it is a point of comparison, relating him to other men, to animals and plants and to his own state as changed by age, climate, time of year or disease.[62]

The ideal temperament, or eucraton, is, mathematically, one in which the proportion of the activity of each element is exactly equal with respect to all the others. This is what Sylvius, in the tables prefaced to the 1549 edition of de temperamentis, calls equal temperament by weight (pondus). But there is another form of equality or eucrasia mentioned by Galen, which Sylvius calls eucrasia ad justitiam. This form of ideal temperament is measured by the perfection of action of the species or individual: "Indeed, we say of whatever plant and animal, that they are best when they best perform their function". Any creature which is so tempered as to best fulfil its function, to perform the actions appropriate to it as perfectly as possible, is of equal temperament. And there is a kind of overall equality of temperament among all such well-tempered animals and plants, even though the actual balance of qualities appropriate to each will differ, being measured, so to speak, on many separate scales.[63]

Temperament, then, is a theoretical means of comparing, and of explaining, the individual and specific differences between animals and plants. In particular, it is a means of explaining the differentiation of the simple parts: blood, flesh, ligament and the rest. Galen devotes considerable attention in de temperamentis and several other works (for example Ars medica, de sanitate tuenda, de usu partium) to a discussion of the temperament of the simple parts. He places them along two intersecting scales, one from hot to cold, and one from wet to dry. Flesh and other 'blooded' parts are the most warm and moist; bone, nerve and other 'bloodless' parts are similarly most cold and dry. The midpoint on both scales is the skin: especially, Galen tells us, the skin of the hands, and in particular of the fingertips. This is appropriate, since it relates the 'hot' and 'cold', 'wet' and 'dry' of temperament to the actual physical qualities we distinguish when we touch an object. Other creatures distinguish objects by touch, but less perfectly: for of all animals, man is the most perfectly tempered, the most perfectly 'in the middle' of all the sensations he seeks to interpret. Thus the touch of human fingertips is the instrument best suited for distinguishing temperament.[64]

This is important in a medical context, since the temperament of a patient is a clue to his health and similarly to the treatment which is suitable. This medical side of the theory of temperaments takes up most of Galen's attention in de temperamentis and other works in which it appears[65], but it does not concern us especially here, apart from a major point which Galen makes in his short treatise On the substance of the natural faculty. The doctor, says Galen, is faced with a variety of conflicting answers to the question of the substance of the soul and its faculties, none of which he can reasonably accept as more than uncertain probabilities. But medically speaking the answer to this question is irrelevant: whatever the soul's substance, and however it is introduced into the body, the latter serves it according to its own nature, which is the result of the temperament of the elements composing it. As long as the body's natural and equable temperament is maintained, the association of soul and body will be preserved; in any case, it is impossible to consider the soul 'in itself', but only in its association with the body. It is not necessary, Galen tells us, to understand the essence of the soul in order to cure disease, to preserve health – or even to lead a moral life.[66] Temperament, then, gives the doctor all the clues he needs.

In addition to equal temperament, Galen distinguishes eight types of unequal temperament, four based on a single dominant quality and four on a pair of qualities. A similar set of temperaments exists for the principal parts, heart, brain, liver and testis: this is a major theme of the Ars medica, for example. Temperamental variations are distinguished in practice, not entirely by touch, as Galen suggests, but by matching temperament to a series of constitutional types or "signs of temperament". This Galen sets out in Book II.[67] In a small separate work, Galen adds to these examples of unequal temperament in the body as a whole or its principal parts the case of "unequal intemperatures", a pathological state in which one part of the body is afflicted by unnatural inflows of a dominant quality from another part. This aspect of the theory of temperaments is particularly easy to confuse with Hippocratic humoral theory.[68]

In addition to flows of temperament from place to place, a man's temperamental state changes over time. Life is a gradual running down from the hottest and moistest stage, that of the embryo in the womb, to the cold and dry of old age and the eventual extinction of heat in death. To each stage of life and of temperament, there is an appropriate kind of action; as one might expect, the hotter and moister stage is more fully active.[69]

This brings us to another Galenic distinction between element and temperament, and the most important. Elements, as we have seen, are used to explain the structure of a body; temperament, to explain its function. The more perfectly tempered an animal is, the more perfectly it is able to perform its life-functions. Turning to this question again in Book III, Galen asks, "What is the power or activity of hot, cold, wet, dry?" The answer to this is striking: for Galen tells us that four basic powers or faculties common to all parts of the body depend upon the prime qualities:

> Certainly, we say that there are four faculties belonging to the whole body, one attractive of what is most suitable, another retentive of it all, a third alterative, and a fourth which is segregative of what is alien. And these same faculties are the result of the total substance in

> whatsoever part of the body, which we also say to be composed from hot, cold, wet and dry mixed among themselves.[70]

Elsewhere, he tells us that in order to understand all the actions of the human body, it is first necessary to understand the proper mixture of wet, dry, hot and cold. What does it mean to say, as Sylvius notes in a marginal scholium, that "the essence of faculty is in temperament"?[71]

What it appears to mean is that if we list the 'actions' of cold, hot, wet and dry within the body, we arrive at what Galen, in the book of that title, calls the natural faculties, and that the seat of these actions or faculties is the 'total substance' of the simple or homogeneous parts of the body. To follow this argument in more detail, it is necessary to look briefly at On the natural faculties and On the substance of the natural faculties, both of which Galen saw as companion works to the one on temperaments.

The natural faculties, as Galen defines them in Book I of the former and larger work, are the cause of the natural actions of living creatures, actions of their nature or φύσις rather than of their soul or ψυχή. They are the actions which animals and plants share; the capacity to grow, to nourish themselves and to reproduce their kind. Galen defines a faculty (δυναμις) as the capacity to perform some action, a capacity deduced from that action (ἐνεργεια) which in turn produces a certain effect (ἐργον). Thus the blood-making faculty shows itself in the activity of blood-making and in the result, blood. Galen lists faculties and sub-faculties appropriate to each of the three areas of natural activity, but he also argues for four general natural faculties possessed by each organ, the attractive, retentive, alterative and expulsive. These are the four mentioned in de temperamentis (and in de facultatum naturalium substantia) as the basic capacities associated with the primary qualities. They can be seen in action most clearly in organs such as the stomach and uterus, but must also be possessed by each organ or part of the body, so that it can attract and assimilate nourishment. But Galen does not associate these faculties with the organ as a whole, but rather with the simple or homogeneous parts – which he describes as distinct 'coats' or 'fibres' – which make it up.[72]

The association of the natural faculties with the simple or homogeneous parts of the body provides one link between the natural faculties and the temperament. The other link is provided by the basic principles of Aristotelian natural philosophy, as applied by Galen to the action of living things:

> As for the scientific proofs of all this, they are to be drawn from these principles... – namely, that bodies act upon and are acted upon by each other in virtue of the Warm, Cold, Moist and Dry. And if one is speaking of any activity, whether it be exercised by vein, liver, arteries, heart, alimentary canal, or any part, one will be inevitably compelled to acknowledge that this activity depends upon the way in which the four qualities are blended.[73]

The four qualities, for Galen as for Aristotle, are the basis for the "motions" of alteration and transmutation: and it is just this type of qualitative action which Galen is describing in the Natural faculties. Just as the temperaments should not be thought of as substances but as a balancing of contrary motions, so the qualities themselves are not static but dynamic, each trying to overcome its opposite, and producing certain alterations as a result. If we think of temperament as a balance of activities, as "the mixture of the motions" of the four qualities, it is easier to see how Galen can say that "every part functions in its own special way because of the manner in which the four qualities are compounded."[74]

The principle is most strikingly proved, for Galen, by disease: for example, the disturbance of function by fever (a disproportion of heat) suggests that normal function must be associated with eucrasia, the proper blending of the qualities.[75] As we have seen, he argues in the fragment de facultatum naturalium substantia for the existence of the same four general faculties, and for their association not with the (unknowable) substance of the soul but with the temperamental constitution of the body. Such a temperamental cause is appropriate to a level of action shared by the simplest plant: the ability to discriminate between substances appropriate as food and those which would be harmful.[76] Plants as well as animals have a temperament, and this in turn is important for pharmacology. The application of the theory of temperament to the action of drugs takes up much of Book III.[77]

For Galen, then, temperament provides us with a means of comparing the constitution of species and individuals; of distinguishing the simple parts and their functions; and of explaining the activities of living things by their qualitative makeup. On the medical side, it allows us to diagnose constitutional types and their susceptibility to disease; to classify disease varieties; and to prescribe appropriate treatment.

Of the four commentaries on Galen's de temperamentis, those of Sylvius and Montanus are very short. Sylvius' is an eight-page introduction, or Isagoge, to the 1549 edition, plus twelve pages of tables setting out the varieties of equal and unequal temperament as they apply to medicine. The title page tells us that he also provided marginal scholia for the edition. His introduction, "some most useful comments and categories in the books of temperament" is useful as a summary, clarifying and bringing out the aspects of Galenic temperament theory that were of most interest to sixteenth century medical teaching. He begins with a definition which reflects those of Aristotle and Galen:

> The four elements...principles of all things which generate and corrupt, being themselves thoroughly blended as to their prime qualities, constitute things in this way or that according to the mode and differences of their mixture...now transmuting the total substance of a thing, as happens in nutrition; now by the formative power in nature, as in the generation of animals: for indeed faculty is like a craftsman, and forms the parts suitably to the character of the soul. We do not attribute the cause of the forming of the parts to the qualities alone, since these are only the instruments of conformation: the true agent is different, whether proximate, as the seed; or of a certain more remote and diviner origin.[78]

The qualities are thus instruments in the activities of generation, producing their effects through their capacity of alteration and transmutation. When a quality alters to its opposite – hot to cold for instance – not only the quality, says Sylvius, but the 'species of the essence' of the substance in which the quality exists is transformed. And so the simple parts differ from one another "from the properties and differences of their temperament". It follows that there are just as many possible variations in the contexture of the parts as there are possible variations in their temperamental balance. In simplest terms, nine combinations: eucrasia, plus four simple and four complex imperfect temperaments, are possible, just as in Galen. But each temperament can be further subdivided, as Sylvius illustrates in his tables. In particular, he adds to the Galenic refinement of four 'grades' of each quality the notion of the latitude of each temperament. A person can thus be hot and dry, for example, not in one mode alone, but over a whole range of possible degrees of hotness and dryness. Sylvius suggests that the range of possible combinations is countless, if not actually infinite; and that this in turn helps to explain the subtle differences to be found even in the simple parts of various animals and plants.[79]

In a similar way, it helps to explain their differences of function. For, says Sylvius, "since the diversity of action follows the mode of temperament, as many as are the diversities of temperament, so many are the differences of function".

The notion of a countless variety of temperaments, covering the whole range of plant and animal structure and activity, is implicit in Galen's discussion of the comparative nature of temperament. But Sylvius makes this conclusion explicit, and where Galen emphasises the basic identity between, for example, the natural activities of a plant and a man, Sylvius uses the comparative theory of temperaments as a means of explaining, and of calculating, the superiority in their action of some animals to others and of animals to plants. Plants are colder than animals and thus less perfectly active, while among animals it is similarly the hotter and moister which are most perfect in their action. Man must clearly be most perfect of all, and as Sylvius explains, this perfection, of body, soul and temperament, is one and inseparable:

> Now since man is the wisest..and..most perfect animal; as his form is a rational soul, most noble of all: so his body is..the best constituted of all, and the most healthy, being most perfectly tempered of all in its simple parts, in the balance of heat, cold, moist and dry.[80]

Man's most tempered constitution implies superior conformation of body, leading in turn to his superiority of action. In conformity with this is the superiority of man's rational soul and his intellect. The argument is, to say the least, somewhat circular.

This grading of plants and animals according to the assumed variations in their temperament is the most striking addition by Sylvius to the basic Galenic pattern. In expanding Galen's eight temperamental types into a galaxy of individual and specific types, he provides a model for the abundant diversity of the natural world, based on change in the relationship of only four qualities. But at the same time, his "latitude of temperament" and other complications which he introduces in his tables subdivides Galen's system of temperamental types to the point of near uselessness. Such an unmanageable explosion of categories is clearly not Sylvius' intention: he sets out his ten pages of tables of the "various divisions of temperament and intemperament" with all the earnestness of someone who is doing his best to be instructive. The result is not quite countless, but he does present the hapless medical student with a total of 770 varieties and sub-varieties of temperament, plus 40 distinct ways in which a subject can be described as hot/cold/wet/dry.[81] Such a passion for classification and sub-classification is by no means absent from Galen's own work: the

Natural faculties is a good example. But Sylvius' tables show something of the lengths to which it could be taken by a dedicated follower.

Montanus' commentaries – his Methodus on the three books of the temperaments and the short commentary on the natural faculties which follows it – are published as companions to his commentary on de elementis. Each is short, but illuminating in the way in which it sets out the main arguments of the Galenic treatise for the use of students. Book I of de temperamentis, Montanus tells them, deals with the nature of temperament; Book II with the signs of each temperament in turn; and Book III with the determination of temperaments from their signs and the way in which temperament can be altered, for good or ill, by outside influences such as food, medicines and poisons. In considering the nature of temperaments, Montanus takes up the question, are there four temperaments (or four pairs of qualities), corresponding to the four elements and the four seasons, or only two (as Averroes, and to some extent Aristotle, argued) due to the incompatibility of the pairs hot and moist, and similarly of cold and dry? Here again we can see a conflict between the system of de natura humana, in which four pairs of qualities are chosen to correspond to the four seasons of the year and of human life; and the active/passive dualism of Aristotle's natural philosophy, which gives, as we have seen, two pairs: hot and cold (active) and moist and dry (passive). Montanus, like Galen, rejects any reduction. Having decided on the number of unequal temperaments, it is necessary, says Montanus, to refer these excesses to some standard and posit "an equal temperament, in which no quality exceeds throughout." We can assume the existence of an equal temperament logically from the existence of unequal ones; conversely, the notion of equal temperament provides a standard to which excesses can be referred.[82]

Book II deals with the signs of temperament, in the body as a whole or part: that is to say, with their actions. In simplest terms, these signs are distinguishable by touch: hot and cold can be determined by touch, and dry and wet can be recognized as hard and soft respectively. On a more sophisticated level, the student is taught to relate certain constitutional types to the action of a certain temperament: hairiness, for example, being a sign of hot temperament, smoothness of a cold one. Once again it is easier for the beginner to learn to distinguish the varieties of unequal temperament, before moving on to the ideal or equal temperament; but in all cases the sign must be related to the action, or the structure, of the individual or part:

> And in the first place, he [Galen] teaches us to recognise the unequals, afterwards the equals, and he puts forward no surer sign than that which is taken from their individual actions. And in the second place from structure..[83]

So temperaments are known by the actions of the qualities composing them, and the actions and 'virtues' or powers of each temperament provide us with a means of dividing patients into constitutional types. But what are the actions of the qualities within the body, especially the active qualities of heat and cold? Broadly speaking, says Montanus, the action appropriate to hot, calida, is "to be changed"; that of cold, frigida, is to "remain unchanged". These two sorts of action are dealt with more fully in the Methodus to the third book.

The purpose for which the student needs to determine the temperament of a patient is, of course, the medical purpose of choosing the appropriate remedy or regimen. This aspect of the theory of temperaments is dealt with in Book III – a book which, Montanus says, is "divine, and worthy to be committed to memory". The modern reader, less dependent on understanding the Galenic theory of drug action than a Renaissance medical student, is likely to be less enthusiastic. But one aspect of Montanus' commentary is of interest, and will be more so in the context of the next chapter. This is the question of the various ways in which we can describe the actions of hot and cold. In the natural world, says Montanus, we say that a thing is calida if it catches fire easily, frigida if it does not, or even extinguishes fire. But this is not quite enough, for we can imagine a substance which does not burn – such as hot iron – but which is nevertheless hotter, and stays hot longer, than, for example, burning straw. It should be possible to apply the same sorts of distinctions to calida and frigida substances acting and being acted upon within the body, such as foods or medicines:

> From this general similarity we can learn which things heat and which cool our body, and that there is in general the same proportion between our heat and the heat of fire.[84]

If a substance is changed by the heat of our bodies – which here seems to be compared with fiery heat – and is "led to act", it is a hot substance, if it is not "led to act" it is a cold substance. The way in which this works out in practice is clearer from Montanus' examples. Food, for example, is a body altered by innate heat in its substance; medicine, in its quality. 'Hot' poisons are also altered, but not conquered; 'cold' ones neither altered nor conquered. So that when an external substance is added to the body, we can suggest a whole range of ways in which it may react with the innate heat, according to its temperament and that of the body.

When we come to examine the relationship between the parts as instruments, their actions, and their potentiae or faculties, where should we begin? Does the structure of the organ determine its function, or the other way around? Galen makes it very clear, says Montanus, that we must begin from the faculties, and the natural faculties in particular, if we are to understand the actions of the body and the structure of its parts. As we have seen, the natural faculties are associated by Galen with the qualities, and with temperament. Moreover, Montanus tells us, what Galen means to say, in de temperamentis and other works, is that all actions – vital and animal as well as natural – result from a moderate degree of heat. He attempted to prove this in a short treatise, "De potentiarum naturalium substantia".

> First he proves it by induction, showing these faculties to operate so long as the heat remained moderate, and this will be revealed by the course of individual actions. For where heat is either weakened, or extinguished, or increased or corrupted, by many disease causes: so also the operations are diminished, removed, or corrupted.[85]

The identity between a qualitative state, that of moderate heat, and the ability of the body to perform its function is proved by the fact that when the degree of heat is altered (as in fever) the function of the body is impaired. This is the same argument which Galen uses in the natural faculties II viii: because the normal activities arise from the proper blending of heat and cold, i.e. eucrasia, we can observe that "the primary impairments of these activities necessarily arise from its derangement."[86]

In his commentary on the Galenic treatise on the natural faculties, Montanus provides us with a definition which unites temperament with faculty. We have already learned that we recognise temperaments, and their equality or inequality, by their actions: the same is true of faculty. We can tell one faculty from another, and determine which one is operating, by observing what it does – just as we can tell one temperament from another, and determine which qualities dominate the balance, by observing what it does, and how.[87] The identity of temperament and faculty is that both are explanations drawn from the same physical evidence: the structure and function of the body in health and disease. Montanus sets out these overlapping systems of explanation concisely, and lucidly, to provide us with a useful Renaissance interpretation of Galen's meaning.

We may remember that in Book III of his De elementis, Contarini defined the prime qualities as "in a way the elements of temperament, for indeed temperament is made up of these just as a mixture is of elements."[88] Of these prime qualities he distinguished caliditas as most important, being most active and nearest to the nature of celestial heat.

In Book IV, Contarini turns to the consideration of certain specific mixtures, among them the simple or similar parts of animals and plants, and to their generation and corruption, the first by natural heat, the second by alien heat. The operations of hot, cold, wet and dry, and of other physical qualities deducible from these, such as frangibility and malleability, are clearly important to any discussion of natural objects and their origins. This leads him to a study of temperaments and their types, to Galen's work de temperamentis, and to a definition:

> Temperament, a secondary quality, is put together from the mixture of the prime qualities, because the mixed body acts with a certain action and is acted upon according to a certain capacity. For as temperament arises from the tempering of the prime qualities, some active, such as heat and cold, the two remaining, namely moisture and dryness, passive, it is necessary that the nature of them all should show forth in the temperament which is made up of all of them. Because of its complexion, therefore, and by this tempering, the mixed body both acts in a certain way and is acted upon in a certain way.[89]

Several points can be drawn out from Contarini's definition.

First, he combines the Hippocratic notion of four equal qualities and the Aristotelian one, taken further by Averroes, that there are two pairs, active and passive. Later in Book IV he suggests that these two positions are not irreconcilable, since Averroes and Galen (following Hippocrates) were speaking of different aspects of qualitative action.

The second is the requirement that the four qualities, active and passive, be combined in such a way that the nature of each is perceptible in the temperament which results. And the third, which we have seen already in Galen, Sylvius and Montanus, is that temperament gives a body certain specific capacities for acting and being acted upon, according to its particular makeup.

Following Galen, Contarini distinguishes nine genera of temperament. Like those of Sylvius, however, each is very broadly defined (latissimum) and contains innumerable grades within it. Of the nine, one is temperata or equal temperament. Contarini, like Galen and Sylvius, interprets this in two ways: first, the temperament in which the four qualities are reciprocally equal; and second, that which is most suitable to the species involved and most perfectly adapts it to its action. In the first sense, there is only one possible state in which all four qualities are equally balanced (which Contarini too identifies with temperament of man). But there are many temperaments which fit the latter sense, as many as there are appropriate natures for things.

Like Sylvius, Contarini calls this latter case equality according to proportion or distributive justice,and gives a similar operational definition of it: "thus, therefore, we say that a complexion is temperate which is suitable to the activity and to its species." His definition of the other type of equal temperament is in a way operational as well. For it depends on an equal balance of the "activeness" of the four qualities, rather than equal amounts of each. Since according to Contarini fire and air are much more active in proportion to their bulk or weight than water and earth, the temperament most nearly approaching a state of balance is one in which the powers and degrees of the prime qualities are equal, or nearly so. In this case, the bulk (moles) of the passive elements will greatly exceed that of the active – as for Contarini it does in all compounds anyway.[90]

The addition of the idea of degrees (gradus) of quality allows Contarini to speak of the balance of qualities in more precise terms. It is a notion which was used by late medieval and Renaissance natural philosophers in their speculations on the nature of heat in bodies and the way in which heating and cooling took place.[91] Hot is the most active quality and the one responsible, according to Aristotle, for the maturation and preservation of mixtures, including the homogeneous parts. The association of heat with the coming together of a mixture, and of cold with resistance to this concretion or concoction, presents a problem: how do cold mixtures come to be, and how are they maintained? The question has bearing on the generation of the simple parts, some of which, such as bone, are temperamentally cold, and Contarini attempts to answer it in two ways.

The first is by distinguishing the action of elemental heat (the basis on which we determine the temperament of a thing) from what he calls 'natural heat'. Montanus too speaks of natural heat, and for him it seems to be the same as innate heat, i.e. the heat which distinguishes living things from non-living ones. For Contarini, as we saw earlier in the chapter, all heat has some connection with aetherial or celestial heat, which he in turn identifies with the magnum seminarium of the Platonic philosophers. It is not the celestial heat itself which permeates all things and secures their preservation, but natural heat, i.e. "the accident and quality flowing from the celestial body, and penetrating all things."[92] Despite its divine origins, this natural heat appears to act in all mixed bodies and not, as in the case of innate heat, in living things only. Furthermore, it is not distinguished from elemental heat in the way that innate heat usually is. Rather than saying that in bodies there is an elemental heat which is responsible for temperament and a natural or celestial heat which is responsible for concoction and preservation, Contarini says that in addition to the power of heating which there is in heat, there is another power, the power of combining like things, by which concoction and the generation of mixtures take place. In the generation of the simple parts, natural heat is the agent or instrument of the celestial qualities, which are in turn the seminarium of all mixtures.[93]

But if there is one quality of heat, with two distinct powers, it is still difficult to see how it can act in two opposing ways: how a mixture, "compacted by heat", can then take on a temperament dominated by cold. How does the heat act "out of its grade" (praeter eum gradum) to form the mixture in the first place? Contarini here presents a complicated argument based on the intension and remission of forms in difform substances. A certain quality, such as hotness, is assumed to exist in intense form in certain particles of a substance which is not uniformly hot. These particles, though a small part of the whole, can transfer their intenser qualitas to the surrounding particles, and so combine it as a mixture, without changing the fact that its overall temperament is dominated by cold. This interpretation, which I have not found in any of the purely medical writers, seems again to arise from the tradition of late medieval physics.[94]

Whatever its origin, this is a solution which distinguishes two levels of presumed action, one macrocosmic and one microcosmic. The alternative solution which Contarini proposes also forces him to distinguish two levels of action, a 'lower' one of the prime qualities and temperaments and a 'higher' one of faculties:

> You can bring in still another cause, if we choose to rise above the prime qualities and temperament, namely to the natural faculties and those which are proper to mixed bodies … These natural faculties have the greatest power of action, and they can incite the inferior qualities to action. For it is not surprising that the lesser power is contained within the greater and more excellent. Thus we say that mixtures act according to their natural or specific faculties, with a heat or cold very different from, and in addition to, that with which doctors have to do.[95]

This 'other' heat, which sets in motion the operations of qualitative or temperamental heat – the heat with which doctors are concerned – is what Contarini calls "animal heat" (calor animalis). It is similar to the aethereal warmth and seems to be different from the natural heat we have already met.[96] The latter, as its name implies, operates on the natural or elemental level as a kind of 'value added' to the primary action of elemental heat. The former, or so it seems, is confined to living things only and operates on a higher, facultative level.

This is not a distinction which we have seen in the two previous authors. Montanus does not so much distinguish innate heat from elemental heat, as identify them. And both Sylvius and Montanus have followed Galen in identifying the level of action of quality with that of, at least, the natural faculties. But our final example comes closer to sharing Contarini's position, represented by this second alternative at least.

Fernel's definition of temperament, in the third Book of his Physiologia, is based on Galen's, but describes the relationship of the four primary qualities with greater complexity and precision. He agrees that temperament has to do with the elements and their combination, "being as it were the offspring and progeny of mixture."[97] Mixture, the coming together of elements in more or less equal proportions – equal, that is, not in their bulk or weight (for there is a greater quantity (copia) of earth in us than of the other elements) but in the strength (vis) and efficacy of their powers, takes place within certain definite boundaries or limits, and it is these limits which give us the concept of temperament:

> Which coming together of prime qualities therefore occurs within these mean limits, so that mixture thus and at once brings forth temperament.[98]

Fernel conceives of temperament by a mathematical analogy, represented as the mean between the two pairs of contrary prime qualities, or an approximation to that mean. He makes his meaning somewhat clearer by an analogy with harmony in music:

> Moreover, temperament is not mixture itself, but the ratio of mixtures. This can be defined as harmony and concord of the principal qualities which are in elemental mixtures. As with singing and the lyre, sounds separated by large intervals, or indeed conjoined but unequal and not separated by a determined interval, do not at all produce harmony, unless the first being tempered with the last, and the mean sounding equally with the rest, a sweet sound pleases the ears: just so the elemental qualities, divided or united, do not indeed form a temperament, unless moderated and brought together in a kind of mean, they are united in one another. Considered in themselves and absolutely, therefore, the primordia of things are not themselves temperament: nor a mixture of them, but the comparison of the mixture... Those who separate temperament from mixture by this distinction, that the latter approximates more closely to the elements, the former indeed to the qualities, come closest.[99]

This is the most sophisticated definition of temperament which we have seen. Temperament is not any random relationship of qualities within a mixture, but those which produce a certain harmony, analogous with that of music.

In the next chapter, Fernel discusses the basic physical problem posed by the theory of temperaments. How are contrary qualities, e.g. hot and cold, each present in its highest degree of intensity, reconciled to produce a mean (mediocritas)? He discusses the solutions given by the Arab natural philosophers, Avicenna and Averroes in particular (to whose arguments he is almost always hostile) and their 'family', and those of certain neoterics, proponents of theories of qualitative change of the sort discussed by Contarini. Fernel here shows his familiarity with the details of contemporary physical speculation on reaction and reduction of heat. He distinguishes between the intension and extension of heat in a body, mentions the doctrine of remission of qualities and the system of degrees of quality, but he introduces this 'numbers game' only to show its futility in solving the problem. In the end, he rejects the arguments of physics in favour of proof by touch: the most reliable judge of temperament available to the doctor. How do you obtain a mean by combining extremes, tepidity from hot grade 4 + cold grade 4? Mathematically, the problem of combining summae is a difficult one: but experimentally, says Fernel, the proof can be obtained by adding equal quantities of hot and cold water. Such 'experimental proofs' are a part of the physical debate on questions such as the effect of mixing a given proportion of hot and cold[100], but Fernel's argument belongs also in a medical context, and in the context of his own 'resolutive' method, discussed above. In chapter v, "that human touch is the rule and judge of distinguishing temperaments", he pursues this point. The portion of each element in a mixture is not knowable to us by reason: the true simple powers of things are hidden from us by the infirmity of our reasoning. What certain knowledge we have comes from the senses: "for what we distinguish with the mind (animo) all takes its origin from the senses". And the senses are competent to judge of their own proper objects.[101] Thus, just as for Galen, human touch is the proper judge of temperament, Man's nature is the "medium omnis exuperantiae", and hence his touch can distinguish the ἐνεργεια of hot, cold, wet and dry where his reason fails.

This brings us to the question of the comparison of temperaments (chapters iii and iv) and the temperament of the simple parts (chapter vi). For Fernel as for Galen there are eight unequal temperaments, four simple and four composite, and one equal one, which can be defined in two ways. This 'most temperate temperament' can be understood as a kind of point from which the others are measured: as Fernel describes it, it is easy to imagine a kind of graph, with an innumerable set of coordinate points:

> What is most temperate is single and without any latitude, and can be understood as a kind of point: the others which recede from it are extended in length and breadth: for not all that we call hot is at the same distance from the mean, nor are cold or wet or dry one and all in the same line. Further, in the composite temperaments the prevailing qualities are either equally removed or one exceeds the other... From which it can be understood that while there are eight unequal temperaments, yet from the greater or lesser extension of each of them the divisions are innumerable, from which the properties both of the simple parts and the infinite natures and properties of bodies arise and are composed.[102]

Like Sylvius, Fernel imagines an innumerable variety of 'latitudes' which correspond in their variations to the variety of natures in the physical world.

He turns to the simple parts in more detail in chapter vi. Here too we can imagine four ordines or series: those of the hotter, colder, wetter and dryer than average parts, with the temperament of the skin forming the central point from which the other temperaments diverge. The structural differences of the simple parts are the result of variations in their temperament, then; the same can be said of the variations which separate one animal from another or an animal from a plant. Temperament, for Fernel as for Galen, is comparative, with living creatures ranged along a scale from hot to cold. In comparison with another point on this scale, a given nature may be comparatively hot or cold: man's temperament is cold compared with that of a lion, hot compared with that of an ant. In fact, as we have seen, man is in some sense the midpoint of the scale.[103]

Fernel gives no space to the central feature of Galen's medical doctrine of temperaments, the listing of signs by which various human temperaments can be recognised. This may reflect his emphasis on touch as the key, but it is curious in a medical textbook. And apart from repeating Galen's dictum that equal temperament ad justitiam is related to the functioning of the individual as well as to its health and appropriate nature, Fernel never associates temperament with function, as we have seen Galen and others do: a point to which we shall return.

He devotes the last four chapters to complications which make it impossible in practice to judge the temperament of the parts (especially the internal parts) from touch alone. For temperament is not completely constant: its appearances can be changed by a number of interfering factors. The first of these is what Fernel calls adventitious or external heat, and which he distinguishes from both elemental and innate heat. This heat is produced in the heart, and transferred to neighbouring internal organs, making them artificially hot. In addition, there is the effect of the non-naturals in changing the indications of temperament. Wine, for example, produces yet another kind of adventitious heat. Finally, there is the change produced in an individual's characteristic temperament in the course of his lifespan: from its hottest and moistest in extreme youth, to its coldest and dryest in old age. The solution, according to Fernel, is to calculate the mean: the temperament of a man at the middle point of his life is his 'proper' temperament from which the others are deviations. To allow for these complicating factors, the help of reason must be added to that of sensation.

The notion that the balance of hot, cold, dry and wet varies in the course of a man's life comes, as we have seen, from the Hippocratic work de natura humana. But Fernel is careful to distinguish the notion of temperament from the doctrine of the humours which appears in the same work. Temperaments arise from the elements and from their qualities in particular, not from the humours. The latter, for Fernel, are superfluous substances within the body (which explains their importance in causing disease), not true, solid, or containing parts; they are within the body but separate from it, 'contained'. Humours can derive from external causes, such as diet; temperament is 'built in' to the congenital nature of the parts. Thus humours do not determine the temperament of the body, and cannot be used as a guide to temperament.[104] Fernel is here grappling, as I suggested at the beginning of this chapter, with a confusion arising from two different interpretations – those of Aristotle and Hippocrates – of the four-quality doctrine. At least he succeeds in distinguishing their results.

I have already said that unlike other Galenists Fernel does not associate the function of the simple parts with quality or temperament. With what, then, does he associate it? The answer is suggested by Contarini's appeal to a "higher level" of facultative action and appears in Book V of the Physiologia, "On the Faculties of the Soul."

Function – even the functioning of the simple parts, and those most basic activities which Galen attributed to "nature" or φύσις, requires a soul, according to Fernel. This soul is inaccessible to our senses, and can be perceived only through its works, i.e. the activities of a living body. And since the study of the doctor is the living man and his actions, "we must turn our thoughts, not to the body alone, but to the man, who consists of soul and body."[105] This position, that the understanding of the soul is an essential part of medicine, is in striking contrast to that of Galen in a work such as de facultatum naturalium substantia. What Fernel has done is to transfer even the simplest of vital functions from the realm of natural or qualitative action to that of the faculties of the soul. The difficulties faced by Contarini in explaining the action of quality "out of its degree" suggest one reason for this move; the association of the Aristotelian-Galenic anima with the Christian soul may provide another.[106]

The soul can be perceived only by actions within the body: hence its definition as "the principle and cause of function of a living body". Furthermore, it is the species of the living body, and hence the ordering of natures, and of souls, proceeds ex vitae functionibus: from a consideration of the differences in their vital functions, assumed to reflect differences in the nobility of their souls.

What then are the faculties of the soul? According to Fernel, they are not its "parts", since the soul is a single, simple entity, but are related to the soul as accidents to substance. Thus just as we can describe an apple by its smell, taste, or the fact that it is a Granny Smith, we can describe a soul by saying that it possesses natural faculties alone (i.e. plant souls); natural plus vital faculties (animal souls) or natural + vital + animal faculties (human souls). In effect, we define the soul by the number of things it can do; and this number is equal to the number of its faculties.

It is the natural faculties with which we are particularly concerned in the study of generation. In the chapters of Book V which deal with the actions associated with generation, growth and nutrition, Fernel gives much the same list as Galen in the natural faculties. He differs, however, in not associating the natural faculties with either the simple parts or with the qualities and temperaments.

The most striking piece of evidence which Fernel gives for his distinction of a facultative and an elemental level of action (he distinguishes several other levels as well, but we can ignore these for the moment) is the obvious difference between a dead body and a living one. The elemental structure, the temperamental balance of qualities are the same in the corpse as in the living man, but the function of the body is completely absent. Therefore, Fernel argues, function must derive from some extra-elemental and extra-temperamental source: which in the Hidden Causes he calls the total form and in the Physiologia the soul.[107]

In De abditis rerum causis he also tells us, in the person of Eudoxus, why temperament is too limited a concept to be useful in explaining the function of a body. Temperament inevitably reflects the dominant qualities of the elemental mixture; this in turn gives it a range of action limited to that of these qualities. The role of temperament, as Fernel explains it is as one of the three preparatory "organisations" which fit unformed matter to receive its total form (or anima). Matter, temperament and form are accordingly the three "principles of medicine", but the two former are subservient to the latter, as it were its instruments.[108]

One might assume that for Fernel all aspects of faculty, all actions of the body, must accordingly be related to the nature of the soul: that there is no room for subsidiary agents, such as the Galenic qualities. But when he comes to discuss the four 'basic' faculties (attractive, retentive, alterative and expulsive), whose action Galen based within the simple parts, Fernel adopts a similar, and much less lofty, explanation: the principle of similitudo. For example, in the case of attraction, "similitude indeed moves to amity, amity to desire or appetite, appetite to attraction". Despite terms such as "amity" and "desire", these faculties are present not only in living things – but, in the case of attraction and repulsion at least, also in things which we would not normally want to call 'ensouled': magnetic rocks for instance, and purgative medicines.[109]

We have seen the difficulties which Contarini found in attributing life-actions and even the preservation of a compound to quality. In associating all function with ensoulment, however, Fernel has led himself into another difficulty – a situation in which anima has so broad a reference as to be nearly meaningless. The difficulty arises in part because of his attempt to identify the Christian soul with the Aristotelian concept of form; and Aristotle's notion, adopted for medical purposes by Galen, of form as the determinant of function places Fernel in yet another difficulty. For a follower of Aristotle and Galen, there is a fine line between what determines the structure of a body and what determines its function: a living thing is as it is because it has a certain function to perform. But for Fernel, as we have seen, the nature of the simple parts derives from their temperament. This is different from the form of the simple parts, which, as he argues in the Hidden Causes, has nothing to do with their elemental construction. And the functioning of parts and of the whole organism derives neither from its temperament nor from the form of the parts, but from its possession of a soul or total form.[110] What, then, of the structure of the animal or plant as a whole – as distinct from its total form?

The temperament of the whole, as Fernel explains in chapter vii of de temperamentis, cannot derive from the temperament of the simple parts blended together, since each part retains its distinctive temperament within the whole at the same time. It requires, then, yet another level of action. Accordingly, the role of integrating the simple parts into an organic and temperamental whole (not simple, like the total form, but "made up of many things coming together", ex multorum concursione conditum) is given to the innate heat and spirits. Fernel is careful to tell us that his innate heat is different from temperamental heat and does not arise from it. Spiritus, similarly, which in De abditis rerum causis is defined as "the substance of calor" is not the substance of elemental heat. Just as in the case of the anima (of which innate heat and spirit are the instruments) the distinction is bolstered by the argument that only a living body possesses innate heat (and the capacity to function), whereas elemental heat is present in a dead one, and in all other 'inanimate' compounds. In the same way, the activities of heat and spirit also serve to explain how plants and animals which are temperamentally cold can live, grow and nourish themselves – the same problem with which we have seen Contarini was concerned.[111]

The innate heat is an important element in Renaissance theories of generation, and this will be the subject of the next chapter. For the moment it is sufficient to point out that while Fernel distinguishes innate heat and spiritus from elemental fire and air in their origins, he considers that in some sense they are a special and more perfect form of "air" and "fire" and hence can operate within the elemental and temperamental world in a manner similar to Contarini's "celestial heat". Thus, the temperament of the whole arises neither from innate heat and spirit alone nor from elemental quality, but from a combination of the two: "perfect temperament is, properly speaking, that which is made one from the primary combination of the elements and from the hot and fervid spiritus."[112]

We are left, then, with three layers of action to account for a range of phenomena which Galen explains by one. The structure and diversity of the simple parts, whether those of an individual, species or genus, are the result of temperamental differences, which thus become responsible, at one level, for the multiplicity of natures in the physical world. At another level, these various natures are an effect of the differing species of the creatures, i.e. of differences in their souls or animae. The anima is for Fernel totally responsible for the functioning of the parts and of the whole, and for specific areas of action such as generation. The integration of these parts into a capable-of-functioning whole is the activity of the innate heat and spiritus: instruments of the soul within the body and ones which play a particularly important part in the operations of generation.

***

We have now reached a stage where we can begin to ask the question: how important are the elements and temperaments in Renaissance theories of generation, and what part do they play? The elements, as we saw, are associated with the basic structure of the parts, the temperaments with their function. More detailed answers can be suggested from a reprise of the arguments discussed above.

It seems that in the main the logical basis of the Renaissance theory of the elements and temperaments lies in the natural-philosophical works of Aristotle, and their development in a medical context by Galen. In these works Aristotle makes the qualities the basis for his distinction of the simple or homogeneous parts. Galen, following this argument, makes quality the basis of the actions which animals share with plants, among them generation. In fact, he makes the understanding of temperament, i.e. of the actions of quality, the foundation of certain medical knowledge. Knowledge of temperament is based on sensory discrimination of a kind for which man, by his nature as a mean, is particularly well fitted. And it does not require the doctor to enter into speculation on the substance of the soul, a subject given over to contention and at best incapable of certain proof. Temperament allows the doctor to differentiate part from part, individual from individual and species from species, and to distinguish healthy from morbid function in each case. Temperament is the level at which the doctor can enter the system with some hope of understanding it. Elements determine the structure, temperaments the actions of the natural part of the body. Both are thus basic to an understanding of generation.

What does it mean to say, as in Sylvius' Galenic tag, that "the essence of faculty is in temperament"? The answer is most clearly given in the companion work on the natural faculties. Here, Galen associates these faculties not with the form of the organ in which they operate, but with its simple parts or fibres and their qualitative makeup. Thus, it is not because the uterus is of a certain size or shape, is an organic whole, or is an instrument of the soul, that it has the ability to attract the semen, and retain, 'concoct' and eventually expel the foetus. Rather, it is because of the δυναμίς of hot, cold, wet and dry: capacities for action which make up its simple parts. Insofar as generation is the result of the natural action of the uterus and of other organs whose actions are considered under the natural faculty, it is the result of quality or temperament. This can be proved most clearly when abnormal temperament accompanies the failure of function as in Galen's explanation of sterility and other 'diseases' of the uterus.[113]

Sylvius, as we have seen, agrees: but he is much more concerned to see the Galenic qualities as instruments of the ultimate cause or soul, or of the more proximate cause, the seed. We can guess from his emphasis on these two causes that the seed and the soul will be important foci of Renaissance speculation on generation. Sylvius' commentary, like Montanus', is written with a view to its usefulness in medical education. This helps to explain his fondness for classification (which reaches a high point in his tables or divisiones). Classification may also be the basis for his development of the notion of the latitude of temperaments, which allows one to distinguish "innumerable" variations in texture and function. It also provides the basis for a comparative 'ladder' of species or natures, distinguished by degrees of excellence which are ultimately traceable to differences in the nature of their souls: with man, of course, at the top of the ladder.

Montanus, a somewhat earlier commentator, places his chief emphasis on the setting out of a method which can be used by students to distinguish temperaments. He is much closer to Galen in his reliance on qualitative action, determined by sensory perception, as a key to the understanding of temperament and of medical theory in general. Temperaments are distinguished by their actions, and these actions are discernible by touch, with some assistance from reason. Faculties, too, are recognised and defined by their characteristic actions. And, says Montanus, according to Galen all actions – not just natural ones – require, and hence stem from, a moderate degree of heat. This heat is analogous with elemental or temperamental heat, and its importance to function is shown by the fact that "these faculties operate so far as is the extent of moderate heat" and by the association of abnormal levels of heat (e.g. fevers) with disturbances of function.

Contarini, following Aristotle rather more closely, agrees that generation is a process which depends on natural heat. Further, he considers that the four prime qualities, hot, cold, wet and dry, are precisely those qualities by which physical bodies differ in their ability to act and be acted upon. The precise capacities of a body will depend upon the balance of all four qualities operating within it: a point on which his definition of temperament differs from that of Fernel, for instance. This association of quality and action is strengthened by his definition (following Galen) of eucraton in operational terms: i) as the most perfect action of a given individual or species ii) as an equal balance of the activity of the four qualities comprising the temperament. He adds a certain precision to the notion of latitudes of quality by the use of the term gradus and by certain other arguments borrowed from contemporary physical debates on the nature and action of qualities, especially of hot and cold. But by associating heat with the composition, cold with the dissolution of a mixture (a peculiarly Aristotelian and natural-philosophical interpretation of their "action", foreign to the notion of temperament as elaborated by Galen) he faces a problem in explaining the existence of temperamentally cold mixtures. This is an argument which he can answer, on a qualitative level, by explaining the action of mixtures not by the dominant qualities alone, but by the action of all four constituting qualities; or, in more detail, by a complicated (and ultimately not very satisfactory) physical argument based on the intension and remission of forms in difform substances. He can bypass the difficulty altogether by appeal to a 'higher' level of action, that of soul, or form, or faculty; or similarly by distinguishing between the elemental action of heat (which determines the temperament of a thing) and its 'celestial' action (which brings it into existence).

Jean Fernel provides us with a complex definition of temperament by analogy with a harmonic mean. His solution of the physicists' problem, the combination of two opposing summae qualitates to produce mediocrity, by appeal to the evidence of touch makes it unlikely that he will accept Contarini's argument based on the same tradition. In any case, he has already rejected the notion that the balance of all four qualities in a mixture determines its action, in favour of the dominant qualities alone. This in turn makes it more difficult for him to explain the function of the body by appeal to the level of temperament. Instead, he associates the structural differences in the simple parts with temperament, the integration of the whole structure with innate heat and spiritus, and the form and function of individual parts and of the whole with the soul, which can, however, only be observed and understood through its actions.

For Fernel, then, in direct opposition to the Galen of de substantia naturalium facultatum, it seems that an understanding of the soul is basic to medicine. In one sense the difference is in words alone, since whatever the level of explanation – temperament, faculty, soul – it is ultimately related to the one bit of evidence the doctor has at his disposal, the actions of the living body in health and disease. Galen bypassed speculation on the nature of the soul by linking these bodily actions with the qualitative actions of the Aristotelian universe. This had the further advantage of basing diagnosis in a set of differentia which, in theory at least, could be distinguished by the senses, in particular the sense of touch. But it also created confusions of terminology (such as the multiple meanings of the verb 'to heat'). These were not of great moment to those (such as Montanus and Sylvius) who were chiefly concerned to interpret the Galenic texts for medical students: they could afford to be uncritical of such difficulties, emphasising instead methodology and the importance of touch as a diagnostic tool. But they presented greater difficulties for those (such as Contarini and Fernel) who approached the problem on a more sophisticated philosophical level, and attempted to reconcile Galen's simple qualitative explanation and the increasingly complicated theory of the Aristotelian universe put forward by contemporary natural philosophers. We can divide these approaches into 'Galenic' and 'Aristotelian', but such labels confuse more issues than they clarify. What is clear, or will be I hope from what follows, is that Galen's attempt to divorce theoretical medicine from speculative philosophy – to link the former with some form of sensory proof and to maintain an agnosticism on the issues of the latter, was, broadly speaking, a shrewd attempt to keep clear of the more contested and tangled philosophical issues of his day. It is a pity that the historian of Renaissance medicine cannot do the same. The reintroduction of the soul into theoretical medicine – as seen in the work of Sylvius, Contarini and Fernel – is a complication which was not without its uses. In enables Fernel, for instance, to distinguish between a living animal and a dead one in terms not possible at the elemental-qualitative level alone. But unlike temperament, which can be explained by touch as well as by its actions, and related to the assumed structure of the rest of the physical world at the level of the elements, soul is an unknown substance. It can be understood only by observing its actions within the body, and proved only by positing its necessity from the existence of these same actions. Thus its use advances physiological understanding very little, and indeed complicates it by the addition of levels of explanation which Galen was able to do without.

Notes

1. For a compact if somewhat overschematic treatment of early single-element theories, cf. Hall, op. cit. 1969; I, chapters 3, 4, 7, 9.
2. Fernel, op. cit., 1542; Gasperis Contareni de Elementis & eorum mixtionibus libri quinque… Paris, 1548; Jacobus Sylvius, in Galeni de Elementis; Lyons, 1550; In Hippocratis Elementa Jacobi Sylvii Medici Commentarius, p. 105; Joannes Baptista Montanus, Methodus..de Elementis, Vienna, 1554. See also Thorndike, History, V, pp. 550–562, for an account of the element theory of Fernel, Contarini, Antonio Luiz and Walther Hermann Ryff; Nardi, Saggi, p. 240–248 for the De elementis of Alessandro Achillini, which considers the question of generation of living things in the context of the natural-philosophical doctrine of forms. For medieval commentaries de elementis, see Richard C. Dales, "Anonymi 'De elementis' from a twelfth century collection of scientific works in BM MS. Cotton Galba E IV" Isis, 1965, 56: 174–89; and "Marius 'On the elements' and the twelfth century science of matter", Viator, 1972, 3: 191–218.
3. Galen, and most modern scholars, consider that this work was not by Hippocrates, but (probably) by his son-in-law Polybus. Renaissance medical writers, however, had no qualms in attributing it to Hippocrates himself, and it is their practice which I shall follow.
4. The major sources for Aristotle's theory of the elements are De Generatione et Corruptione, especially Book II; De Caelo, Books III and IV; Meteorologica, Book I, chapters 1–3 and Book IV, especially chapter 12; and as background, discussions of matter, form and motion in the Physics.
5. Hippocrates, De natura humana ed. cit.: the mechanical separation of qualities in the embryo is more fully described in De semine.
6. Op. cit. note 2."Hippocratis librum de Elementis" is the work De natura humana. It is worth noting that both the Galenic treatise and that by Sylvius are described in this edition as commentaries on the Hippocratic work: p. 1, De elementis, describes it as "Galeni de Elementis ex Hippocratis sententia" a description which really applies to Book II.
7. Ibid. pp. 2, 26–8, 31–3.
8. Ibid. pp. 16–18, 34–37, 50–1. The argument is taken from Aristotle, e.g. G & C II ii & iii, viii; Meterologica, IV i 378b-379a. Both Aristotle and Galen support this with evidence that all four elements are taken in as food.

9. Ibid. p. 56: "Is ex primis simplicissimisque his sensibilis elementis constat, quae similaria appellantur: nempe fibra, membrana, carne, adipe, osse, cartilagine, ligamento, nervo, medulla atque etiam omnibus aliis, quorum universae particulae eandem omnino formam retinent, quae et ipsae quoque ex aliis quibusdam ipsis proximis gignuntur elementis, ex sanguine nimirum, & pituita, & bile utraque, pallida & atra, quae item ab illis ortum habent, quibus vescimur, quaeque potamus, at haec ex aere, igne, aqua et terra gignuntur.."
10. De elementis, pp. 64–5: "particulae horum utrounque in minima infringuntur, moxque inter se mutuo agunt & patiuntur, & eo facilius sibi invicem qualitates impartiuntur, quo fuerint in minutiora redacta.."
11. Ibid. pp. 38–42, 54, 56, 59–60.
12. Ibid. 66–7; 80–3; 68–9: "Ecquid ergo ex eadem omnia procreata sunt substantia? Vel potius sagax rerum opifex natura, quo primum tempore ex sanguine ab ipsa matre in uterum delato foetum procreabat atque efformabat, quod in eo crassius erat ad solidiora corpora constituenda attraxit; quod vero tenuis, ad molliora; & simili etiam modo, quod calidius erat, ad calidiora, & etiam quod frigidius, ad frigidiora. Et sanè mihi videtur quidem longe esse magis naturae consentaneum, & statim ab ipsis initiis formari foetum, ac subsequenti etiam tempore universo unamquamque particulam suum habere alimentum augmentumque ex propria peculiarique materia. Nam sanguis, ut lac etiam, una quaedam res videtur, ratio tamen docet illum non esse unum, sicut nec lac."
13. Ibid. pp. 84–5; 87–91 (i.e. chapters vii, viii, ix, x).
14. Ibid. p. 105: "Sperma autem tam viri quam mulieris, & sanguis maternus, nostrae generationis principia, illud [c]eu forma, hic tanquam materia, ambo ex elementis quatuor conflata, sed interventu humorum quatuor sanguinem constituentium." Galen (De elementis 48–9) also distinguished between a principle and an element: a principle is not necessarily of the same genus as the thing of which it is the principle; an element is. Neither Galen nor Sylvius make this distinction clear in practice. For Fernel (see below) "principle" seems to be identified with the formal, "element" with the material aspect in the context of generation.
15. Ibid. pp. 105–7.
16. Ibid. pp. 107–9.
17. Ibid. pp. 111–113; "Haec ipse divinus calor semini animalium atque plantarum insitus per se molitur, sed divinum addidi quod calor, qua calor est, heterogenea quidem illa separat ac dissipat, & homogenea cogit, qua vero divinus est, cerebralem in semine portionem sursum distribuit, & cranii portionem ex eodem semine secretam, ipsi cerebro circumponit: cordis portionem, in medio pulmone & thorace, mediam ex ipso semine secrevit, pulmonariam vero thoracicamque in semine portionem cordi circumdedit, eademque iustissima ipsius oeconomia, in reliquis partibus effulsit... Idem calor in semine divinus, seu spiritus ille calidus, & nostri formator, venas, arterias, canalesque reliquos, & cavitates sensu manifestas, atque etiam obscuras..fistulat, easque quibus maxime ad actionem, aut usum convenit, applicat & adaperit.. Quid non facit? idem quibusdam, aut anima, aut primum ac principalissimum animae instrumentum

censetur ob vires in gignendo, formando, vivendo praecipuas. Haec vero effecta, licet a calore illo diviniore non impedito sint inseparabilia, tamen illius non sunt propria, cum reliquae ignis & aliorum omnium elementorum qualitates, in ea cooperentur, ut mox dicam, sed tamen igneus ille vigor, cuius caelestis origo est, primas partes in hac formatione sustinet totus ubique actuosus."

18. See the chapters on innate heat (Chap. 15 below) and ancient theories of generation (Part I).
19. Sylvius, op. cit. p. 114.
20. Ibid. p. 118, 121.
21. It is possible to speak of "higher" and "lower" elements in an Aristotelian sense, as well as in the sense discussed by Dr. Walter Pagel and Marianne Winder (note 70). For example, in Book IV of De caelo, Aristotle 'rates' the elements according to their degree of lightness or heaviness, fire and earth being absolutely light and heavy respectively, air and water relatively so. Earth's natural place is lowest, then come, in order, water, air and fire, forming concentric circles. And Aristotle adds, "We hold further that that which surrounds is on the side of form, that which is surrounded is on the side of matter (IV iv 312a 12–15). In other words, the elements which are physically 'lower' or 'higher' in the Aristotelian universe, are also 'lower' or 'higher' along a scale from pure matter to pure form. Sylvius and Contarini develop this assertion in various ways. Though I have identified transmutation according to rare and dense with the Stoics, and transmutation according to sensible qualities with Aristotle, it is clear that Aristotle's own position is ambiguous. See also Meteorologica, I iii 341a 5–9: "whenever a particle of air grows heavy [ατμίς] the warmth in it is squeezed out into the upper region and it sinks, and other particles in turn are carried up together with the fiery exhalation."
22. Sylvius, op. cit. 125–6, 137–9.
23. Ibid. 139–40, 142: "Humectare autem multo magis quam hybernus, hinc deprehendas, quod germinationi & animalium generationi est aptissimus: quae duo sine larga humiditate non fiant, & largiore quam seminis calor nativus, aut Solis possit depasci. Si enim totam depasceretur calor alteruter, neque generatio fieret, neque per nutritionem auctio. Aer igitur omnium penitissime & citissime & potentiss[i]me humectat calido adiutus, & tenuitate, levitate raritate quibus qualitates contrariae, aquam penitus & potentibus humectare prohibent."
24. Ibid. pp. 143–5, 144: "Haecque cum terra materiae plurimum corpori tribuit, cuius continentia fere sunt aquea simul & terrea, atque etiam contenta praeter impetum facientia, seu spiritus quae aerea magis & ignea sunt: ut igitur terra ex aqua subacta fit lutum, habile ad formam quamvis suscipiendam, sic nostrae generationis primordia semen viri & mulieris, & sanguis maternus, ambo ex quatuor elementis composita: sed sanguis aqueus & terreus magis, semen vero, quamvis initio aereum magis, ac igneum est, dissipata non multo post horum elementorum magna vi & substantia, quae ex illo fiunt partes, aqueae & terreae, evadunt."
25. Ibid. pp. 147–9, 151–2.

26. The title and the dedication by Martinus Stopius, dean of Medicine at the University of Vienna, suggest that it was based on Montanus's lectures. a ii v – iii r. "Etenim humani corporis elementa, temperamenta, humores, spiritus, facultates, cuiusque, actiones, naturae, functiones, adeoque omnium partium constructionem (quam anatome docet) ad unguem nescient, quomodo (cedo) incognitis naturalibus, eaqu[a]e praeter naturam obtingunt, afficiuntque, explorata habebunt?"
27. Ibid. 1 v – 4 r.
28. Ibid. 11 r – v.
29. Op. cit. fol. 3v.
30. Ibid. title page and 1 r (Dedication).
31. Especially I, chs. ii, iii; and IV. In his work as a whole, Aristotle gives two alternate explanations for the activity of the elements, one based on their natural motions (up and down) and the other on their qualitative alterations. It is the former explanation which Contarini seems to take up (cf. e.g. Meteorologica Bk. I iii 341a 5–9) and which we have also seen in Sylvius' commentary.
32. Contarini op. cit., Bk. I, fol. 3v – 4r, 5r–9r. He devotes several pages to an analysis of the argument in the Timaeus, because, he says, the original is unclear.
33. Ibid. 11r–11v. "Haec elementa inferioremque mundum sese habere veluti materiam, corpora vero coelestia ac superiores mentes esse causam efficientem, quae scilicet imprimant motum suo lumine, & tepore illo, qui assecla est luminis, diverso quodam a calore elementari & igneo, quem jure appellare possumus teporem aethereum. Hunc Platonici & antiquiores philosophi putaverunt esse corpus quoddam aethereum tenuissimum omnia penetrans, quod etiam magnum naturae seminarium nonnulli eorum vocare consueuerunt apud Aristotelem, quoniam corpus nullum quamvis tenuissimum, quin etiam mathematicum aliud corpus penetrare potest, utique nullum tale est corpus permeans inferiora haec elementa, sed coeleste corpus aethereum est, ab elementis separatum, cuius naturalis motus est circularis: ab hoc tamen corpore in elementa refluit lumen motus, & hic quem diximus tepor aethereus, non corpus sed aetherea qualitas, quae omnia penetrat & permeat elementa. huius teporis meminit in libris de generatione animalium, quem dicit ibidem non esse calorem ignis, sed proportionatum elemento stellarum. Cum ergo elementa sint materia quae formatur a corporibus coelestibus eorumque mentibus, nulli est ambigendum quin multa fiant in elementis quorum causam referre non possumus in naturas formasve elementorum proprias: sed referatur omnis eorum causa necesse est in coelestia corpora, quorum vim & actionem elementa recipiunt."
34. Walter Pagel and Marianne Winder, "The higher elements and prime matter in Renaissance Naturalism and in Paracelsus." Ambix, 1974, 21, 93–127 esp. 102–3, 123; – "Die Konjunktion der himmlischen und irdischen Elemente in der Renaissancephilosophie und im echten Paracelsus" in Paracelsus Werk und Wirkung, Festgabe für Kurt Goldammer zum 60. Geburtstag, Wien, 1975, 187–204. Paracelsus, however, in at least one work (Philosophia de generatione et

fructibus elementorum, I iv) uses higher elements to equal fire and air, lower to mean water and earth, in much the same sense as Sylvius. (pp. 193, 200).

35. Contarini op. cit. fol. 11v, 13v-14v.
36. Ibid. Bk II 23r-27r: "Ideoque quemadmodum universa haec elementorum sphaera se habet veluti quaedam materiae massa & moles, quam corpus coeleste effingit & format: ita caetera tria elementa, si igni comparentur, habent se tanquam materia: ignis vero forma eorum est & figulus, qui ex illis mixta omnia efficiat" (23v-24r).
37. Ibid. 28r. "corpus quoddam calidum erit, praecipueque humidum: igne densius, sed longe rarius aqua & etiam vapore aqueo: quod neque ita efficax est in agendo, sicut ignis, humiditate etenim calor obtunditur.."
38. Ibid. 30r-31v.
39. Ibid. 38v. "Et quemadmodum animalis corpus ubique fovetur spiritu, si haec terra plurimum intra se continet aetherei vaporis, & igneae exhalationis, quibus immixta generare & nutrimentum praestare omnibus potest."
40. Ibid. iii, 43v-46v; 48v-49v.
41. Ibid. 60v.: "Non omnino dissimilem esse generationem corporum mixtorum a generatione animalium plantarumque, quamvis in plurimis dissideant. Nam sicut in generatione animalium est quaedam certa materia in qua inducitur anima (quae primo inexistit) participaturque a calido naturali, sibique proprio, ac per eum membra format: ita in generatione mixti corporis est forma quae inducitur in sua materia, & primo inest calido naturali[s]. Item quemadmodum in generatione animalis praeter calidum in quo est anima, existit etiam semen in quo est vis generandi animal a parente deducta: Ita in generatione mixtorum omnium est qualitas coelestis, in qua veluti in quodam magno seminario est vis generativa mixtorum omnium deducta a corpore coelesti & a mentibus illis coelorum motricibus ac vi veluti quadam arte, a suprema illa arte derivata, & dispensantur elementa in mixtorum generatione & forma mixti inducitur." I am grateful to D.P. Walker for corrections to my translation.
42. Ibid. 61r-v.
43. Ibid. 54r.
44. D.A.R.C. [p.7]
45. Ibid. Bk I [p.5], II, p. 143; De Elementis p. 69.
46. Claude Bernard, Introduction a l'Étude de la Médecine Experimentale, Paris, 1865, is a much later example, but in his division of medicine into physiology, pathology and therapeutics, his acceptance of a hierarchy of causes, and his insistence that reason,not empiricism, is the validating principle in science, he resembles Fernel.
47. D.A.R.C. I [55, 57], 58–9 (summary); II, 150.
48. Ibid. [p. 56]: "PH: Inexplicabilis est, mea quidem sententia, temperamenti modus. EU: Id in conformatione longe magis dicas: nam hoc aut illud temperamento esse calido satis intelligitur, eo quod calidum & igneum elementum caeteris elementis in permistione praepolleat. Itemque tactu frigidum si senties, frigidum elementum dominari: sed quanam ratione factum sit, ut haec aut illa fuerit huius aut illius rei figura, id vero est, quod nec dicere nec mente consequi

possis. Plurima enim cernimus quae cum sint eiusdem temperamenti, dissimiles tamen figuras acceperunt rursumque aliis consimiles esse figuras, quorum sit dissimilimum inter se temperamentum: quod ex medicorum sententiis plenius postea & uberius probabitur."

49. Ibid. p. 43. "Partium quae in composito sunt singularum formas, esse veluti quasdam praeparationes inducendae totius formae, & eas esse multiplices."
50. Ibid. p. 45. "..quo idipsum unum sit, & quo vitam agat."
51. Ibid. II 157, 160, 162, 163, 167–9 (Galen); 159, 169 (Hippocrates); 153, 161–2 (Plato); 160–2 (Aristotle).
52. Ibid. Bk ii, Praefatio, 143–4; 150, 153, 160–1.
53. Medicina, De Elementis iii-v (pp. 74–76).
54. Ibid. 78–9; D.A.R.C. p. 46.
55. De Elementis ch. vii, p. 80: "Hae [substances and qualities] quidem nequaquam dissipatae aut extinctae oblitescunt, sed repressa…referuntur."
56. Ibid. ch. viii, pp. 81–82.
57. Ibid. Bk IV, De spiritibus et innato calido, develops this argument more fully. See below, Chap. 15.
58. Meteorologica IV i and iv, v, viii; ch. x, and xii 390b: "Now heat and cold and the motions they set up as the bodies are solidified by the heat and cold are sufficient to form all such parts as are the homogeneous bodies. For they are all of them differentiated by the various qualities enumerated above, .. all of which are derived from the hot and the cold, and the mixture of their motions." Also cf. De caelo, II, ii and iii.
59. Galeni de Temperamentis Lyons, 1550, I pp. 23–4. "ex calidi, frigidi, sicci, & humidi temperatura conflari corpora dicunt, de iis, quae summo gradu sic se habent, ipsis scilicet elementis, aere, igni, aqua, terra intelligendum aiunt."
60. Galen, de naturalium facultatum substantia. In addition to Kühn, IV 757–766 I have used two Renaissance editions: Galeni de Naturalibus Facultatibus.. Additusque est de naturalium facultatum substantia liber, … Lyons, 1550; and Galeni … Libri Tres. Primus, de facultatum naturalium substantia [and other works] Paris 1528. The work was evidently relatively well known to Renaissance readers; apart from its appearance in the collected works, the Bibliothèque Nationale's catalogue Galien lists Paris editions of 1528 and 1547, Lyons 1550, 1552 and 1560 and Toulouse 1554, by three different translators. "Quod quidem omnia, quae hoc terrarum orbe corpora continentur ex quatuor elementis totis per tota contemperatis, non ut Empedocles putabat in minutissimas particulas redactis, constant" (1528 edition, used throughout for quotations).
61. Ibid. (1528) 3r: "Probabilius..mihi videtur, temperamenta pro qualitatum ratione constitui".
62. De temperamentis, I 43–5, 45–6, 47, 48–52, 58–60.
63. Ibid. (Sylvius) pp. 10–11; (Galen) 53–57. "Quippe animal, ac stirpem quamlibet, tum optime se habere diximus, cum optime suam functionem obeunt".
64. Ibid. 64–5; 68–9; 73; 97–100 especially.

65. E.g. Ars medica, de sanitate tuenda, de naturalium facultatum substantia, de naturalibus facultatibus, de usu partium – all of which are cited by one or another of the four commentators.
66. De naturalium facultatum substantia (1528) 2r and v, 3r
67. De temperamentis, especially I 30–1, 60–3, 70–3; ii, 97–118, 120–6.
68. De inequali intemperie.. pp. 175–6.
69. De temperamentis I 37–43; II 78–86.
70. Ibid. iii p. 134, 141: "quatuor nimirum dicentes totius corporis proprias facultates esse, unam idoneorum tractricem, alteram eorum omnium retentricem, tertiam alteratricem, & quartam quae alieni sit segregatrix, easdemque facultates effectus esse totius in quovis corpore substantiae: quam etiam constare ex calido, frigido, humido & sicco inter se mixtis, dicimus."
71. Ibid. ii p. 86, iii p. 141 ("In temperamento est facultatum essentia, ut in natura diximus".)
72. Galen on the Natural Faculties, with an English translation by Arthur John Brock, M.D. London, 1928 (Loeb Classical Library) Book I, i; ii 6–7; vi–xi (especially ix) for faculties and sub-faculties; III, especially i, vi, ix for general faculty (e.g. vi 159–60; III viii 173, 175 for attractive and expulsive fibres; xi for retentive ones; iii for uterus as example of retentive and expulsive faculty.
73. Ibid. II ix par. 126–127.
74. Ibid. viii 118; also I iii.
75. Ibid. II viii 121.
76. De facultatum naturalium substantia (1528) 1v-4r
77. De temperamentis Book III, esp. 141–151, 156–160, 162–170.
78. Ibid. In libros temperamentorum commentationes partionesque aliquot utilissimae. [Sylvius] 1549. p. 3–4. "Elementa quatuor ignis, aër, aqua, terra rerum omnium generabilium & corruptibilium principia, per se tota a suis qualitatibus summis confusa, pro mistionis modo & differentia, res alias atque alias constituunt…" nunc tota transmutandae rei substantia, ut fit in nutritione: nunc formatrice in natura virtute, ut in animalibus generandis: quae quidem artifex facultas est, & particulas animi moribus accommodas effingit, ne solis qualitatibus formandarum partium causam tribuamus, cum hae formationis tantum sint instrumenta: formator alius, tum propinquus, ut semen: tum remotior, divinioris cuiusdam originis."
79. Ibid. p. 4 and p. 8: "quarum omnium magna est latitudo, innumerabilesque sunt differentiae, majoris et minoris ratione." Galen's distinction of four grades (gradus) of each quality is found not in de temperamentis but in the pharmacological work De simplicium medicamentorum temperamentis ac facultatibus (Kühn XI 375–892, XII 1–377).
80. Ibid. pp. 6, 8: "Cum autem homo sit animal sapientissimum, & ..perfectissimum: ut formam omnium nobilissimam rationalem animam: sic corpus estsortitus omnium optime constitutum, idemque sanissimum, quod est similaribus cunctis partibus temperatissimum, in calidi, frigidi, humidi, sicci symmetria.."

81. Ibid. pp. 10–22, "Variae Temperamentum & intemperamentum divisiones." Here he allows four grades of each of the eight varieties of unequal temperament.
82. Montanus, op. cit. (Libri Primi de Temperamentis Methodus) 5v-6r. "Temperaturam aequalem, in qua nulla penitus qualitas excedit".
83. Ibid. (Libri Secundi de Temperamentis Methodus) 7r v: "Et primo nos docet inaequalia cognoscere, deinde aequalia, ac nullum signum efficacius ponit, quam quod a singulorum actionibus sumitur. Secundo a structura."
84. Ibid. (Libri Tertii de Temperamentis Methodus) 8r-v. "hac similitudine omnino docemur, quae calefaciant, & quae refrigerent corpus nostrum, ac eandem esse omnino proportionem inter calorem nostrum, & inter calorem ignis." (8r)
85. Ibid. 9r-v: "Primum probat inductione, ostendens tam diu operari has facultates, quam diu fuerit calor temperatus, & hoc per singulas actiones patebit discurrenti. Ubi enim calor vel resoluitur, vel extinguitur, vel augetur, vel corrumpitur ob multas morbificas causas: ita etiam minuntur, auferuntur, corrumpuntur operationes."
86. (Ed. cit. p. 121) It is possible that in speaking of 'moderate heat' Montanus is thinking not of eucrasia but of a moderate degree of innate heat. However, his argument is very close to the Galenic one, and here eucrasia is clearly meant.
87. Montanus, op. cit. Methodus de facultatibus naturalibus, 10v.
88. Contarini, op. cit. 49r. "Nam qualitates primae sunt veluti elementa quaedam complexionis, ex illis etenim conflatur complexio sicuti mixtum ex elementis."
89. Ibid. Bk IV 62r, 68r: "Complexio secunda qualitas est conflata ex mixtionibus primarum qualitatum, quia mixtum corpus agit certa actione, & patitur certa passione. Nam cum temperamento primarum qualitatum fiat complexio, nonnullae vero activae sint ut caliditas & frigiditas, duae quae supersunt passivae, humiditas scilicet & siccitas, necesse est ut harum omnium natura reluceat in complexione quae ex ipsis conflatur. Per complexionem ergo mixtum corpus & certa actione agit per cuiusque temperamento, & certa item passione patitur." This resembles Aristotle's description of qualitative action in Meteorologica IV 5 382a.
90. Ibid. 70r; 68r-v. "Sic ergo dicimus complexionem temperatam & aequam quae operationi cuiusque speciei conveniat" (68v).
91. See also above, note 79. Gradus never seems to be used to describe the intensity of wet and dry, only of hot and cold. In the sense that hot and cold gradus cancel one another out, it can possibly be best imagined as "positive" and "negative" degrees. Thus a Galenic "thermometer" might look something like this:

For examples of use of the term by late medieval and early Renaissance natural philosophers (many of them medically trained) in physical speculation, cf. Marshall Clagett, Giovanni Marliani and Late Medieval Physics, New York, 1941, especially Chapter II, pp. 34–35 and note 13 for Galenic origins of gradus. For an early thirteenth century medical account of degrees, in the context of a commentary on the Articella, see Michael McVaugh, "An early discussion of medicinal degrees at Montpellier by Henry of Winchester", Bull. Hist. Med., 1975, 49: 57–71.

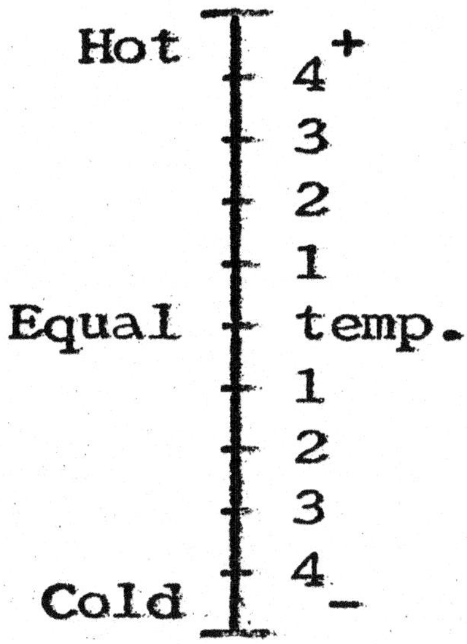

92. Contarini 70v–71v; "non ita tamen ut ponamus seminarium hoc esse corpus aethereum omnia permanans, sed accidens qualitatemque a corpore coelesti defluentem, omniaque penetrantem."
93. Ibid. ("Item diximus praeter vim calefactionis quae est in calore, inesse etiam in calore vim cogendi homogenea..") What the celestial power adds, according to Contarini, is not so much a greater power of acting but a greater power of resisting the action of cold, even when the latter is present in a much higher degree. So the power of action of a quality is assumed to be proportional to its gradus, but not necessarily its power of resisting. The language which Contarini uses is that of natural-philosophical speculation on the nature of qualitative action. For example, the notion of a body's resistance to qualitative change was invoked by natural philosophers such as Marsilius of Inghen and Albert of Saxony to explain the phenomenon of reaction: i.e. the cooling of a hot body by the cold one which it is at the same time heating. (Clagett, op. cit., esp. p. 46, note 41; he also notes that this 'Parisian' theory had considerable influence in Italy).
94. Contarini, 71v-72r; Clagett op. cit., especially ch. ii (on reaction) and iii (reduction).
95. Ibid. 72v-73r. "Alia etiam causa adduci potest, si vellemus ascendere supra qualitates primas & complexionem, ad potentias videlicet naturales & proprias…hae etiam naturales potentiae vim habent maximam etiam agendi, eaque inferiores qualitates agere possunt. Quid enim mirum est si inferioris virtus contineatur in superiori & excellentioribus. Dicemus ergo quod mixta illa per naturales potentias seu proprietates agunt, cum calorem seu frigorem multaque alia praeter haec de quibus medici tractant." Contarini here appears to agree

with Galen that temperamental heat is all that doctors have to concern themselves with.

96. Ibid. 73r.
97. Physiologia De Temperamentis Lib. III ch. i p. 84: "quae quodammodo propago & soboles permistionis est."
98. Ibid. "Quae igitur intra mediocritatis huius limites inciderit principum qualitatum concursio, ut mistionem sic & continuo temperamentum progignit."
99. Ibid. "Est autem temperamentum, non ipsa mistio, sed mistionis ratio. Poterit & id earum quae in mistis elementis sunt principum qualitatum harmonia & concentus definiri. Ut in cantu & fidibus soni longis locorum intervallis distracti, aut conjuncti quidem, sed dispares & non pro rata portione distincti, concentum haudquaquam efficiunt, nisi primi cum ultimis temperati, & medii utrisque respondentes aequabiliter dulci sono aures compleant: ita neque diuulsorum elementorum qualitates, neque coeuntes illae quidem temperamentum efficient, nisi moderata & concordi quadam mediocritate sese mutuo complectantur. Non igitur ipsa rerum primordia per se & absolute perpensa temperamentum sunt: neque illorum mistio, sed mistorum comparatio.. Non desunt qui temperamentum a mistione hoc discrimine seiungunt, quod haec maxime elementorum propria sit, illud vero qualitatum."
100. Ibid. ch. ii p. 86; cf. Clagett, op. cit., pp. 43, 47, 71–2, 76 for appeal to experiment in controversies concerning reaction and reduction.
101. Ibid., ch. v, p. 90. "Quicquid enim animo cernimus, id omne originem traxit a sensibus."
102. Ibid. ch. iii p. 88: "Temperatissimum quod est, individuum & omnis expers latitudinis quasi punctum intelligitur: caetera quae ab hoc recesserunt, longe lateque fusa sunt: neque enim quae calida dicimus, omnia pari a mediocritate intervallo distant, neque aut frigida, aut humida, aut sicca, uno eodemque omnia sunt ordine. In compositis porro evincentium qualitatum aut par est recessus, aut alias alia antecellit..Unde intelligitur octo cum sint impara temperamenta, uniuscuiusque tamen ex maiore minoreque recessu innumerabilia fieri discrimina, ex quorum proprietatibus tum similium partium, tum infinitae corporum naturae & proprietates consurgunt atque constant." If we made a diagram from this description, it might look something like this:

Equal temperament is of course the point at the junction of the coordinates; T_1 represents a simple temperament, cold in the first degree (say); T_2 a composite temperament, hot and moist. The number of possible co-ordinates is clearly unlimited.

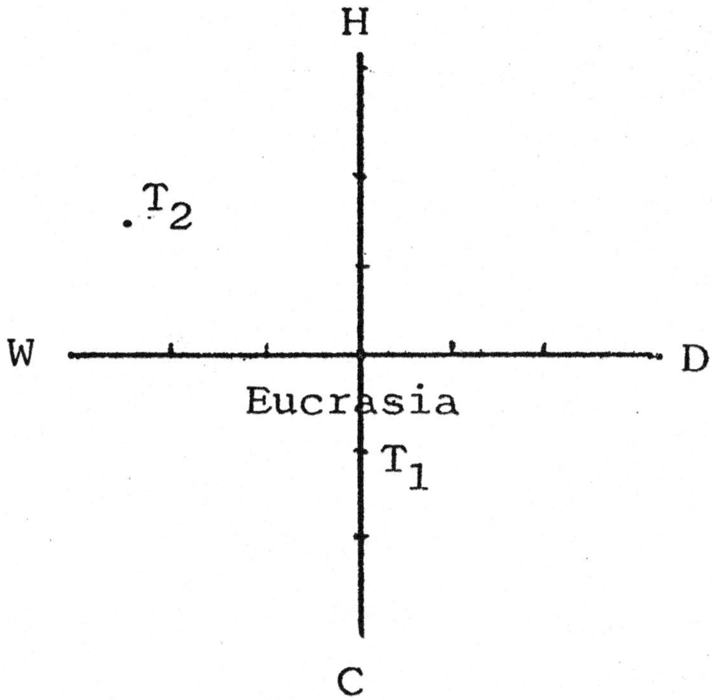

103. Ibid. p. 91. In De abditis rerum causis (I iii 24–6) Fernel adds to this the notion that the simple parts differ in essentia (and not only in quality) because they differ in function. Temperament thus becomes a sign of these essential differences; though elsewhere in the same work (e.g. pp. 36, 160) it is described as purely qualitative, and distinguished from the form of the simple parts: which, like function, is supra-elemental and derives ultimately from the soul. In fact, the distinction between structure and form is a difficult one.
104. Ibid. pp. 89, 95–101 (chapters iv, viii–xi).
105. Ibid. De Animae Facultatibus Lib V., ch. i, p. 122 "At quoniam non in solius corporis, sed in hominis qui ex corpore & animo constat, meditatione versamur."
106. For a more detailed discussion of particular problems concerning the soul and generation, see below, ch. 4.
107. Ibid. ch. i, ii; 122–5; D.A.R.C. ch iv, p. 45 for definition of total form; Physiologia V ch i for definition of soul.
108. De abditis rerum causis II i pp. 145, 147, 150. In Book I he gives two further reasons why form cannot arise from temperament: i) temperament is qualitative, form substantive: no substance can arise from a quality (p. 36) ii) the simple and opposing motions of the Aristotelian elements, which are in effect their form, are insufficient to explain the total form of a living thing (pp. 41–2).

109. Physiologia V iv p. 130 for range of 'ensouled' substances; v p. 131 for activity on basis of similitudo: "similitudo quippe amicitiam, amicitia desiderium seu appetitum, appetitus attractionem movet."
110. Ibid. V i 122, ii 123–5; De abditis rerum causis II chs ii, iii, iv.
111. Physiologia III (De temperamentis) vii pp. 93–5; also in Book IV, "On spirit and innate heat" (and see below, ch. 3). For a definition of spiritus, De abditis rerum causis II iv 168.
112. Physiologia III vii p. 95: "id proprium perfectumque temperamentum est, quod ex prima elementorum confusione & ex fervido calentique spiritu integratur."
113. cf. e.g. De locis affectis IV 5, "De uteri affectibus".

Chapter 15
Spirits and Innate Heat

Abstract The *spiritus* is a characteristic feature of Fernel's physiology. It is an important part of his explanation of generation. And it is an aspect of his physiology which called forth comment from contemporaries. The first section of this chapter gives an account of innate heat and *spiritus* as Fernel explains them in the *Physiologia* and *Dialogus*, the second deals with the action of *spiritus* in generation and the third considers contemporaries and successors who wrote commentaries on heat and spirits and commented on Fernel's own system. Those considered are Joannes Argenterius, Jean Riolan the Elder, Zabarella, Paparella, and Caimo on innate heat, and Delphinus, Bertacchius and Bronzerius on spirit.

Fernel was in a minority on the question of the celestial origin of *spiritus*. In his own terms, despite the flaws in his argument pointed out by later writers, Fernel makes a good case, in his own terms, for the necessity of a *spiritus*, and for its association with a power beyond the elements.

We saw in the last chapter that for Jean Fernel the elements and temperaments did not suffice as an explanation of the structure and function of the living body, and that he appealed instead to a higher level and ultimately to the soul. We also met the innate heat and spiritus, with which this chapter will be concerned. The spiritus is a characteristic feature of Fernel's physiology, and is of interest to us for two particular reasons. First, it is an important part of his explanation of generation. And secondly, it is an aspect of his physiology which called forth comment from contemporaries. The first section of this chapter gives a fairly detailed account of innate heat and spiritus as Fernel explains them in the Physiologia and Dialogus, the second deals with the action of spiritus in generation and the third considers contemporaries and successors who wrote commentaries on heat and spirits, and in particular commented on Fernel's own system.

L. Deer Richardson, *Academic Theories of Generation in the Renaissance*,
History, Philosophy and Theory of the Life Sciences 22,
https://doi.org/10.1007/978-3-319-69336-1_15

The Spiritus and Innate Heat

A spirit is traditionally a body so subtle as to be almost, but not quite, immaterial, a substance which links the two worlds of body and soul, animate and inanimate, by having a nature midway between them. In its standard medical usage, a spirit usually meant one of the three grades of spirituous substances within the body, as described by Galen and elaborated by Arab and later commentators. These were usually distinguished as the animal, vital and natural spirits. Each took its origin from one of the principal organs of the body, brain, heart and liver respectively. And each was responsible for a certain area of bodily function, corresponding to the areas of the three major faculties. Galen defined faculty as "the cause of an action" and the spirits accordingly came to be considered as the instruments by which these actions were accomplished.

But the spiritus of Fernel is not any one of these standard medical spirits, and its relation to the level of action represented by faculty is not at all clear. It seems to have both a different origin and different functions from the medical spirits of Galen.

The genealogy of the spiritus is complicated, and I do not propose to go into it here in any more detail than Fernel himself provides. Instead, since my own interest in the spiritus concerns the uses which Fernel finds for it, and particularly its application to the problems of generation, I have chosen to attempt to explain spiritus in terms of its function in Fernel's physiology, rather than in terms of its origins.[1]

Fernel's spiritus first makes its appearance in Book IV of the Physiologia which is entitled "Of spirits and innate heat". Despite the plural title, only one spirit – which we call the spiritus to distinguish it from the three classical medical spirits – is mentioned for most of this book. Book IV represents a turning point in the Physiologia. In the first three books, Fernel was concerned with the structure of the human body: in Book I, its visible, anatomical structure; in Books II and III, its elemental and temperamental organisation.

The rest of the work is concerned, broadly, with the functioning of the body as distinct from its structure. Book IV discusses the role of vital heat, Books V and VI the facultative causes of activity for both body and mind. And the final Book is concerned with one special activity of the plant or animal body, that of generation.

Everyone agrees, says Fernel at the beginning of Book IV, that all living things, plants as well as animals, require heat to accomplish their activities. The 'natural' activities which plants and animals share, i.e. growth, nutrition and procreation, depend on this no less than the specifically animal activities of sense and motion. But this vital or innate heat, which is the efficient cause of plant and animal function, cannot be the same as the elemental or qualitative heat which is one of the causes of structure in living things. For one thing, structure, and therefore elemental heat, are present in inanimate objects, which were never alive, as well as in a corpse, in which life and activity are lost. There is something in a living man 'by which he is more excellent than a dead man' as Fernel puts it, and which is the cause of his life and activities. This something cannot be related to the temperaments, because even those animals, such as snakes and salamanders, and plants, such as mandragora, which are temperamentally of a cold nature, still live. But since in them the elemental quality of cold dominates that of heat, they must live by virtue of a separate, vital heat, which is superior in its origins – in fact, divine – and which 'does not stink of the crasser nature of elemental fire'.[2]

The notion that all forms of vital activity are the result of the presence within the body of a heat of a certain kind was a natural inference which could be tested by touch in the case of man and the higher animals. Philosophically, it was an idea especially associated with the Stoics. The closest analogy to vital heat, in the Stoic system, is not the destructive heat of earthly fire, but a "universal fire" which is the source of all life and preserver of the entire universe. This "hot and fiery principle" is itself possessed of will and conscious intelligence, and is the source of all motion and consciousness in living things. Of visible fires, the closest approximation to vital heat is the heat of the sun and other stars, which are "created out of the purest and most active elements of the aether, with no admixture of any baser substance". Neither can be compared with elemental fire.[3]

This parallel between the celestial heat of the body and that of the sun is one which is to be found also in Fernel. But such ideas were also current in the medical tradition. Though Plato and Hippocrates had distinguished imperfectly, if at all, between ordinary fire and that within the body, both Aristotle and Galen stated the distinction clearly. Aristotle, in the Generation of Animals, wrote that "the heat which is in animals is not fire, nor does it take its origin from fire". And Galen spelled out the reasons for this distinction in his work On Respiration. Ordinary fire acts 'by no means rationally', destroying what it feeds on, while the innate heat has the startling ability to organise its own nutrition, as well as the other functions of the animal body, and thus to replace its 'fuel' as it uses it. Thus, argues Galen, "we have to consider the indwelling heat and the nature of the organism as one and the same thing."[4] This definition is one which we shall find, at least by implication, again in Fernel. In it, the innate heat comes very close to being identified with the organising principle of the living body, corresponding to the Aristotelian form and the Christian soul.

I have devoted quite a lot of time to the vital heat, but it is important as a central, and unorthodox, part of Fernel's system. By claiming that the efficient cause of vital actions lay in a heat whose origin was "different from, and more excellent than that of the elements", Fernel introduced a host of problems which could not be solved by recourse to the system of medical spirits and humors derived from Galen. Elemental heat could quite naturally perform its actions through and by means of the elemental and temperamental composition of the body of which it was a part. But a celestial heat required a different, and in some sense 'nobler' base for its actions. It was for this reason that Fernel introduced, in chapter ii of Book IV, the concept of the spiritus, which is intimately bound up with the vital heat in all its actions.

What is the spiritus, and why is it necessary? It is necessary first of all, says Fernel, to enable the vital heat to act within the body. Because heat is a quality and not a substance, it must be 'carried' by a substance within the body acting as its substrate or vehicle. But an elemental, humoral substance is not suitable for this:

> I am of opinion [says Fernel] that it was necessary to contain this [i.e. heat] in some fluid and freely-flowing body. But no humor was suitable or fitting to pass through all bodies with such speed in this manner; wherefore it was necessary that the heat should subsist in a very tenuous material, capable of extremely rapid motion, which at the same time would be similar and compatible with the fostering heat.[5]

Such a substance is available in the material which the ancients called spirit, or aether, and which they associated closely with heat. The doctrines of the ancient philosophers, in particular the Academics or Neo-platonists, provide guidance also as to the "necessity and substance" of this spirit:

> In order that the necessity and substance of the spirit may be more fully shown, we must revisit and recall the doctrines of the ancient philosophers. The Academics were the first to suppose, when they realized that two entirely dissimilar natures cannot be associated together without the interposition of a suitable mean, that our soul, created by the supreme maker of all things, before its emanation and immigration into this thick and solid body, put on as a simple garment a certain shining, pure body like a star, which, being immortal and eternal, could never be detached nor torn away from the soul, and without which the soul could not become an inhabitant of this world. Then they surrounded the soul with another body, also fine and simple, but less pure, less shining and splendid than the first, not created by the supreme maker, but compounded of a mixture of the finer elements, whence it is named aerial and aetherial.[6]

Dr. Walker points out, in his article on the astral body, that "the theory of the astral body was … not generally considered safe or respectable" in the sixteenth century and it remains something of a puzzle why Fernel should have used it in this context. Other writers, for example the Aristotelian commentator Alexander Aphrodisias, also considered a spirit of this sort to be the suitable medium between conflicting natures, such as those of the body and soul. And the notion of a mean between extremes which is of an intermediate nature was in itself an extremely common one.[7]

Arguments for the spiritus can also be taken from anatomy. The ventricles of the heart and brain contain no visible substance, but they must have been made to contain something, as nature does nothing in vain. That we breathe is a further argument for the existence of a tenuous, spirituous substance within us. Just as the coarser elemental substance of the body renews itself by taking in food which is also a mixture of the elements, so the spiritus requires a 'food' similar to and suitable for its nature:

> Wherefore it is necessary to draw in and attract air, by which substances similar to something in our nature may be joined to it, as though converted into food. If nothing in us were a tenuous and spirituous substance, scarcely anything would really force us to the necessity of breathing.[8]

Many arguments, therefore, prove the spiritus to exist. Its quintessential nature is appropriate to its importance in the functioning of the plant and animal body, just as the celestial nature of the vital heat itself fits its role as the efficient cause of all vital actions.

But in turning from the nature of the innate heat and spiritus to their substance within the body, Fernel descends abruptly from this celestial plane. If the fire in living things is compared with ordinary fire, it now appears that there is no essential difference. All things which burn, Fernel explains, do so simply because they contain a certain oily material which is able to support flame. This oil is the analogue of the sulphur of the chemists, by which they explain the combustible nature of certain minerals. It is the same in all living and non-living things, and can be extracted from them mechanically by burning or by chemical distillation. Fernel might seem to be saying no more than that there is a substance within the body which makes it subject to ordinary fire. But no: this oil is present, like the vital heat itself, as the necessary companion to all life: "In short, nothing in the animal or plant genera can possess life, which cannot release 'oil' from itself."[9]

> In the following chapter this analogy is made more explicit: the spiritus which we have demonstrated to dwell within us, because it is a tenuous and aetherial substance perfused by indwelling heat, corresponds by analogy to what we properly call flame.[10]

We thus have an altered definition of the spiritus. It is the flame of the fire within us, neither the celestial heat alone nor its aetherial vehicle considered separately from this heat, but the two in association and acting together. Pursuing the analogy, the 'primigenial moisture' which is the subject of chapter four corresponds to the 'oil' which is the fuel of ordinary flame. If we write the comparison out in the form of a ratio, we have elemental heat : flame : oil :: celestial heat: spiritus : primigenial moisture. We thus have a three-level system similar to that in the analogy of the vestments of the soul which I quoted earlier. In that analogy, the radical humor, or primigenial moisture, can be seen as the final link with the world of substance.

For the radical humor, to give it its Arabic name, is one of the humors of the body, and shares its elemental nature. In addition to the four constituent humours which make up the similar parts, there are three further humours – added to the Hippocratic-Galenic system by Arabic commentators – which preserve each of the similar parts. Among them is "that humor which we call radical, inherent to life and primigenial".[11] It is this which Fernel considers most suitable to be the substance of the innate heat within the body: an indwelling, permanently available form of sustenance which lasts as long as life itself and provides the means by which the heat of the body is preserved and maintained.

In what way does the radical humour differ from the spiritus? It is, as we have seen, a humour, that is one of the material substances of the body. It is elemental: Fernel tells us that it is formed of aery and fiery elements. And it acts as a permanent source of "fuel" for the "fire" within the body.[12] In short, it has none of the celestial and superelemental attributes which Fernel earlier tried to persuade us were so important in distinguishing the heat of living things from ordinary fire. But it is to this material substrate, not the nobler celestial heat itself, that Fernel turns his attention in the rest of Book IV.

The distinction between the various levels, celestial, aetherial and material is further blurred because they have, according to Fernel, a common origin. All are connate, and take their origin from the semen. Spiritus, indwelling heat, radical humor and the solid or seminal parts of the body all arise from this same source in our generation. Acting together, the vital heat and primigenial moisture contained in the semen of the parents work to form a kind of network of fibers, which are then literally 'fleshed out' with material blood.[13] Their common origin in the semen – which itself has divine aspects for Fernel, as we shall see – serves to underline the special nature of all these substances and to explain the complex of relationships linking heat, spiritus and radical humor.

A further source of confusion appears when Fernel attempts to describe the relationship between the spiritus as we have presented it here and the three medical spirits of standard Galenic physiology. He tackles this important question in a puzzling passage at the end of Book IV. It is puzzling because, if the implications are drawn out, it seems to deny most of what he has told us about the spiritus in the rest of the Book. The spiritus, he says, occurs in two forms, one 'indwelling' in the similar parts of the body, the other flowing through the parts. (This distinction is found also in Galen, for instance in De formatione foetuum.) On the basis of earlier chapters, one would have assumed the flowing form of the spiritus to be that responsible for distributing the innate heat to all the similar parts. But instead Fernel divides it into three kinds. These sub-spirits correspond exactly in their origins, and so far as one can see in their functions, with the animal, vital and natural spirits of medical theory. But they are much more difficult to reconcile with the spiritus as it is described elsewhere in Fernel's work, in the Hidden Causes as well as in the Physiology. For one thing, only the noblest of the three, the animal spirit, is aetherial in nature. But it is the air-like vital spirit, and not the animal spirit, which is associated by Galen, and Fernel, with the distribution of heat to the various parts via the arteries. And finally, these spirits are and must be elemental in nature and origin. They are, after all, successively elaborated from the blood, or rather from "an exhalation of the purer portion of the blood" as Galen tells us in the Use of the Parts.[14] I think the best that can be said for this passage is that it is an understandable attempt to reconcile, or rather conflate, two parallel systems each featuring a substance labelled spirit. But it does seem that the spirits of the two systems differ radically in their nature and origins. And as we shall see they differ in their function as well. Such a reconciliation thus seems to be plainly impossible, and Fernel's attempt is not convincing. As we shall see, commentators such as Argenterius and Caimus distinguished clearly between ordinary medical spirits and the spiritus of Fernel, and I think we must do the same.

The Spiritus in Generation

We now have a somewhat clearer idea of what the spiritus is, according to Fernel. But what does it do? In the most general sense, as we have seen, it is associated with all the actions of living things. It is the vehicle of the vital heat. In particular, it is associated with growth and procreation. And in practice it is only in considering this latter function of living things, in Book VII of the Physiologia and in the Dialogus, that Fernel really puts the spiritus to work. In this he follows Aristotle, who also invoked a spirituous substance, closely linked with the vital heat, at several points in De generatione.

The most important of these passages is that on the 'element of the stars' (G.A.II iii 736b-737a) which Fernel quotes, or refers to, repeatedly in both his works when the question of the spiritus is introduced.[15] Now, exactly what relationship Aristotle meant to suggest between the power which acts in fertile semen and the element of the stars or quintessence is not nearly as clear from this passage as either Aristotle or Fernel would imply. But there is a section of the Hidden Causes which may help us to understand what this text meant to Fernel. In Book II, chapter seven Eudoxus uses this same quotation to attempt to persuade the sceptical Brutus that there is a heat different from and superior to that of the elements, and that it is associated with a spirit "for the most part aetherial". Philiatros, (who can be depended upon to ask the right question), is puzzled. "Tell us briefly what you mean by the appellation 'divine' " he asks. "As Aristotle said, corresponding by analogy to the element of the stars" replies Eudoxus. Philiatros pursues: "What do you mean by 'the element of the stars'?" Eudoxus' reply gives us perhaps the clearest and most concise answer to the question of how Fernel sees the workings of the spiritus in generation:

> Just as we call the four elements of simple nature those which come together in all solid bodies, and from which material things proceed, so we should rightly call elements celestial, or stellar, which might bring to bear and direct such powers. If you consider in the first place the procreation of animals, which is from semen and maternal blood, you will observe that the blood is constituted from the four elements which are contained in the food, just as the material of the semen is put together from the purer and more useful blood. But you should remember that the nature, spirit and heat which the semen contains enclosed within itself was produced from other sources. You understand that their arrangement and powers are above those of the elements, nor can their superiority pertain either to these lowly and transient elements nor to the properties of the temperaments. But the faculties, spirits and indwelling heat arise from this spiritus of the semen, in which both heat, natural properties and divine powers reside.[16]

I think we miss the point here if we consider the heat and spirit of which Eudoxus speaks as a sort of superelement, the Renaissance version of helium or hydrogen as it were, or even, really, as a substance at all. What Fernel is drawing attention to here, and elsewhere in the work, is not the substance of the spiritus but its extraordinary powers and activity. Certain aspects of our makeup, he seems to say, are indeed divine, but this divinity is revealed not so much in their composition as by their powers. It is to this part of our nature, this class of 'hidden causes', that the spiritus belongs.

As Fernel distinguishes them, the activities of the spiritus in generation are of two sorts. In part, they arise from its assumed properties as a spirit, that is, a tenuous, air-like body capable of swelling, ebullient motion. For the rest, they derive from powers which it is assumed to have in a more abstract and philosophical sense.

Aristotle had argued that evidence for the existence of the pneuma or spiritus could be found in the 'swelling' of the male and female sexual parts – most obviously the testes and female breasts – at puberty. (G.A. I, xx, 728b) Fernel draws on this example, and the parallel argument that it is the motion of the spiritus in these organs which is responsible for sexual desire, to argue that semen is produced by both parents and not the father alone as Aristotle thought. The swelling motion of the spiritus, by contributing to the pleasure of both parents, is a direct impetus to intercourse and hence to generation itself. Thus in some sense it is the spiritus, or the vital heat working through it, which is directly responsible for perpetuating a species.[17]

These 'works' of the spiritus, and, similarly, the spurting forth of semen or of arterial, spirituous blood, are argued from observation and from the assumed physical characteristics of any aetherial body: its air-like ability to 'puff up' a substance, its motion, its ability to instantly transmit a physical effect.

There are other, less visually apparent characteristics which are the basis for the other works of the spiritus in generation. It is the spiritus, 'diligently mixed' with the material part of the semen, which is responsible for making it fertile. It is analogous, in this role, to the anima or soul itself, which Fernel had earlier described as the 'principle and cause of functioning' of the living body. In the same way, the spiritus is the principle and cause of functioning of the living semen. Fernel's argument runs along lines similar to those in which he argued earlier for a heat 'different from and more excellent than' that of the elements. There is living (fertile) semen and dead (infertile) semen. Both have the same elemental construction, and even the same temperament, but one is capable of producing a new human being and the other (e.g. the semen of a diseased person) is not. So there must be something present in the one, and missing from the other, which is responsible for this difference in capacity. This something is the spiritus, which is neither elemental nor temperamental in its nature.

Here again, Fernel's arguement draws on Aristotle's – and at the same time goes beyond it. In G.A. II i, Aristotle argued that heat and cold could produce the secondary qualities of hard, soft, tough, brittle but not the logos or organising principle, 'by virtue of which flesh is flesh and bone is bone', which he referred to the 'movement' transmitted from the generating parent. Fernel echoes this passage, but changes it to emphasise the superiority of the soul and its instrument, the spiritus, to the world of the elements:

> But the individual plan and essence, by which a thing is what it is said to be, in no way arises from the elements. For in the same way as the reason and essence of man, which is the soul, has a diviner origin than that from the elements; so also the reason and essence why flesh is flesh or bone is bone derive not at all from these dense and concrete elements, but from the wholly celestial element of the stars: and this consists in a spirit, as it were celestial, moreover this spirit is in the seed by which all things generated are ordered. In almost the same way, the nature not only of all other animals, but also of plants takes its origin from the heavens. They possess more hidden and profound powers, in no way brought about by elemental temperament, which are also called celestial.[18]

Again, it is necessary to turn to the Hidden Causes in order to find a fuller explanation of this passage. In Book I, chapter four, Eudoxus distinguishes three 'preparations' of the material which is to form a new individual, necessary before it can receive its final form. The first is that by which the similar parts are formed from the 'primordial mixture' of elements and given suitable temperaments. In the second these similar parts are organised into systems to form the organs and other 'dissimilar parts'. And the third is that by which "a certain spirit", in which is contained a beneficial (salutaris) vital heat, is diffused throughout these organs, bringing them together as a whole and preparing them to take on the nature of a new living creature.[19]

This brings us to the final and greatest work of the spiritus in generation, its part in transmitting, via the semen, parental qualities and the power of organisation to the new individual. The spiritus is fitted for this by its nature as intermediary between the soul and the body. And Fernel develops two parallel analogies to explain how its activities in this sphere may be understood. Both derive ultimately, like much else of his argument, from passages in De generatione. In the first, the spiritus is made responsible for preserving and mediating the motion of the celestial heat within the developing embryo. This is similar to its activity as Fernel described it in Book IV. This heat is not fire, as we have seen, but that "heat of the sun and of animals" without which no generation can take place. In the second analogy, the spiritus is seen as an instrument for the transfer of motion from the soul of the parents to the embryo via the semen. The image which Aristotle had used (G.A. II, i, 733b.e.s) and which Fernel takes up is that of the spiritus as one of the intermediate wheels in a machine or automaton. This image is particularly well-chosen for explaining the successive and regular development of the parts in the embryo. A description of this development takes up most of the latter half of Book VII.[20]

Fernel summarises the process which takes place in chapter v, "Of the nature, excellence and powers of the semen, and that spirit and faculty are the cause of all". The vital heat generates semen from material attracted to the male and female testes, mixing with it the vital spirit which flows through every part of the body. The temperament of the resulting mixture is elemental; but it is this heat, in some way celestial or divine, in which all its life consists.[21]

So much Fernel tells us of the work of the spiritus in generation. But he is not always clear in distinguishing his terms, and some points need further clarification. What, for instance, is the relation of the spiritus to the semen itself? In the chapters on the formation of the semen, he implies that it has a dual nature, partly material and partly facultative. The material part of the semen is formed from the 'ultimate residue' of the food, transported to the testes by the veins. Similarly, in the arteries associated with the testes, the spiritus "remains and is caught as in a net." The mechanism of elaboration is exactly similar in both cases, although we are told that two entirely different classes of substance are involved.

These two components, material and spirituous, are then 'diligently mixed' in the testes to produce the fertile, life-giving substances which is emitted by both parents. Fernel usually describes the spiritus as 'contained, or enclosed, or immersed' in the semen, suggesting a somewhat tenuous connection between the two. But on the other hand, he emphasises that the power or activity conferred by the parental soul is inseparable, or one with, the spiritus (una cum spiritu) and that neither the spiritus nor the soul itself can function apart from the body of the semen.[22] But if neither the spiritus nor the soul of which it is the instrument can act apart from the particular material in which they are contained, if they cannot function without it or beyond the limits of its capabilities, then Fernel's careful distinction between elemental and celestial powers seems to have come unstuck.

This is a point to which I should like to return. But first I should like to look briefly at some other opinions on the action of heat and spirit in generation, beginning with an important commentary by a near-contemporary.

Innate Heat and Spirits in Generation: Contemporaries and Successors

Johannes Argenterius

Argenterius, an Italian medical writer who was born sixteen years after Fernel, has received more attention from medical historians as a critic of the Galenic doctrine of spirits than for his opinions concerning his contemporaries.[23] But the section of his work On Sleep and Waking which deals with spirits and innate heat gives considerable space to a discussion of Fernel's spiritus. To devote so much time to a contemporary in a general medical treatise is somewhat rare. And Argenterius' commentary is valuable and interesting for several other reasons. First, it is early: On Sleep and Waking was published in 1556, two years after the appearance of the Medicina. Secondly, it combines great respect for Fernel as a doctor and scholar with vigorous opposition to his ideas. Finally, it is intelligent: and thus gives us a perceptive and often illuminating look at the spiritus through contemporary eyes.

Argenterius' own preference, as he tells us in the course of the work, is for a single spirit rather than the traditional Galenic three. He is especially concerned to question the animal spirit of Galen – chiefly because there are no anatomical structures in which he feels it could plausibly be elaborated or confined. In anatomy, the major Galenic argument is founded on the supposed existence of a network of vessels, the retiform plexus or rete mirabile, at the base of the brain. This is supposed to be the place in which the animal spirits receive their final elaboration. But, says Argenterius, the anatomists tell us that this plexus does not exist in man, whereas logically it should be largest in the noblest of the animals. Moreover, the heart – which is the spirit-maker favoured by Argenterius – does not show a similar plexus formation. He goes on to argue from this first that the animal spirit cannot be proved to exist, since the plexus itself does not; and then, somewhat contradictorily, that therefore a spirit of the kind which he has in mind does not require a plexus for its elaboration. Concerning his predecessors, he concludes somewhat tartly:

> What has been said, and written, by our ancestors, and afterwards accepted and published by us without reason, concerning so many kinds of spirits, is nonsense, stories, and the invention of the Stoics.[24]

Argenterius' own spirit comes closest in its details to the vital spirit of Galen. It is elaborated in the heart from the purer portion of the blood and distributed throughout the body by the arteries. Like Fernel's spiritus there is a passive form which resides in the individual parts of the body and an active form which flows through the body and is responsible, among other things, for the mechanism of sleep and waking. Again like Fernel's spiritus, it is the vehicle of vital heat. A single spirit, flowing through the body and distributing heat to all parts, they have in common: but Argenterius' spirit is derived entirely from the elements which make up the body, and it is this difference which forms the basis of his criticism of Fernel.

As he had done with Galen, Argenterius devotes part of his attention to the logical inconsistencies to be found between the details of Fernel's system. But his opposition runs much deeper than these logical niceties. It reveals the most fundamental assumptions of his own position, and reveals, too, the basis of his uneasiness with Galen's doctrine of medical spirits as well. For Argenterius, the basic danger represented by Fernel is that his celestial heat and the non-elemental spiritus with which it is associated may usurp the rightful place of the soul in physiology; furthermore, any confusion between the nature and powers of the spiritus and those of the soul, such as Fernel seems to imply, is a logical absurdity. The soul, in Aristotelian terms, is the form of the body, and not any part of it; the spiritus, however, can only be a substance acting within the body. It is "unable to be a form". Properly speaking, the innate heat is only the "first and common instrument" of the soul. To confuse the two is to make the basic error of confusing the form of a thing with its instruments:

> To assert heat to be our nature, and that which directs solid bodies, and the substance of the soul, is not to know that the form of a thing is distinct from its instruments.[25]

Furthermore, to "assert heat to be our nature … and the substance of the soul" is to come dangerously close to the suggestion that the soul may be mortal.

Elsewhere Argenterius summarises and refutes Fernel's arguments for the existence of a spirit

> whose substance is not corporeal, nor from the elements, or anything else which bears the mark of this inferior world, but divine, and celestial … an aetherial body, the abode of heat, link between the body and soul, and first instrument of all action.[26]

The arguments which he summarises are those which we have already discussed: (1) the doctrine of the 'Academics' concerning the astral body; (2) the 'element of the stars' passage from De generatione; (3) the anatomical demonstration, that the cavities of the body must contain something, but something so tenuous as to be invisible, therefore a spirit; (4) the argument that inspired air serves to nourish the spirits within the body, and (5) a further argument from the body's inflammability, implying the presence of a 'thin and pure air' to support the flame.

Argenterius' opinion of the Academics and their astral body matches his scorn for the 'Stoic invention' of a plurality of spirits. Anybody who could quote the Neoplatonists as an authority could believe anything. But his reply to the heart of Fernel's argument is to agree that there is indeed 'something divine' and 'not composed of the elements' which directs the actions of the body in life. But this 'something' is the soul itself, which as we have seen is for Argenterius neither a body nor a part of the body.

> There is in the living body something divine, truly, the soul: but it is not therefore so that either the entire body, which is alive, or the parts of the body within it are formed of celestial matter.[27]

Instead, they are formed from semen and spirit, which are in turn generated from blood and aliment, both clearly elemental substances.

Argenterius suggests that he believes Fernel actually identifies the spiritus which the celestial heat 'puts on' in order to work within the body with the aetherial and aerial vehicles which the Neoplatonic soul 'puts on' before it can enter the earthly body. If so interpreted, the analogy does come dangerously close to considering the soul as a kind of vital heat, and hence presumably mortal. Moreover, Argenterius supposes Fernel's analogy to require two spirits, one finer and more celestial than the other, to correspond to the aetherial and aerial garments of the soul. They are thus similar to Galen's three spirits, and Argenterius argues against them in the same way, suggesting that they would inevitably mingle within the body and so contaminate one another. He also notes that, in the final chapter of Book IV, Fernel had spoken of the spiritus as having 'species'; i.e. that it differed in its nature and substance in the different similar parts – bone, nerve and so on. This too seems to Argenterius a senseless multiplication of spirits. Why not have each of these parts constructed of a different set of elements? he asks sarcastically. It would make as much sense as a spirit in bone which differs from that in nerve or in muscle.[28]

But the basis of Argenterius' practical argument against a celestial heat and its spiritus is the necessity which both have to take in some form of nourishment. This squares very badly with their supposedly divine status. If spirits are truly divine and celestial, and do not consist of elements – what element could possibly nourish and conserve them? Moreover, why should they need to be nourished? It is as though we said that the stars had to have food. But in fact they are nourished, says Fernel: both by the air we breathe and by the 'oily humor' within the body. From this, they should surely be no more than air-like; or should consist of a mixture of elements, like all the remaining substances in the body.[29]

Argenterius' final comment combines respect for Fernel's obvious gifts and for his learning with disappointment at the uses to which they have been put:

> These and others like them are the words of those men, [that is, Fernel and his followers] indeed most learned, but (as I judge) too much influenced by inveterate opinion and the authority of the Greeks, they have in no wise been able to free themselves from this common error.[30]

Lest he should seem to be condemning such distinguished men without cause, Argenterius hastens to put forward his own opinion, arguing, as we have seen, for a single spirit, which is generated and nourished by elemental substances within the body, particularly the blood; and which is not an instrument of faculty but simply of vital heat, with which it is closely identified. Since neither heat nor spirit are the vehicles of any faculty – as Galen and Fernel would have them – it is not necessary to assume that there are as many different spirits as faculties, that is, different kinds of possible actions, operating in the body. Instead, Argenterius considers that he has proved

> that spirit does not at all contain faculties of the soul within itself, nor is it their vehicle, but that of heat itself, by which the soul exercises its power and functions in the body.[31]

Spirit is in no sense 'the substance of the soul'. It is merely an instrument of an instrument: the instrument of heat, which is in turn the instrument, (and not the analogue) of the body's governing soul.

Jean Riolan

In his commentary Ad Librum Fernelii de Spiritu et Calido Innato, Jean Riolan the Elder takes a much less critical attitude to Fernel's doctrine of heat and spirit, although in the end he too rejects the notion that the innate heat and spiritus are celestial in origin. In the first two chapters of the commentary, he quotes Fernel's arguments and examples, and his definition of spiritus as "an aetherial body, the seat and bond of faculties and of heat, and first instrument in performing their functions" is clearly based on Fernel's own.[32] At the same time, he wrestles with the problem of a celestial heat forced to act on an elemental plane:

> This pure, uncombined heat must mix [elemental things] for mixture is necessary to change, and change to mutual action and suffering. Celestial things do not suffer by acting, because they have nothing in common with the elements … therefore, some say, why are there not two kinds of heat in living things, one celestial, one elemental? Because, as the philosophers tell us, entities are not to be multiplied without reason. However, a heat single in number, varying in its conditions, can produce as many different effects as required … [Thus] dry heat condenses, moist expands, immoderate heat destroys, moderate preserves.[33]

Riolan distinguishes three kinds or species of fire – lux, flamma and carbo – which differ not essentially, but only in their effects. Lux or light corresponds to celestial heat and illumines without burning; flame, which both lights and burns, to elemental. The chief difficulty which he, like Argenterius, sees in the identification of innate heat with celestial heat is that the former clearly requires nourishment, yet celestial heat does not:

> Celestial things are not nourished, for nourishment is by making [something] like [oneself]; but elemental things cannot be made like to a celestial one, and nourishment is proper to mortal things .. yet native heat decays for lack of food: how then is it celestial?[34]

Forced to distinguish between vital and elemental heat in the case of living things of cold temperament (like the salamander, which can live happily in the midst of fire) Riolan admits the possibility of two forms of heat, one elemental and one "in origin and essence celestial, which yet in the world of elements in some way becomes elemental". But he adds, still not altogether happy, "For the rest, it is not easy to remove this scruple, that celestial heat should be nourished."[35] Later, Riolan argues, somewhat fatuously, that by comparing innate heat with flame (which is 'like' elemental heat) rather than with light (which is 'like' celestial heat) Fernel himself suggests that he does not intend the celestial origin of vital heat to be taken literally:

"the innate heat is elemental, celestial only by analogy."[36] Clearly this is Riolan's own position, and he supports it somewhat more effectively in his commentary on chapter six, where he looks at the supposed origin of the vital heat and spiritus. Everybody agrees, says Riolan, that we take our origin from two sensible and conjoined elements, semen and maternal blood. But how can the heat which is in the blood and semen be of two different kinds? The semen is made from the purer portion of the blood, elaborated by native heat: who then could say that the heat in the semen is celestial, that in the blood elemental? It is 'celestial' only by courtesy, as it were, being formed from the purer and more concocted blood. The argument is neatly made, and effectively undercuts one of Fernel's arguments for the special nature of vital heat.[37]

Despite his pedantic approach, certain major inconsistencies in Fernel's explanation are pointed out by Riolan, for whom Fernel's "celestial heat" has become virtually elemental in name as well as in fact. It is only a matter of decades from the time of Riolan until Descartes will claim that the heat which is in the heart of living animals has nothing special about its nature, but is exactly similar in all respects to the heat which is to be found in damp hay.[38]

Riolan's commentary suggests two of the major difficulties which helped to contribute to this shift. One is the difficulty of maintaining in practice a system of physiology in which an intermediary substance (the aether or spiritus) is used to link two components of contrary nature, especially when only the lower, elemental nature can be adequately described in terms of current theory. (Though as Descartes' own work suggests, this method of argument was by no means extinct a century later.) The second is the focussing of attention, strengthened by the work of anatomists in Fernel's time and later, on the linked problems of nutrition and respiration. If spirits are incorporeal, how can they be nourished? Moreover, how can they be 'starved' for want of nourishment? If they are corporeal, in what respect do they differ from the crasser elemental substances of the body? Paradoxically, the fact of respiration, which was for Fernel one of the best proofs for the existence of tenuous, spirituous substances within the body: ("If nothing in us were a tenuous and spirituous substance, scarcely anything would really force us to the necessity of breathing") has become for Riolan one of the major stumbling-blocks which casts unhealthy doubts on the notion of spirit itself.

De calido: Zabarella, Paparella, Caimo

Innate heat does not feature, on the whole, in our 'compendia' treatises.[39] But it was a topic of interest to natural philosophers, as we saw in the case of Cremonini; some of whom, including the three listed above, treated it in its medical context.

Jacobus Zabarella's solution to the problem of distinguishing celestial from elemental heat is both lucid and intelligent, and helps to explain Charles Schmitt's description of him as one of the most distinguished (and from the historian's point of view, most interesting) natural philosophers of his day. It comes in chapter xi,

"Quis sit calor coelestis distinctus ab elementari" of the treatise on celestial heat in his De rebus naturalibus. Both Aristotle and Averroes distinguished two heats: one prolific and celestial in origin, the other elemental and destructive. Yet in Meteorologica I iv, Aristotle explained that all elemental heat was produced by the heavens. Zabarella deals with this apparent contradiction, and with the question of whether heat is a quality or a substance, in an ingenious manner. He begins by distinguishing two classes of form: the anima or soul, which is the form of living things; and the elemental form of mixed, inanimate bodies. The former is 'superadded' to the elemental mixtio, as we have seen in Fernel, but both classes of form are able to be 'drawn out' of matter by celestial heat. So heat serves as a preparation of matter for the drawing out of form but once drawn out, it is an accident of the form itself. For instance, the heavens by their light and motion produce heat, and by means of heat generate fire; but once generated, fire generates more fire by means of this same heat. So heat always acts as secondary, or as instrument, both to the heavens and to the form of elemental heat which we call fire. A parallel process takes place in the generation of living things, which have within their seed a vis prolifica or principle of generation. This, like fire in the previous example, is made apt for generating a new creature by the instrumentality of heat.

Zabarella's solution has a number of virtues: it explains, quite plausibly, a puzzling contradiction in Aristotle's theory, and his equally puzzling 'element of the stars' passage; it provides a model for distinguishing between the generation of mixtures and of living things, while identifying a single heat in the two cases; and it explains how 'celestial' heat can have a role in elemental substances, which still can act according to their own form or essence.[40] Perhaps Zabarella's solution was too clever; it does not seem to have found many followers.

Sebastian Paparella's De calido is a much less subtle, encyclopaedic work, based in Aristotelian physics. Book I considers fiery heat, which he distinguishes into aetherial and elemental. However, these are different only in their rarity and density; the form of heat has great latitude, and these two sorts are only species within it. Furthermore, it is "our material fire" which is taken in by the body as food, and goes on to make the principles of generation. Paparella's reasoning here is like that of Argenterius and Riolan: the proper "working sphere" of elemental heat is in elemental substances.[41]

The second book, on natural, innate, or connate heat, is the one which chiefly concerns us; Book III treats fevers or praeternatural heat. The innate heat in living bodies is not, according to Paparella, a medium between soul and body (as the Platonists, and Fernel, claimed); rather, it is a mixed body, the offspring of fire, but differing from ordinary fire in being tempered or temperate. The source of this position, which Paparella says he will follow "without swerving a nail's breadth", is the Peripatetics.

Paparella quotes Fernel's definition of innate heat as "primigenial moisture, with indwelling spirit, perfused with heat", but much of what Fernel had to say about innate heat is in his judgment, "useless": for radical humour and spiritus are both formed from the elements; the latter is of the nature of air and fire, the former of air "well tempered with moisture".

Paparella distinguishes seven species of heat in the animal body: three, seed, blood and geniture, active in forming it, and four active once it is formed: spiritus, blood and the simple parts, which together make up the fourth, "heat so called".[42] Natural heat is thus elemental and corresponds to flame; but it is at the same time the ratio vitae, distinguished from fiery and feverish heat by its 'tempering' and suitability for the functions of the body.

Pompeio Caimo, or Caimus, we may remember as Cremonini's opponent, and like Cremonini the full title of his treatise De calido innato contains a hint of its purpose. Caimus's aim is to explain the nature of the innate heat as it is understood by medical writers, to give "solid arguments" for their position, and to defend Galen against the attacks of recent writers. Caimus tells us in his dedication and preface to the reader that after 25 years at Rome lecturing on philosophy and the theory and practice of medicine he returned to his native Padua and was raised to the first chair of theoretical medicine. Innate heat is a topic of immense controversy in academic circles, and one of importance to both doctors and natural philosophers; his purpose is to shed light on the subject, "making the implicit explicit, distinguishing the confused, spelling out the compressed, illuminating the obscure". In fact, his aim, as he explains it in chapter x of Book III, is not so much polemic as conciliatory: having taught both medicine and philosophy, he sees his role as smoothing the differences of opinion between his ancient authorities.[43]

The authorities which he seeks to explain and if possible to reconcile are the familiar four of Champier's Symphonia: Plato, Hippocrates, Aristotle, Galen. In the first chapter of Book I, he outlines the differences of position. For Hippocrates, the heat which was proper to humans was the instrument of the human soul; for Plato, similarly, it was "almost an instrument of God", infusing forms into chaos, active and creative.

Celestial heat was thus active in the sublunar world, and Hippocrates and the Stoics emphasised its power in procreation, generation and growth of all kinds. The Peripatetics, however, refused to accept the "Platonic, and Pythagorean" dogma that celestial substances could operate in elemental situations; but they too admitted the power of heat. Their conclusion was that there were two forms of heat: one active and celestial, one passive, subject to change and elemental.

The heat in mixtures is, as we have seen, temperamental, arising, says Caimus, from the mutual action and passion of the contrary qualities in the mixture.

But when in chapter three he turns to the nature of the heat in living things, Caimus concludes, with Fernel, that this is not temperamental heat (arising from the elements) but arises instead from the seed and menstrua, or the corresponding principles in the case of lesser living things. So the material of living heat is different from that of temperamental heat: and is distinguished by its heat and moisture. Living things require the presence of moist and 'oily' substances; the heat of inanimate mixed bodies can be dry.

Averroes, Argenterius and Fernel all argued for a distinction between vital and elemental heat, and one might have thought from the argument so far that Caimus would join them. But in chapter five he argues that innate heat is celestial not in its essence but simply in its origin and name. The essentia of innate heat is in fact elemental; but it differs from the heat of mixed bodies in its association with a particular moist and oily matter, the seed and menstrua. The innate or seminal heat is thus – as for Paparella – not 'fiery' and uncontrolled but tempered or 'moderated'; and it acts only in its own, proper subject, seed or its 'oily' equivalent. In chapter six, he uses these arguments, and his own interpretation of the element of the stars passage to set out Fernel's arguments for an innate heat, celestial in essence, and refute them.[44]

In Book II, Caimus applies his notion of innate heat more specifically to generation in man. He argues, citing Galen and Averroes on the importance of the testes as generative principles, that these organs, in male and (to a lesser extent) female are seats of innate heat. So, as we have already seen, is the seed itself, together with the menstrua.

The nature which is in the seed parallels, in its actions, the influence of the celestial intelligences on sublunar life; this analogy is the explanation of the 'element of the stars' passage:

> For just as life is communicated to inferior things by the celestial intelligences, so living things are brought forth by the seminal spirit, whose power is generative.

But the spirit is in effect instrument both of the seed, and of the heavens: "for to the power proper to the seed is added celestial virtue".[45]

In this chapter, and also for example in his explanation of the animation of the seed, Caimus seems closer to Fernel's position. But there are differences, and in some ways Caimus's interpretation, though different from Zabarella's, is similarly more logical than Fernel's and more effective in solving the problem with which Fernel himself was faced. It is based on careful distinction of two points:

(1) heat is a quality, and therefore must act in and through a substance; innate heat is distinctive because its substance, the seed, is distinctive. (2) the analogy made by Aristotle in the 'element of the stars' passage is an analogy, and not an identity: that is, it compares two different processes: the action of the heavens in influencing life and motion in general in the sublunar world; and the action of the nature and spirit in the seed in bringing forth particular living things, resembling the parents from whom the seed came.

De spiritibus: Delphinus, Bertacchius, Bronzerius

Spirits, unlike innate heat, were primarily a medical doctrine, and a fruitful source of controversy and confusion. The statements of Galen and Hippocrates on the number of spirits, their nature and uses, were cryptic and often contradictory, and Arab commentators complicated the picture still further. And moreover there were aspects of the opinions of Hippocrates, Plato and Galen on the subject – not to mention the Stoic writers and the pseudo-Aristotelian treatise De spiritu – which worried some medical writers in the Renaissance, as we have already seen in the case of

Argenterius, because they seemed to make it difficult to distinguish between spirit and soul.[46]

Julius Delphinius's short commentary on spirits, which forms a section of his Quaestiones Medicinales (1559) contains three questions: what is the substance of spirit? how many are there? are they animate, or not? To the first, he provides a standard definition, based on Aristotle: spirits are hot, active bodies and instruments of the soul. In his view, spirit is elemental; but it approaches the divine because of its special purity (as Galen also testified, in De usu partium VIII vii) and because, as Avicenna suggested, it is the most finely tempered of substances.

This suggests the answer to the third question: spirits (the animal spirit is the example he chooses) are not soul, but the instrument of soul, as Aristotle argued in de motu animalium. The example he gives is the transfer of voluntary motion from the (self-moving) soul to the muscles via the nerves. In his commentory on the Ars medica (1557), he similarly defines spirits as the immediate instrument of the soul and the mean between soul and body, and uses the examples of local motion and of the effect of maternal desires on the foetus to illustrate how "the anima is moved by a spiritual motion, the spirit by a corporeal, that is local, one."[47]

The second question is less straightforward, largely because the authorities are not clear. Galen distinguished three faculties: animal, vital, natural; for which Avicenna suggested three corresponding spirit-instruments, based, like the faculties, in the brain, heart and liver respectively. In de placitiis, however, Galen also suggested the existence of a generative spirit; and while brain, heart and testis all have structures suitable for the elaborating of spirit, the liver has only veins, suitable for producing blood. Delphinius offers several possible solutions to this problem: perhaps the heat from the heart is the instrument of the nutritive and alterative faculties as well as the vital one; or perhaps, indeed, the liver is the "principle of nutritive heat" but with blood, rather than spirit, acting as the instrument of the soul in nutrition. Delphinius does not make his position entirely clear, but he seems to associate generative spirit with the testis rather than the liver.[48]

Delphinius is interesting as an indication of the puzzlement, or perhaps the cheerful agnosticism, which could be created by Galen's spirits as complicated by Arabic tradition. Unlike Argenterius, he seems not to have questioned the anatomical evidence for the elaboration of spirits in the brain; and the fact that in the end he allows the possibility of a natural spirit, on authority rather than anatomical grounds, suggests that (as for Argenterius) anatomical arguments were not conclusive.

In the Preface to his De spiritibus, Dominicus Bertacchius tells his Candid Reader that he has spent the major part of his life in teaching and practising medicine. Exegesis on the question of spirits is needed because of the nature of the sources: Hippocrates is short and cryptic, Galen wordy but often tangled and involved. The result has been battles over mere words, as well as over ideas. Bertacchius's ancient sources are the standard four; he also mentions the Pseudo-Aristotelian De spiritu and the opinions of Democritus and the Stoics. Speaking generally, Galen is his chief authority, but it is not always easy to distinguish an opinion among the exegesis. De spiritibus is in effect a textbook, and of interest precisely for that reason.

His treatise is much longer, but the questions he asks are basically the same as Delphinius's:

1. What is the substance of spirit: that is, what is its definition?
2. Is it a body, or soul, the seat of soul or faculty, or its vehicle?
3. How many spirits are there, and what are their uses?

Bertacchius tells us that *spiritus* in Latin, or *pneuma* in Greek, has in the past been defined both as a body – subtle, tenuous and invisible, as are air, wind, breath, vapor; and as an incorporeal substance, either resident in a body (as soul) or separate from all matter, as God, celestial intelligences, demons. That's quite a range, but Bertacchius narrows it down: spirit, he argues, is a body, but it is "not air, or wind, or any other thing which in this manner is constituted outside the [living] body." It is thus associated exclusively with living things, and is most necessary to life and health; however, it is not soul.[49]

He takes his argument very largely from Galen, who divided the substances of the body into solid parts, humours and spirits; and as we have already seen in Fernel's summary, put forward arguments for the existence of spirit from anatomy and bodily actions. Bertacchius thus concludes, like Fernel:

> It seems, that spirit is a most tenuous body, and most apt to 'rush about' from one part of the body to another.[50]

Chapters iv and v deal with the argument that spirit is not the same as soul, and with an attempt to prove that Hippocrates and Galen did not really mean it on the several occasions when they seem to have said it was. In chapter vi, he concludes that the spirit is the seat of the faculties of the soul, but not its vehicle. Chapters vii–x deal more narrowly with the substance of the spirit: is it aethereal or elemental; *quasi* aethereal as Galen reckoned; similar to the substance of the heavens, as Avicenna suggested; or more divine than the elements, as in Aristotle? Is it a simple, or a mixed body? Surely the latter, as it arises from "concocted, commixted" bodies, blood and seed. But should it not be placed among the elements of the living body? Bertacchius, like most of our commentators, seems to conclude that it should.[51]

The remainder of the book deals with the number and *species* of spirits, and their uses. Both are the source of controversy, as we have seen, with the natural spirit at the centre of both debates. In Book III, chapters iv, v and x, Bertacchius opposes Actuarius, Altomari and Manardi on the natural spirit and its uses; in chapter xi, he disputes Argenterius on the question of the number of spirits.[52]

On the spirit in generation, and on the position of Fernel, he is less forthcoming; his usefulness to us is as a general guide to the arguments and authorities put forward in the debate.

The last of our selection of authorities, Joannes Bronzerius, speaks directly to the debate on natural spirit and, more importantly for our purposes, to the use of the *spiritus* in generation and Fernel's argument for a 'celestial' heat. The first three chapters of his *De innato calido et naturali spiritu* are used to argue against the existence of a natural spirit. It is not mentioned in Galen, and not necessary either for such actions as nutrition, generation and blood formation; and so Bronzerius argues, like Argenterius (whom, however, he also criticises) for one heat, and one spirit: i.e. the vital spirit.[53] In the remainder of the treatise, particularly chapters iv and ix, he turns to the origin of heat and spirit in the seed and its uses in generating and preserving the human body.

In chapter iv, he argues that the vital spirits establish (constituere) the innate heat in the seed, and in the formed foetus, and that from the three archigones or centres of organ development described by Galen, by means of spirit, blood and seminal moisture, the solid or seminal parts of the body are formed. Spirit, heat and moisture remain in the foetus to preserve it, not dissipating into vapour as Aristotle argued. In chapter vi, Bronzerius defines innate heat and its relation with vital spirit and seminal moisture:

> innate heat is a hot substance, from the mixture of vital spirits and seminal moisture, suitable for the accomplishment of the functions of the soul, when aroused and fostered by the heat, which enters from outside.[54]

Here Bronzerius seems to be arguing for heat as the agent of faculty, and for two forms of heat, which differ in their origins: however, like Caimus, he distinguishes the heat in living things and the heat in elemental mixtures not by their essence but according to the special material with which the former is always associated. He will not dispute, he says, whether innate heat is aethereal or elemental; he has argued 'elsewhere' that it is elemental, and he has the majority on his side:

> In which opinion I find all others zealous for truth, only excepting Fernel and a few lovers of novelties.[55]

What is Bronzerius' own opinion, and how does it compare with those of the other writers, presumably also 'zealous for truth', whom we have examined? He agrees with Fernel in some things: both in arguing (like Caimus) that the substance in which innate heat inheres is special – peculiar to living bodies – and in his recognition of the origin of both the spiritus and the seminal moisture (or radical humour) in the seed. He quotes Fernel's definition of radical humour with approval. He is aware, however, that if both spirit and radical humour are vehicles of heat, there is a problem of redundancy:

> Now if heat is in spirit, and spirit in humour, as the testimonies of Aristotle and Galen agree; I accept without doubt that with humour as the subject of heat, spirit seems to be excluded.[56]

In chapter ix, he refers back to his definition of innate heat, as 'excited' or 'roused' by elemental heat: and to his insistence that elemental and innate heat are not two 'layers', one super-added to the other, as in Fernel, but are one in subject and in form. But the two are not quite the same: the difference arises from the presence, or absence of soul, or form: without soul, the heat of mixed bodies contains only the powers of mixtio with which it is provided; it is not 'vivifying'. When soul, or form, is added, it requires an instrument or material corresponding to it; this is not, however, aetherial but the blending of spirit, seminal moisture and menstrua, to make up 'the substance of heat': a point which reminds us of Paparella.[57]

In fact – as Bronzerius tells us in chapter ix – the name 'innate heat' has been transferred to spirits and to blood, from their presence in the body in these 'primordia of generation' and from the intense and powerful heat which the primordia contain.

Unlike Argenterius, Bronzerius is not reluctant to associate this spirit with faculty, though not with its essence; as he concludes,

> one must conclude that vital spirits are indeed a cause, without which the power indwelling in the parts could not exercise its proper functions; but which does not extend to the major principle of operation, or to the constituting of the essence of that same faculty.[58]

Seed is the instrument of form, and of innate heat; but innate heat is not anima or form. Heat and spirits are a cause, not only of life, but of life-functions and the vigor, as Bronzerius calls it, with which they are performed.

Conclusion

It seems that Bronzerius was right to suggest that Fernel was in a minority on the question of the celestial origin of spiritus. Of the writers we have just considered, none mentions this opinion except to refute it, although other doctrines, e.g. the radical humour, met with more approval. Argenterius was particularly successful in pointing out some of the weak points in Fernel's theory of the spiritus which later writers took up. In effect, he places Fernel in a Morton's fork position. Either the innate heat and spiritus are celestial in their powers, in which case they duplicate the function of the soul and usurp its place, with clearly dangerous implications. Or else they are not – and considering that they are nourished by elemental substances, this seems more probable – in which case they lose their claim to be considered as superior to the other substances of the body.

William Harvey took a similar position almost a century later. For him too Fernel's aetherial spiritus was both unlikely and unnecessary. Writing in the Generation of Animals of 1651, he remarked that

> Fernel and others have invented a spirit of celestial origin and nature, namely a body most simple, most subtle, most fine, most mobile, most swift, most lucid; aethereal and participating in the Quintessence. But nowhere have they demonstrated that there is such a spirit, or that it acts beyond the power of the elements, or that it performs greater works than blood alone could do. We indeed, who in our investigations use sense as our guide, have not been able to find any such spirit anywhere.[59]

But for Harvey, all the powers and excellence which Fernel had placed in the spiritus dwelt instead in the blood itself. Hence his claim that Fernel's spirit can perform no 'greater works than blood alone could do'. As an anatomist, he was concerned to confine his attention to substances which could be perceived by the senses. And it is hardly surprising that he failed to dissect out the spiritus. But for Fernel, it was almost axiomatic that the real causes of both the body's structure and its activities should be hidden. Elements and temperaments, humors and spirit, the soul and its faculties – all these, as he reminds us at the beginning of his section on the elements, can be "discerned only by thought". Such hidden causes, of which the soul is the ultimate example, are 'visible' only through their effects. They cannot be anatomised. Anatomy gives, in any case, only an idea of the structure of a thing. But 'that which makes a living man more excellent than a dead man' is not anatomy. It is not simply a matter of more complex organisation. There is a vital, dynamic component as well.

In his own terms, then, and despite the obvious flaws in his argument as pointed out by later writers, I think Fernel makes a good case, in his own terms, for the necessity of a spiritus, and for its association with a power beyond the elements. If the elements and temperaments are made responsible for the physical form of a body, there is still something missing, which is its life. To explain this, it becomes necessary to bring in a power 'which acts beyond the power of the elements', and which indeed performs greater works than blood, considered as a simple elemental compound, can do.

But as we have seen, Fernel is rarely successful in practice in distinguishing celestial areas of action from material ones. His account of the spiritus leads him repeatedly into ambiguous and contradictory statements. If the spiritus is truly celestial in its nature and origins, how can it be nourished by elemental substances, and why should it need such nourishment? How can it be identified with those medical spirits which are elaborated from the blood? How is it that the spiritus in the semen is elaborated in the same way and by similar instruments as the material part? Why, most importantly, are the spiritus and the vital heat unable to act beyond the capacity of the material substances with which they are associated?

In part, contradictions such as these seem to be a result of the attempt to combine parallel systems, one or another of which becomes superfluous. For example, the liver produces both venous blood and bile, or melancholy. The veins which transmit both carry at the same time the Galenic natural spirit. Similarly, the arteries are crowded with arterial blood, humors, vital spirits, and finally the innate heat with its aetherial vehicle. The Galenic system is becoming seriously overloaded.

The concept of spiritus itself suffers from a similar piling on of functions. For instance, Fernel tells us that the semen is endowed with a divine and celestial heat, which inheres not in the material semen itself but in an aetherial spiritus 'mixed' with it. But this spirit is also the seat and first acting instrument of the sensitive and nutritive souls. Unless the celestial heat can be identified with these aspects of the soul, the spiritus is instrument and carrier for two powers at once: literally doing two jobs at the same time.

One of the most interesting questions raised by Fernel's discussion is that of the relation between the vital heat and its vehicle and the soul with its faculties. To critics such as Argenterius, the image of the astral body seemed to imply a confusion between the Neoplatonic soul and its aetherial 'garment' and the celestial heat with its aetherial 'vehicle'. It is a debatable, and perhaps unanswerable, question whether Fernel would have accepted such an interpretation of his use of the astral body. In the Medicina especially he is rarely so philosophically precise or dogmatic. Perhaps the celestial heat was intended to pass as a 'likely story', sanctioned by the authority of Aristotle and other classical writers, with which to confront the mysteries of life and generation.

However, there are aspects of Fernel's argument which suggest that the vital heat must logically be regarded as an analogue of the soul in some sense. Its activities in generation are exactly similar to those of the anima, except for the separate infusion of the mind. In all other areas, the anima tends to be pushed out of the picture in its favour. And since the anima and celestial heat share the same 'first instrument', the spiritus, they are in a sense doomed to perform identical actions. This redundancy is a serious problem in Fernel's work. He never resolved the contradiction, and it was left as a thorny question to plague his successors, philosophers, doctors, and ultimately even historians.

Notes

1. Fernel cites, among others, the stoic concept of pneuma, the 'astral body' of the Neoplatonists, and the medical spirits of Galen. But his notion of where and how the spiritus operates seems to derive very largely from Aristotle, despite its eclectic sources. For a comprehensive account of the origins of spiritus in Stoic, medical and Neoplatonic thought, see G. Verbeke, L'évolution de la doctrine du Pneuma du Stoicisme à S. Augustin, Paris, 1945, especially chapters I, II and IV.
2. Physiologia IV i pp. 102–3.
3. Verbeke, op. cit. chapter 1; quotes from Book II, 23–26 of Cicero's The Nature of the Gods, (pp. 132–133 of the Penguin Classics edition) an account of Stoic philosophy which may have been one of Fernel's sources; another was clearly Aristotle.
4. Galen, De respiratione, Kuhn, VII p. 674; translation from R. E. Siegel, Galen's System of Physiology and Medicine, Basle and New York, 1968.
5. Physiologia IV ii p. 104: "fuit opinor necessarium hunc corpore aliquo fluxo et profluente contineri. Caeterum nullus humor ad hoc aptus erat et habilis, ut tante celeritate corpus omne traiiceret: quocirca necesse fuit calori materiam substerni tenuissimam, pernicitate velocem, quae simul fovendo calori familiaris esset ac amica."
6. Ibid. "Quo autem spiritus necessitas atque substantia plenius demonstretur, repetenda sunt et replicanda veterum philosophorum decreta: quorum primi Academici cum perspicerent fieri non posse ut naturae quammaxime dispares, nullius idonei interjectu societatem inirent coirentque, animum nostrum a summor e rerum opifice conditum, priusquam emanaret immigraretque denso huic et concreto corpori, censuerunt corpore quodam illustri, puro et astro simili tanquam simplici veste indui: quod immortale et sempiternum nunquam ab animo absolui divellique posset, et sine quo non fieret huius mundi incola. Ab hoc deinde alterum corpus animo circumjecerunt, tenue illud quidem et simplex, sed tarnen impurius, minus illustre & splendidum superiore: non ad summo id opifice procreatum, verum elementorum praesertimque tenuiorum permistione concretum, a quibus nomen inveniens aereum et aethereum appel-

latur." See also Walker, "Astral body", for the translation and an analysis of this passage.

7. For Fernel's references to Alexander, see Ibid. p. 105 and D.A.R.C. II vii p. 187. The latter also contains a definition of spiritus in terms not of its substance but its powers: "Quaecunque substantia intelligitur corporis expers, nostrosque sensus effugiens, ea. suas vires nobis caducisque corporibus provehente spiritu impartit". (Ibid. p. 186).
8. Physiologia IV ii 105: "quare necessum est. aerem hauriri et attrahi, ut finitimae & cuipiam in nobis sui generis substantiae jungatur, eique velut in pastum cedat. Si nulla in nobis esset tenuis et spirituosa substantia, vix ulla profecto nos ad inspirandum necessitas impelleret."
9. Physiologia, IV iii, pp. 106–7. "Nihil denique in stirpium aut animantium genere, vitam habuit, quod not possit oleum de se profundere."
10. Ibid. iv p. 108 "spiritus quem in nobis inesse demonstravimus, quia tenui est. & aetherea substantia insito calore perfusa, ei proportione respondet quae proprie flamma appellatur."
11. Ibid. IV vi 112: "id humidum quod radicitus inhaerens vitale et primigenium appellavimus.."
12. Ibid. chs. v and vi.
13. Ibid. vi 112.
14. Ibid. IV, pp. 119–121.
15. Ibid. VII v 228: "Nam (inquit Aristoteles) omnis animae sive vis sive potestas, alterius cuiusdam corporis particeps esse apparet, eiusque divinioris quam quae elementa appellantur. Verum quemadmodum nobilitate obscuritateve animae inter se discrepant, ita & natura eius corporis differt. Continet enim in se semen cuiusque, foecunditatis suae causam, nempe ipsum calorem, qui igneus minime est., neque id generis facultatem aliquam aemulatur: sed spiritus qui in semine spumanteque corpore coercetur, & natura quae in eo est. spiritu proportione respondet elemento stellarum. Quamobrem ignis nullum animal generat, neque constitui quicquam densis, vel humidis, vel siccis videtur. At vero solis calor & animalium, non modo qui in semine continetur, verumetiam si quid excrementi sit, quanquam diversa a natura, tamen id quoque principium habet vitale. Caeterum calorem in animalibus contentum, nec ignem esse, neque ab igne originem ducere, apertum ex his est."
16. D.A.R.C. II vii, p. 193. "Ut quatuor simplices natures elementa vocamus quod in rerum omnium concretionem confluunt, illisque cedunt in materiam: ita coelum ac stellas quod suas etiam illis vires adhibeant ac impertiant, elementa jure appellemus. Si in primam animantis procreationem, quae ex semine, ex maternoque sanguine sit, animum cogitationemque figes, animadvertes sanguinem ex quatuor elementis qui in cibis inerant, constitui: ex eisque materiam seminis, quam purior utiliorque sanguis contulit. At naturam, spiritum, & calorem quos semen in se comprehensos continet cum aliunde prodiisse memineris, intelligis supra elementorum vires & ordinem esse, neque posse illorum praestantia vel ad caduca haec et sordida elementa, vel ad temperamenti proprietatem pertinare. Partium humani corporis subjectam materiam ab elementis, & ab tempera-

mentis: at facultates, spiritus, insitum calorem, & figuram, ab eo seminis spiritu, in quo tum calor, tum natura, visque divina insidebat proficisci."

17. Ibid. VII, chs. iv-vi.
18. Ibid. especially ch. vi p. 229; paraphrasing Aristotle, G.A. II i 734 b. "At propria cuiusque ratio et essentia, qua idipsum est. quod esse dicitur, nequaquam ex elementis emergit. Quemadmodum enim hominis essentia atque ratio, quae illius est. anima, diviniorem quam qui ex elementis est. ortum habet: sic et essentia ratioque carnis qua caro est., aut ossis qua os est., neutiquam ab elementis his densis atque concretis, sed ab elemento stellarum plane coelestem naturam sumit: & haec in coelesti quodam spiritu consistit, hic autem spiritus est. in semine, per quod genito toti impertitur. Ad eundem prope modum tum reliquorum omnium animantium, tum stirpium propria natura coelesti ortu profertur. Hinc facultates quasdam obtinent occultiores & reconditas, quas etiam coelestes appellant, nequaquam ex elementorum temperie profectas." For Fernel's definition of soul as "principium et causa functionum viventis corporis", cf. Book V, i p. 122.
19. D.A.R.C. I iv pp. 45–50.
20. Physiologia VII v 227 for the role of spiritus in transferring the 'motion' of the parent to the foetus; chs. ix-xi, pp. 237–244, for order of development; also ch. xiii, 245–250, for consideration of debate on order in which faculties are developed in the foetus, and whether the male seed forms part of the embryo or not.
21. Ibid. VII v 226–229.
22. Ibid. 227–8; also ch. xiii, p. 247: "Hic [i.e. the spiritus] a corpore inseparabilis est., nec per se potest solus constare, neque anima cuius est. sedes.."
23. Walter Pagel, Paracelsus: An Introduction to Philosophical Medicine in the Era of the Renaissance. Basel and New York, 1958, pp. 301–4; Temkin, Galenism, esp. 141–2, 150–151. Argenterius' works include: De Erroribus Veterum Medicorum, Florence, 1553; In Artem Medicinalem Galeni commentarii.., Paris 1553, Monte Reggio 1556; The Commentario in Hippocratis Aphorismam primam, secundam & quartam was published as part of the Opera of 1556. So was De Febribus tractatum singularum: & primi libri ad Glauconem praeclaras explanationes. The short appendix on native heat which follows this treatise is almost the same as the corresponding chapters in De somno et vigilia. His earliest publication, Varia opera de re medica, Florence, 1550 was reprinted in 1557 as De Morbis Libri XVI. Other treatises on therapeutics include De Consultationibus medicis sive de collegiandi ratione, Florence 1551.
24. Joannes Argenterius, De somno et vigilia.. Florence, 1556; 298–302. p. 302: "Sunt ergo nugae, fabulae, & Stoicorum commenta, quae de tot spiritibus dicuntur, et literis mandata a maioribus nostris, ac a nobis postea absque ratione suscepta, & celebrata."
25. Ibid. pp. 58–9. "Calorem asserere esse naturam nostram, et illam quae regit solida corpora, animaeque substantiam, est. ignorare formas rerum distinctas esse a suis instrumentis.."
26. Ibid. Bk II, vi, 273: "..Fernelius, quum admirabiles viderent spirituum effectus, eorum substantiam non corpoream, neque ex elementis, aut alio quod habeat

inferior hic mundus, notam, sed divinam, et coelestem esse, ..corpus aethereum, caloris sedem, animae et corporis vinculum, primumque actionum instrumentum.."

27. Ibid. p. 276 "Est in corpore viventium aliquid divinum, nempe anima, sed propterea non sit, ut corpus universum, quod vivit, aut pars corporea, quae in eo est., ex coelesti materia sint confecta."
28. Ibid. pp. 288–9.
29. Ibid. p. 275 ("Quo pacto quod coeleste est., nutrimento eget?")
30. Ibid. p. 289 "Haec et similia illi doctissimi quidem viri, sed (ut arbitror) inveterata opinione, & Graecorum authoritate nimis persuasi, a communi hoc errore se nequaquam liberare potuerunt:"
31. Ibid. II, xi, p. 318: "..probavimus, spiritus nullam in se facultatem animae continet, nec vehiculum est. illarum, sed ipsius caloris, quo anima in corpore suam vim, et functiones exercet."
32. Ad Librum Fernelii de Spiritu et Calido Innato, Jo. Riolani Commentarius. Paris, 1576. chapter 2, 6v. "Sit ergo spiritus, corpus aethereum, caloris ac facultatum sedes & vinculum, primumque obeundae functionis instrumentum.."
33. Ibid. chapter 1, 3v-4r: "Iste calor immixtus miscet, nam mistilia prius alterari necesse est., alterantur actione & perpessione mutua, coelestis non repatiuntur agendo, quia nihil habent commune cum elementis..Ergo, dicet quispiam, duplex in viventibus calor, unus coelestis, alter elementaris, quidni? quia ut philosophi loquuntur, non sunt entia sine necessitate multiplicanda, unus autem ad numerum calor, varie affectus, potest effectus quantumvis differentes,... Calor siccus densat, humidus laxat, immoderatus corrumpit, temperatus conservat.."
34. Ibid. 4r: "Coelestis non nutritur sit enim nutritio assimilatione, assimilari autem non potest elementare coelesti, estque proprium caducorum nutriri...atqui nativus calor inopia pabuli marcessit, quomodo ergo coelestis?"
35. Ibid. 4v: "Est igitur in viventibus calor duplex. Unus agens & formalis, qui elementarem varie permiscet: non tantum analogia...sed origine & essentia coelestis, qui tamen in elementari mundo quodammodo fiat elementaris...alioquin non est. facile istum scrupulum, quod calor coelestis nutritur, eximere."
36. Ibid. chapter 3, 7v: "calor nativus revera erit elementarius, coelesti tantum analogus."
37. Ibid. chapter 6.
38. For an account of Cartesian and later explanations of innate heat, see June Goodfield, The growth of scientific physiology, London, 1960.
39. Although it is discussed by, for instance, Montanus (Medicina universa, ch. xxi, pp. 21–22; Donatus Antonius ab Altomari, "De Alteratione..ex Hippocrates et Galeni sententia Methodus", pp. 20–21 in Opera Omnia, Lyons, 1565; and Cardanus, Contradicentium Medicorum, I i chs. 10 and 11. For a valuable survey of sixteenth century opinions on the relation of elemental, innate and praeter-natural heat, based on treatises about fevers, see Iain Louie, "Fever pathology in the sixteenth century: orthodoxy and heterodoxy", a paper presented at the Wellcome Symposium on Theories of Fever from Antiquity to the

Enlightenment, 13 June, 1980; to be published, with other symposium papers, as a supplement to Medical History.

40. Jacobus Zabarella, De Rebus Naturalibus, Cologne, 1590, "Liber de calor coelesti", ch. xi, pp. 522–3; Charles Schmitt in D.S.B. XIV, pp. 580–2; Temkin, Galenism, pp. 143, 148.
41. Sebastian Paparella, De calido, Perugia, 1573, I xii and xiii, 14r-17r.
42. Ibid. II, xxvii-xxxi, 62r-71v.
43. Pompeius Caimus, De calido innato.., Venice, 1626: Dedication (unpaginated); Preface, pp. 1–2; III x 408–432.
44. Ibid. ch. iii, pp. 35–39 especially. Caimus here takes his evidence from the Galen of De temperamentis, De semine and De sanitate tuenda, and the Hippocratic generation works; later (p. 45) he also cites the Collectanea of Averroes on the distinction between innate and temperamental heat. Chapter v, pp. 61–79; vi pp. 80–94. Quoting the Greek text and his own translation, and citing Plato and Plotinus for authority, he argues that analogous substances are in fact different. He also cites Averroes, XII Metaphysics 18, for evidence that seminal or innate heat is not fiery because "non est. absolutus, et immodicus calor, sed temperatus..."; the same argument that we found in Paparella.
45. Ibid. II, viii, pp. 249–262 (testes as subjects or seats of innate heat); xi [error for ix] pp. 262–283. "veluti enim a caelesti intelligentia communicatur inferioribus vita, sic a natura contenta in spiritu seminali, quae vis est. generatrix, progignuntur viventia…ad proprium seminis virtutem accedat etiam virtus caelestis." (pp. 267–8).
46. The other major authority, especially for the actions of spirit in generation, was, as we have seen, Aristotle, in particular the 'element of the stars' passage from De generatione. On ancient pneumatology see, in addition to the references by Peck, Temkin and Verbeke already cited, Friedrich Solmsen, "The Vital Heat, the Inborn Pneuma and the Aether", Journal of Hellenic Studies, 1957, 77: 119–123.
47. Julius Delphinius, Quaestiones Medicinales, Venice, 1559; "De spiritibus", 80v-81v, 86r–v; In III Galeni artis medicinalis..Venice, 1557, p. 19: "quod anima spirituali motu moventur, spiritus vero corporali, idest locali.."; p.20 for means by which "signa in foetibus appareant propter matris appetitum." Once again a mental alteratio causes a physical one, through the medium of the spirit and the blood.
48. Ibid. (Quaestiones) 82v-85r.
49. Dominicus Bertacchius, De spiritibus libri IV, Venice, 1584; ch. ii, 1v: "nec aer erit, nec ventus, nec alia, quae extra eiusmodi viventia sint constituta." As with the innate heat of Caimus, Bertacchius' spirits occur only within living things.
50. Ibid. ch. iii, 1v-2v: "videtur, spiritum tenuissimum corpus esse, et ad excurrendum ab una in alia corporis partem maxime idoneum" (2v).
51. Ibid. chs. iv-vi, fol. 3v-5r.; vii-x, 5r-8r.
52. Ibid III iv-v, x-xi. Joannes Actuarius (fl. 13th c) was the doyen of commentators on the spirits: his treatise, which was published in the Renaissance, deals mainly with animal spirits. Argenterius, as we have seen, argued for a single

spirit, corresponding to Galen's vital one; Bertacchius argues for the traditional three, including the natural spirit added by Arabic commentary. The controversy (or confusion) as to its uses was outlined by Delphinius.

53. Joannes Bronzerius, De innato calido et naturali spiritu disputatio. Padua, 1626, chs. i-iii, pp. 1–33. The place and date of publication, mutual advocacy of Galen and other details of the argument suggest a link with Caimus, but I have not established the connection. Bronzerius' other treatise, De principio effectivo semini, is considered in the next chapter, although their subjects, and their arguments, overlap to some extent.
54. Ibid. ch. iv, 34–49, vi, 55–59: "calidum innatum sit substantia calida ex mixtione vitalis spiritus, atque humidi seminalis, animae functionibus exequendis accommodata, si ab alio superveniente calore nativo suscitetur, et foveatur." (p. 56) The native heat and innate heat here seem to be distinguished; for some indication of the complications which the definition of heat called forth, see Lonie, "Fever pathology", op. cit.
55. Ibid. p. 56: "In qua sententia, uno excepto Fernelio, et paucis novitatum amatoribus, plerosque alios omnes veritatis studiosos concordes reperio." 'Elsewhere' I presume refers to De principio effectivo.
56. Ibid. ch. vii, p. 67: "Quod si calor in spiritu, et spiritus in humor, concordi Aristotelis, et Galeni testimonio; procul dubio accepto humido tanquam caloris subjecto, spiritus videtur excludi."
57. Ibid. ch. ix, pp. 80–88; referring back to vi, 56–59.
58. Ibid. p. 83: "concludere oportet, vitales spiritus esse quidem causam, sine qua vis partium insita nequeat proprias functiones exercere; sed quae ad principale operationum principium, adeoque ad ipsam facultatum essentiam constituendam non pertineat."
59. William Harvey, De Generatione Animalium, 1651; quoted by Walker in "Astral Body".

Chapter 16
The Soul in Generation and the Animation of the Foetus

Abstract The role of the soul in generation was a subject for controversy, rooted in Aristotle's definition of *anima*. The classic questions concerning the soul in generation, which we find in Fernel, his contemporaries and successors, arise from this definition of *anima*. Put briefly, they are: is the seed animate, or is it an instrument of an "external efficient" as Cremonini called it? If the latter, is this external cause the parents, God, Nature, the heavens? Is the *anima* (as form) the same thing as *species*? And, related to this, how are the faculties or actions of a complex living creature such as a man, composed of parts of different forms or *species*, integrated into a functioning whole?

These questions are discussed by Fernel in his *Medicina* and by others. The *de principio effectivo semini insito* of Joannes Bronzerius asks "what makes semen fertile?" Archangelo Piccolomini's *Anatomicae Praelectiones* is much broader. Scaliger's discusses the soul in generation in his *Exotericarum Exercitationum* (1557). What is clear, in this all too brief and sketchy survey of Renaissance ideas of the soul in generation, is that Aristotle's *De generatione animalium* is probably the most important single authority for the Renaissance debate.

Even more than the nature and uses of the innate heat and spiritus, the role of the soul in generation was a subject for controversy. It was a complicated issue, involving the authority of theology as well as of medicine and natural philosophy. In this chapter, I have concentrated on medical contributions to the debate in relation to Fernel in particular. Philosophically, the controversy was rooted not so much in Church doctrine as in the interpretation of Aristotle's definition of anima, which presented certain difficulties when applied to the phenomenon of generation; and in attempts to reconcile this with the cryptic statements of Plato, Hippocrates and Galen on the subject.[1]

Aristotle's definition of anima as the form of the body implied that it could not exist in the foetus before organisation at a certain level, as we saw in Cremonini's arguments in Part II.[2] The anima rationalis could not precede rational thought, according to this argument (raising the interesting question 'at what point does a child become rational?'); nor, in a more familiar and straightforward example, could the faculty of walking come before feet. Yet Aristotle's definition of anima as the self-moving, organising principle of the body suggested that it should be present in

L. Deer Richardson, *Academic Theories of Generation in the Renaissance*, History, Philosophy and Theory of the Life Sciences 22,
https://doi.org/10.1007/978-3-319-69336-1_16

the geniture from the moment of conception: or that if it was not, some other means of forming the foetus in the early stages would have to be found. In De generatione, Aristotle attempted to get around this problem in two ways: by positing a hierarchy of souls which developed successively in the foetus and suggesting that those souls which were not yet active were present in potentia; and by his notion of instrumentality, according to which the seed transferred 'motion' from the soul of the male parent, acting as his instrument in the formation of the foetus. The contradiction in this argument, between soul as something whose presence signalled a certain level of organisation, and soul as the 'mover' for bringing this organisation about in the first place, reappears in the Renaissance debate. Furthermore, both soul as form, and soul as the origin of self-movement, were difficult to reconcile with the Christian requirement for an immortal soul. The usual solution was to identify the immortal soul with Aristotle's anima rationalis, "entering from the outside". In the Christian version, it was infused into the foetus, usually at the fortieth day, by which time the embryo was supposedly already formed and sufficiently organised to receive it.

The classic questions concerning the soul in generation, which we find in Fernel, his contemporaries and successors, arise from this definition of anima. Put briefly, they are: is the seed animate, or is it an instrument of an "external efficient" as Cremonini called it? If the latter, is this external cause the parents, God, Nature, the heavens? Is the anima (as form) the same thing as species? And, related to this, how are the faculties or actions of a complex living creature such as a man, composed of parts of different forms or species, integrated into a functioning whole?

The basic question: 'is the seed animated?' was the focus of two major debates in the course of a century. In the first of these, Thomas de Garbo, son of Dino, used the final chapters of his Summa Medicinalis, a compendium on the Conciliator model published in 1529, to argue the question 'whether the spiritus existing in the animal body is formally animate'. He did so in the context of generation, and argued, against his opponents, Joannes de Penna and Franciscus de Zanellis de Bononia, that the seed was not formally animated.[3]

Almost a hundred years later a similar debate developed, with Fortunius Licetus and Thomas Fienus both arguing, on somewhat different grounds, against the orthodox theological opinion put forward by Ponce de Santa Cruz, physician to Philip of Spain; Ponce in turn was supported by Louis DuGardin.[4]

Like the debate of female seed, which is somewhat resembles in its narrow focus and the virulence of the combatants, there were more than two possible positions on this basic question. The orthodox position, in which the foetus was assumed to be animate from the fortieth day, was, as we have seen, difficult to reconcile with the Aristotelian definition of anima as a self-moving principle. Fienus mercilessly probes the absurdities to which this argument leads, but his own opinion, according to which the soul enters on the third day (not the fortieth) presents exactly the same problems: what forms the foetus for the first three days of its development?

The difficulties in pushing the point of animation backwards to conception or beyond were theological as well as logical: logically, a human soul required a certain degree of organisation, which was not present in the geniture; theologically, there was the question of the soul's immortality to consider. Neither Licetus nor

Fienus argued that the seed was animate when formed in the body of the parents, or even when emitted – despite DuGardin's somewhat hysterical suggestion that, if we admitted the animation of the geniture, every ejaculation would require baptism[5] – but Fortunius Licetus came close enough to make more orthodox colleagues anxious. His argument was that a unitary soul exists in the seed from the beginning, "contained" in actu within the seminal matter, not in potency: but with its faculties in cold storage, as it were, until needed. This solution emphasised the soul as mover at the expense of the soul as organisation; and it led Licetus to suggest a soul which was not simply a way of talking about the vital, facultative aspects of a living thing: soul as life – but a soul which sat within the geniture (in Fernel's phrase) like the helmsman within his ship.[6]

Thomas Fienus's solution, as we have already suggested, was similar: seed as seed, before its reception of soul in the uterus, has no active virtue, nor powers of conformation, except as materia. This conclusion, he tells us, is a 'paradox', going against the opinions of philosophers and doctors, both of whom allow an active principle in the seed of one, or both, parents. The rational, immortal soul, infused into the conceptus on the third day (the time when the foetal membranes are first observed) both forms it and is its form: another paradox, admits Fienus, but 'most true'.[7]

Another possible solution was to suggest that the soul of the parents organised the early development of the foetus, before it received its own self-driving principle of movement; the powers of the parental soul might either be transmitted through the seed or be present in the seed itself; or the foetus might be considered as forming part of the mother.[8] The question raised by this argument was the same: when did the foetus take on responsibility for its own development? Most writers assumed that the infant's dependence on the maternal soul did not last until birth.

The action of the soul in generation was not a question confined to specifically polemical tracts of this sort. It is the theme of the early treatise, De virtute formativa, (1506) by the medical humanist, Nicolaus Leonicenus. This is a "conciliatory" work, setting out the arguments of the major authorities in the debate, with particular emphasis on the late Greek commentators, which most later Renaissance writers virtually ignore. It was the subject of specialist monographs, such as those of Bronzerius and Schegk, mentioned below; and appeared as a major question in theoretical treatises of generation, in Latin and the vernacular, and in even more general medical works.[9]

With the obvious exception of Leonicenus, these examples are later than the Medicina and De abditis rerum causis, and most of them take Fernel's complex and somewhat unorthodox position into account in developing their own argument.

How, then, did Fernel tackle the problem of defining the soul, and its action in generation? How did he deal with the questions which, we have suggested, were important in this debate:

- is the seed animate, or instrument of an external cause?
- what is the relationship of anima and species, and of both to 'total form'?
- how are the faculties integrated into a functioning whole?

"Not to the body alone, but to the man, who is composed of soul and body, we turn our thoughts" wrote Fernel at the beginning of Book V of the Physiologia, "De Animae facultatibus". In fact, despite his suggestion that we are in for something new, a 'value added', Fernel's definition of the soul is something of a paradox, taking us back to questions – 'something divine' evident in the body's function, the role of spirit – which we have discussed in earlier chapters.

In the first place, as Fernel reminds us at the beginning of Book V, the soul is the ultimate 'hidden cause'; "hidden and unknown to the senses", it can only be understood by its works, which are knowable to the mind and perceptible by the senses.

So Fernel's definition of soul (like his definitions of temperament, and of spiritus) is ultimately based on its effects: "there is something in a living man by which he is nobler and more powerful than a dead man, and by which he performs his functions". The soul is the "cause in us of a certain stable and constant life-activity"; it is, as we have already seen, "the principle and cause of functioning of the living body".[10]

The anima is associated, therefore, with the functions which all living things perform: nutrition, for example, or generation. And to say that a creature is organised (or possessed of organic parts) is another way of saying it possesses anima: since, as we have again seen, anima is for Fernel associated with the highest level of 'organisation', in which the simple and composite parts are 'harmonised' into an organic whole.

Anima is also, as for Aristotle, and Cremonini, an indication of species: as creatures differ in their level of function, so do their animae; and so one can imagine a scale or ladder of animae, with the simplest creatures, barely distinguished from inanimate bodies, at one end, and man at the other. There are three major species of soul: natural, sentient and intelligent: corresponding to Aristotle's nutritive, sentient and rational souls, and to the three great categories of living things, plants, animals and man.

These same names: natural, sentient, intelligent: apply to the parts of the human soul. But Fernel is at pains to explain that despite this verbal distinction, the human soul is a unity; the natural and sentient functions are incorporated in the intelligent soul, just as the figures of a triangle and tetragon fit inside a pentagon.[11] This analogy is recalled at several points in the Physiologia and De abditis rerum causis, with reference to the development of the three souls within the foetus. Together with the heat and spirits, the natural or nutritive and sentient souls (or parts of the total soul) are present in the semen. But they are there only potentially, not in act. As Fernel explains at some length in Book VII, they cannot be the soul of the semen, which is not a living individual, nor the actual soul of the child, which does not yet have organs capable of nutrition or sensation. But following conception, these faculties, which until now have lain dormant, leap into action and show themselves by their powers. The faculties, like the celestial heat, come not from a mixture of the elements, but from the parents, in whom they formerly existed in act. And they are transmitted from one generation to the next by the semen, and especially by the spiritus within it. And so, says Fernel, a man who generates, producing semen, produces at the same time all the faculties of the human body. But he is not responsible for generating the human mind, associated with the highest, immortal portion of the soul. This is infused full-blown into the embryo when its bodily development is sufficiently advanced.[12]

Each of the lower 'souls' is retained as part of the total form (or soul), in the same way as the forms of the simple parts are retained within the organic parts of the body.

The mind, however, has a different origin, and this has caused certain philosophers to argue that the soul cannot be simple, consisting as it does of an immortal and a mortal part;

> The former as it were separate and simple, does not mingle with the body, but at most is present within it as the helmsman within his ship; the latter indeed is thoroughly immersed in the body, from which it cannot be loosed or torn away. The former does not need the assistance of the body; the latter can do nothing without the suitable preparation of the body and its organs.[13]

Two such dissimilar substances, according to the arguments of Plato and Aristotle, cannot be associated in one and the same substance.

Fernel admits the difficulty, but uses his pentagon analogy to insist that, as the lesser faculties are comprehended in the higher, so the immortal soul can perform all the functions of the body or each individually, acting either per se (as in rational thought) or through the medium of the body.

His argument is not entirely convincing, for while arguing for the simplicity and unity of the soul, Fernel sets out a complex three-part system of souls, faculties, and, as we saw in the last chapter, spirits. The relationship between faculty, which Fernel defines as "the force and power by which the soul as it were reaches out from its centre and performs its functions" and spirit is not clear; officially spiritus, and the Galenic spirits which Fernel introduces towards the end of Book IV, are instruments of faculty, but in practice they are closely identified with it (as indeed they are in Galen). Faculties, Fernel insists, are not 'parts' of the soul, but inherent properties of it; in Aristotelian terms, accidents of its essence or substance.

At the same time, he argues that each faculty has its appropriate organ or seat within the body. Anatomy supports this, and so does the specific action which each organ possesses.[14]

All of this seems orthodox enough, based as it is on an Aristotelian-Galenic notion of faculty and similar in many ways to the explanation of natural philosophers such as Cremonini. But though he uses similar terms, Fernel's distinctions do not correspond exactly with those of Aristotle and Galen – which makes for considerable confusion, especially as these two authorities differ between themselves. A table may make this somewhat clearer:

Parts of soul		Faculties (and spirits?)	
Aristotle	Fernel	Galen	Fernel
plants: nutritive/ generative	natural	natural (nutritive & generative)	natural liver (nutritive?)
animals: sentient	sentient	vital (sensation, motion)	vital heart (nutritive)
man: rational	rational	animal (rational)	animal brain (sentient & rational)

The vital faculty, because of its association with innate heat and spiritus, is for Fernel the most important of the three. He associates it not with sensation but with nutrition, the basic life-function; and it, or the innate heat which is its associate, performs the central function of integration, melding the body and its various functions into a coherent whole.[15]

The aspect of Fernel's discussion of faculty in generation which brought most criticism from his successors was a detail: what one might call 'facultative panspermatism'. We saw in Chap. 15 how the uses of the Galenic natural faculty, and the existence of the natural spirit, were topics of debate. The generative faculty was, in Galenic theory, part of the natural one; but Aristotle had argued that the heart, not the liver or testis, was the controlling organ in semen production. Orthodox Galenists had to choose between the testis, as the most obvious organ involved, and the liver, traditional seat of the natural faculty. Fernel attempts to sidestep this problem by suggesting that, as the embryo will possess all three faculties, so all the principal parts contribute to the production of seed. This argument was related to his suggestion, which, as we have seen, Argenterius criticised, that because organs have a specific function, e.g. seed is not produced except in the testes, milk except in the breasts, both their substance and their instruments (humor or spirits) should differ, as species.[16]

Fernel's notion of faculty thus has two opposing tendencies, which he struggles to reconcile. If faculty, or anima, is defined by life and life-actions, it must be specific: not all organisms, or parts of a complex organism, perform the same functions. At the same time, a living thing is integrated: this, and the Christian belief in the soul's immortality, requires that the soul should be, as Fernel repeatedly insists, 'simple', and one.

The explanation of the soul in generation which Fernel gives in the Dialogi is sufficiently different from that of the Physiologia to cause further confusion. Eudoxus puts forward the idea of a hierarchy of forms, in which lower forms are subsumed by the higher – for example, the elemental form by the total form of man. As we have seen, he suggests that three 'preparations' or levels of successive integration fit a living thing to take on its total form. Similarly, the seed undergoes three 'preparations': its elemental organisation results from the 'power of nature'; next the geniture takes in nourishment and acquires an appropriate temperament; and finally, through the agency of vital spirit and heat, it acquires form, which comes from its parents, the proximate agents of generation. Do these three preparations for the reception of a total form correspond to the three souls which succeed one another in the Physiologia account? It appears not, for these are preparations of matter, which place it into proper state – tempered, organised, provided with heat and spirits – for the reception of a form.[17]

A further difficulty arises out of Fernel's interpretation of the idea of potentia. Aristotle contrasted it with act or actualisation – thus the higher animae exist in potentia in the foetus, before they appear in actu. Fernel's interpretation approximates to the Aristotelian use of materia, the opposite of form; a point which emerges only obliquely from a lengthy debate between Brutus and Eudoxus on the question of the animation of the seed.

Brutus puts forward an impassioned argument for animation, that is, for an internal principle which governs the development of each living thing: his evidence, he insists, is based on what he has himself seen:

> Have I not often seen a plant, hanging down from the roof in the open air, working from nothing external to itself, spread itself and bring forth green plants? Or the egg bring forth a chick, only by the gentle heat of an oven?.. for when that chick begins to live and move, from whence does its soul come? whence is it produced? what introduces it? where, outside it, is your opifex which carves its form into the egg? Is it not clear as day that potentia takes its existence from the egg itself, and from this same egg, the universal chick?[18]

Eudoxus, however, disagrees: the form must come from without; the nature which is in the seed is imperfect, and acts simply as one of the preparations for the reception of a form.

What is in potentia is not yet, Eudoxus argues; it is, simply speaking, not; and what is not cannot produce an animal. The seed is not yet an animal, therefore it cannot produce one. And what Aristotle meant, when he said that the geniture possesses soul 'in potentia', was that there were two natures responsible for procreation: one, imperfect, in the semen, which is not yet in act; the other, the true efficient, which is the nature of the plant or animal from which the seed comes.[19] So Fernel's answer to the question 'is the seed animated?' is a decided no. Instead, it is an agent or instrument – but of what?

The obvious answer to this is 'of the parent', and indeed this is the first answer which Fernel gives. Procreation, as he defined it in the Physiologia, is a 'motion towards form': but the seed, although potentially a living being of a certain species, is not responsible for its own development: that power, and the efficient cause, resides in the parents, and, behind them, particularly in the case of the mind, in powers which are 'truly divine'. The true cause of generation – of the reception of a form – is the same in seminal as in spontaneous generation: the heavens. Only in the case of the higher animals, the seed, and a genitor of the same species, act as intermediate steps.[20]

We saw in Chap. 14 that for a straightforward medical writer like Sylvius, faculty could be explained in terms of temperament. This is the position which the rebellious Brutus also takes. For Jean Fernel and his spokesman Eudoxus, however, faculty inheres in a complex of supra-elemental causes. First there is the spiritus itself, which is 'divine' not only in its actions but in its substance, and from which the faculties and spirits emerge. Fernel compares it with the Aristotelian element of the stars, and with the 'aetherial and shining body' which Galen suggested was either the substance of the soul, or its instrument. In addition to the spiritus, there are the three souls – two present in the seed, one entering from without; they too, as Fernel argues repeatedly in the Hidden Causes, are in some way divine in their nature and substance. In fact, Fernel suggests, on the authority of Aristotle and Galen, that this divine substance is the spiritus, and the heat which resides within it as a quality.[21]

This system has its analogue in the remote causes of generation, which, as we have seen, are the heavens, source of motion, constancy, and celestial heat. The divine laws of nature are, says Fernel, administered by the heavens, and there is thus no conflict between the doctrines of Plato and his followers and Holy Scripture. It is the former which Fernel follows, however, in arguing for the existence of a spiritus mundi, which 'brings forth and governs' the bodies of the inferior world; and for the activity of spirit-intermediaries or daemones.[22]

The dialogue form allows Fernel to explore the paradoxes, and the contradictions, in his subject; like his contemporaries and successors, he struggled to reconcile the conflicting positions of his classical authorities with the Christian requirement for an immortal soul.

Fernel's influence on his contemporaries, in this as other areas, seems to have been considerable; fortunately, though, most of them produced systems which were more orthodox, and less complex. One way of comparing other approaches to this question is to adopt the division suggested by Walter Pagel. In Part II of his article "William Harvey revisited", Pagel considers Harvey's place in the Aristotelian tradition and the suggestion that the impetus theory of late medieval physics was applied to the question of the transfer of motion from parent to child via the semen. In line with his distinction between 'immanent' and 'dualist' systems of physiology (Harvey being an example of the former, Fernel of the latter), Pagel argues that there is a clear-cut opposition between those who argue for the seed as animate and those who see it as instrument: of the parents, of God, of Nature.

Those who argued, like Fernel, that the seed was an instrument of external, formative powers, might argue this point in various ways. Pagel mentions Jacob Schegk, who spoke of a formative or spermatic virtue in animal semen. Semen itself

> is thus neither an animate nor an inanimate body, but the operative instrument of an animate body – the parent – the latter being the principal efficient … The instrument is said to be possessed of the moving force (kinesis, energeia) of the principal agent, but not its essence (species, eidos) – just as the art itself is not in the tool of the artist, but his movement (energeia) is … It is by this seminal instrument that the animate body can effect and generate another animate body.[23]

The details of Schegk's argument – his distinction between the energeia of the parents and their species, and the curious analogy which he makes with the transfer of the energeia of the visible object to the eye in vision, are not found in Fernel; but the basic argument for the seed as instrument, and , indeed, the reference to Aristotle's comparison of generation with artistic creation (G.A. II i 734b–735a), are.

The de principio effectivo semini insito of Joannes Brònzerius is, like Schegk's De plastica seminis facultate, a specialised study of the problem, which focuses on the question, "what is the cause of foecundity in semen?" His answer is circular, to say the least: seed is fertile because it is endowed with an efficient principle, or faculty, which is active in generation.

In Part I, Bronzerius considers the opinions of ancient and modern authorities on the role of spirit in generation, including Fernel. His chief quarrel with Fernel is that he makes the seed originate from all the principal parts: Bronzerius rejects this and all other forms of 'panspermatism'; the seat of faculty is in the testes alone. Plato, Aristotle and Galen all suggested that the spirit in the seed was the cause of its foecundity; but Bronzerius argues against those who would divide the semen into an active and passive portion, representing the spirit and the body of the seed respectively. Seed is one, and one cannot say that the soul is 'in' the spirituous part, and not in the material.

Part II concerns the role of the soul in generation. Having rejected any suggestion that spirit is soul, or the vehicle of soul, in the seed, Bronzerius goes on to develop his own argument: the seed is not animate, but, as Plato, Hippocrates and Galen have all suggested, has within it a 'soul-forming principle' (principium animasticon). The true efficient cause, the generans, is external, and the semen receives motion from the generans according to the principle of impetus, as Aristotle rightly argued.[24]

Archangelo Piccolomini's Anatomicae Praelectiones is much broader in its focus than these specialist monographs. As his complete title reveals, he is concerned, like Fernel, with the "whole man". His anatomy is arranged according to faculties, and the lectures in Book I cover topics common to all the faculties, in particular the "cause of our conformation". His technique is to focus on the true cause by eliminating a list of false ones which includes the parents, maternal imagination, the seed, its temperament or innate heat, and either God or the soul considered in isolation. Instead, Piccolomini argues that the true cause is "the natural (i.e. vegetative) soul, prepared and informed most judiciously by God." Sennert names Piccolomini among those who argued for the animation of the seed, and this argument, together with his definition of the action of the whole soul as "life" (more particularly the life appropriate to a particular species) does suggest such an interpretation. But like Fernel, Piccolomini follows the Aristotle of De generatione ii to argue that the heavens confer substantial form on the foetus, for which the parents provide only the preparations which make it receptive. And again following Fernel, he distinguishes the form of the simple parts, whose powers arise from matter, form and temperament, from that of the individual as a whole.[25]

Other 'animationists' on Sennert's list include some familiar names: Foroliviensis, Licetus, the 'compendium' writer Valles, Fernel's critic Argenterius and both Cardanus and his opponent Scaliger. Three of these last four are also mentioned by Jacques Guibelet, in his vernacular treatise on generation.[26] Of these, Scaliger will serve as our last detailed example.

Scaliger's discussion of the soul in generation comes in Exercise Six of his Exotericarum Exercitationum (1557), following a section on the anima mundi and the role of the Platonic Ideas.

He begins with Cardan's definition of the 'generation' of form: "Form, then, is drawn out by the power [potestas] of the agent"; and Fernel's statement, in De abditis I v, that "the drawing out of form from its potentiality [potentia] in matter, is the preparation of matter for receiving a form." This capacity exists, says Scaliger, in the seed itself:

"Form is in the seed of a dog; which is said to be in potency [potestate] because the seed is capable of giving form, which is contained in it." This form must be equally provided by the two parents: one cannot provide form, the other matter.

Form comes from substance, not accident; and from within, not from outside: that is, from the form of the seed itself, which, if it builds the body and causes it to grow, is "equal to soul" because it performs functions which are the work of the soul. There is thus in the seed an anima formatrix, preceding the 'informed form' or soul, and prior to it in nobility and importance as well as time. There is, thus, a soul contained in potestate within the seed, which builds its own body. Scaliger's position here is similar to Daniel Sennert's as described by Pagel: the soul hidden within the seed is agent, and acts on seminal matter.[27]

In chapters 8 and 9, Scaliger considers the objections of those who argue that semen cannot act in this way. One is that nothing acts in itself; soul acts in the body by instruments: spiritus in the case of generation. But spiritus is part of the seed; matter cannot be efficient at the same time. Scaliger's reply is that, first,

> … spiritus is not part of the body which is formed.. It is, I say, in the body of the seed, as active in the passive. It is not separated by location, as perceivable by the senses, but by intellection.

Spiritus is 'in' the seed as fire is in wood. Secondly, Scaliger argues that an instrument is not necessary to the action of form on matter:

> the anima moves [the body] without the spirit as instrument… Does not earth descend without an instrument? Between form and matter no intervening body is necessary.[28]

Two other arguments against the animation of the seed are more practical. The first is that, if the soul of the dog is in actu in the seed, the seed of the dog will be a dog. Scaliger admits this: "it is an imperfect dog, lacking only articulation". Secondly, how can the active efficient cause in a wheat seed remain inactive in the granary? Scaliger's reply is that it is truly 'inactive' (otiosus) only if, when placed in a situation in which it can germinate and 'vivify its matter' (earth in the case of the wheat) it does not do so. In the same way, "the soul of the tree creates nothing in winter."[29]

What, then, is the potentia of seed? It lies in its disposition (habitus, conformatio, dispositio) to receive the first acts of the soul; a disposition which pre-exists in the seed, before it finds material expression in organs. As Themistius (De anima i) put it, "the soul is architect of its own dwelling."

If the seed is the new animal, the term 'generation' takes on a new meaning. A tree generates when it produces its seeds, not when they germinate; and a dog generates when the seed is produced in the parents' body, not when the pup is born. But then, the parent dog will be 'full of souls'. No, says Scaliger: it has one soul, which suffices for the generation of many; and since the soul is 'all in all, and all in each part', there is no reason why the soul of the pup should not be a part of the soul of the parent. As Pagel points out, this sort of argument leads naturally to preformation. It was also potentially theologically worrying. But Scaliger argues that, though the vegetative and sensitive souls arise from the seed as Aristotle argued in De generatione ii, our immortal, human soul arises not from seed, nor from the heavens, but from God as sole author and creator. There are thus three 'generations': from the stars, as in the spontaneous generation of flies and other lower animals (Scaliger is emphatic in rejecting the suggestion that humans can be produced from sidereal forms, impressed on matter, without coition); from the seed, as in the higher animals; from God, in the case of the human soul.[30]

Scaliger's position is very different from Fernel's, and though he was clearly familiar with the Hidden Causes, and derives some points of his argument from it, it is hard to see how Harvey, or Pagel, could lump them together. Both Fernel and Scaliger base their arguments on the authority of Aristotle: but so do Harvey and the 'monist Aristotelians' Pagel cites.[31]

The medical debate on the soul in generation provides us with an opportunity to look for the source of these opinions in the classical texts of Part I, and to compare them with those of the natural philosophers – in particular Nifo, Cremonini and Telesio – in Part II. We may remember that for Plato, seed, or marrow, was clearly animated – an unstable compound of soul and a specially prepared body, the *pneuma*. Echoes of this appear in both Fernel and Scaliger, and both, as we have mentioned, also recall other elements of Platonic doctrine: the world-soul and the intelligences or *daemones* who act as intermediaries between the *anima mundi* and its *corpus*.

More important to all the discussions in this chapter, and in Part II, is the Aristotelian definition of soul: soul as capacity, soul as source of internal movement, soul as 'instrument' in transmitting the motions of the male parent. Equally central to the Renaissance debate is his association of the capability of soul with a substance 'above the elements': introduced to solve two problems which his definition created. First, how can the semen transfer the higher qualities of the soul, which it will never possess in actuality, to an embryo in which they will be realised only gradually? And secondly, with what 'body' can the intellect be associated? We have seen that the natural philosophers in Part II, following Philoponus and Averroes, associated this supra-elemental substance with the generation of the intellect in particular. Fernel, on the other hand, associates it with all forms of faculty.

Finally, Galen, like Aristotle, associates the semen with extracorporeal powers and substances: the spirit in the semen is a self-moving principle of the animal; either soul or its vehicle. His distinction of semen into a self-moving, spirituous portion and a passive, liquid one is clearly a factor in the arguments of 'animationists' like Scaliger, Fienus and Licetus. And his agnostic stance left a way open for speculations which could claim passages in *De placitis* and *de formatione foetuum* as their authority: Fernel is an example of this.

Nifo, Cremonini and Telesio all considered the question of the role of the soul in generation from the point of view of natural philosophy. Nifo argued, against Philoponus and on broadly orthodox Aristotelian lines, for the seed as instrument, and Cremonini's position, in which the seed possessed a *vis animastica* but was the instrument of an "external efficient", was similar to this and similar, also, to Fernel's description in Book VII of the *Physiologia*. But the account Fernel gives in *De abditis rerum causis* is more similar, in some ways, to the position of the 'Aristotelian revisionist' Telesio, particularly in the major role both give to *spiritus* – though they locate it in different parts of the body.[32]

It may be worth mentioning that the topic of the soul, perhaps because of its elusive and controversial character, seems to have furthered the use of a *quaestiones* format among the medical writers considered in this chapter. Apart from the *Dialogi*, Garbo, Scaliger, Fienus, Licetus and Piccolomini all set their work in the form of a series of questions, which are refined successively to reach their final judgment. This may be because all of them, apart from Piccolomini, were involved in controversies of some sort, and all, apart from Fernel, were identified with the somewhat unorthodox position that the seed was animated or in some way self-directing.

What is clear, in this all too brief and sketchy survey of Renaissance ideas of the soul in generation, is that Aristotle's *De generatione animalium* II is probably the most important single authority for the Renaissance debate. And what is perhaps most interesting is the variety of different opinions which scholars, all working within the same framework, could devise.

Notes

1. For the doctrines of Plato, Hippocrates, Aristotle and Galen on the subject, see Part I, above, pp. 62–64, 67–69, 87–90 and 110–115 respectively, plus references. Particularly on the natural philosophical side, much more secondary literature exists than for some other questions discussed in this thesis. My own treatment is therefore brief. For further detail, see for example Giovanni di Napoli, L'immortalità dell'anima nel Rinascimento, Turin, 1963; Walker, especially "Astral body", op. cit. and, especially important for the medical writers, Walter Pagel, "William Harvey Revisited", History of Science, 1969–70, 8:1–31, 9:1–41, particularly Part II. More general discussions of the soul in Renaissance philosophy can be found in the work of Kristeller, Schmitt and Nardi cited in Part II, above.
2. For Cremonini, see Part II above, especially pp. 232–243 and the references in Chap. 8, Note 43.
3. Thomas De Garbo, Summa Medicinalis, Lyons, 1529. Book I, question 80, "Utrum spiritus existens in corpore animalis sit formaliter animatus", fols. lxxxi ro- lxxxvi vo; especially "Reprobationes Joannis de Pena" and "Tractatus comminantium magistri Fr. de Bonon.", fols. lxxxi vo- lxxxiii vo. Lipenius, op. cit. p. 419, seems to have taken his references to the debate from Garbo's account, as I have done.
4. The sequence of events in this debate is complex. Santa Cruz, in his In Avicennam primam primi… of 1622, seems to have attacked two earlier works of Fortunius Licetus, De ortu animae humanae..., Genoa, 1602 and De perfecta constitutione hominis.., Padua, 1616; as well as Thomas Fienus's De formatrice foetus liber..., Antwerp, 1620. Both Licetus and F.ienus argued for positions which were disturbing to orthodox (particularly Catholic) medical writers, and Santa Cruz was supported by the Flemish doctor Louis DuGardin whose De animatione foetus quaestio..., Douai, 1623 was aimed particularly at Fienus's 1620 Liber. Fienus produced a detailed reply, aimed in turn at DuGardin rather than Santa Cruz. Both were attached to Catholic Belgian universities (Fienus at Louvain, DuGardin at Douai) and both dedicated their works to high-ranking medical men in Catholic courts: Du Gardin to Santa Cruz, private physician to the King of Spain, and Fienus to Francisco Paz, "Regiae Catholicae majestatis et … Belgii Archiducam Archiatro..." One suspects a certain professional rivalry. Fienus's reply was the De formatrice foetus liber secundus of 1624, and despite his scornful opinion of the 1623 Quaestio ("videns nihil in eo solidi contineri, praeter derisiunculas, cavillationes, et inania verba..") he still devoted 170 pages to a detailed response (pp. 1–2). The third round came in 1629, with

Fienus's ..Apologia adversus Ant. Ponce Sanctacruz, Louvain, 1629, and Du Gardin's Anima rationalis restituta.. of 1629, also dedicated to Santa Cruz, in which he sets out the sequence of events so far (Epistola Dedicatoria, fol 2v–3r). Licetus replied to Santa Cruz directly, in his De Anima subjecto corpori..., Padua, 1631.

5. Du Gardin, op. cit. 1623, pp. 56–78 esp. 62–64 for baptism issue. His arguments are mostly religious, founded on the blasphemy that 'spilled seed' should have a rational soul, and thus be capable of salvation or damnation. His authorities are mostly Scriptural, but on page 58 he cites Jean Fernel's statement that mola (shapeless masses formed in the womb) did not have a rational soul or require baptism, and on p. 60 he applies the same argument to monsters.
6. Licetus, op. cit. e.g. 1631, xvi p. 51: "Nostram igitur opinionem e formis in materia velut in loco existentibus, cui nullam actum exhibeant.." In his 1616 work, he divides the geniture into an active, spirituous portion, which contains within itself the active principle of generation or substantial form; and a crasser portion on which the former operates (iv, pp. 20–3, v pp. 23–25). He also argues (ch. vi, pp. 26–30) for a variety of panspermatism. For Fernel's use of the helmsman analogy, see below. Licetus himself (1631, ibid.,) uses the analogy of a vessel or container: "Anima in corpore degens ut in vase" to explain how the soul can be in actu within as yet unformed seminal matter.
7. Fienus, op. cit., 1620, Conclusio Septima, p. 91; ch. 7, 124–157, esp. Conclusio Octava, p. 128; ch. 8, 158–220 and Conclusiones Nova (p. 161) and Decima (p. 184).
8. A similar example is in the De Miraculis Occultis Naturae of Levinus Lemnius, Antwerp, 1574. Lemnius argues (I xi 53–55) that the soul is not transmitted in the seed, but newly created and infused by God on about the 45th day. Earlier foetal development is directed by the formative faculty or vis vitalis of the seed, and the facultas uteri. Lemnius's popular anthology, with subjects ranging from the geography of Zealand, sleepwalking, the basilisk, the use of lampreys as torches, the sex of plants and planets, to how to grow a beard, appeared in numerous editions, including an English translation, The Secret Miracles of Nature, in 1658. It is mentioned by Cole, op. cit.: Needham, op. cit., describes it as "without value".
9. Nicolaus Leonicenus, De virtute formativa, Venice, 1506. Generation treatises in which the soul is an important feature include those of Jourdain Guibelet, "Du Principe de la Generation del'Homme", in Trois Discours Philosophiques, Evreux, 1603; Victor Cardelinus, De origine foetus, Vincenza, 1628; Joannes Gallego de la Serna, De principiis generationis..., Lyons, 1634. More general treatments with a medical bias appear in Chrisopher Rumbaum, De Partibus Corporibus Exercitationes, Basle, 1586 and Archangelo Piccolomini, Anatomicae Praelectiones, Rome, 1586. Rumbaum is not a doctor of medicine but of divinity, and his purposes, as he explains, is to demonstrate the works of God in the creation and the unity of truth; and, in considering the question of the generation of the soul, to reconcile Aristotle with theology (Epistola Dedicatoria, ii r–8 r). Piccolomini is considered in more detail below.

10. Fernel, Physiologia V i p. 122: "..non in solius corporis, sed in hominis qui ex corpore et animo constat, meditatione versamur…inesse quippiam viventi quo sit is praestantior atque potentior demortuo, & quo ad obeunda munia se comparet."
11. Ibid. vii 123–4, xviii 160–1.
12. Ibid. VII v 227–9: "Ex quo intelligi potest eum hominem qui emisso semine generat, omnes quoque corporeas facultates una producere. Caeterum animum et eam animi partem quae princeps est, quaeque mens nominatur minime is confert, quia cum animus sit a corporis natura diversus, eius substantiam non comitatur, sed extrinsecus in id illabitur ceu divinum munus." (p. 229; cf. translation and commentary by Figard, op. cit. pp. 200–201). This statement is very similar to that of Aristotle, and almost identical with the position of Cremonini (above, Part II Chap. 11). It varies somewhat, as we shall see, from the discussion in D.A.R.C. I iv–vi.
13. Ibid. V xviii 161: "Illa quasi separata & simplex, corpori non permiscetur, sed ei duntaxat ut navi gubernator assistit: haec vero in corpus penitus immergitur, a quo nunquam diiungi divellique potest. Illa denique, ut cogitet, ut contempletur, & intelligat corporis subsidio non indiget: haec vero nihil efficere potest citra idoneam corporis eiusque organici praeparationem.."
14. Ibid. IV x and xi, 119–121; V ii 125 ("Est enim facultas vis illa & potestas quam anima tanquam de sinu suo promit, & ad muneram functiones profert.."); xii p. 148 for Fernel's statement that,not calor but the anima of which it is vehicle and instrument is opifex in the functioning of the body; xiii, xiv and especially xv pp. 152–157 for the argument that each faculty has an appropriate 'seat' in the body.
15. Ibid. xvi p. 158, xvii 159–60. As we saw in Part I (p. 88) Aristotle himself associated generation both with the nutritive soul and with the primordial heart; the confusion is not entirely of Fernel's making.
16. Ibid. xv pp. 155–6. For Argenterius, cf. De somno, p. 288. Joannes Rogerius, De sede animae, Naples, 1574, is based on this same question.
17. D.A.R.C. I iii 27–37 for Eudoxus on the hierarchy of forms; iv 55–56, 58–59, v 64 (61) for the three preparations preceding the reception of the total form.
18. Ibid. I v pp. 59–66, especially 65–66; the argument continues in vi ("Gignendae formae, ne minimum quidem ante in materia fuisse, neque potentiam quicquam esse formae, neque ad substantiae genus referri"), pp. 68–85: "An non cepam saepenumero vidisti, de tectis aere libero propendentem, per se nullius extrarii ope, sese diffundere, et herbescentem ex se viriditatem proferre? aut ovum solo fornacis tepore pullum excludere?…dum iam vivere ac moveri incipit ille pullus, unde suam accipit animam? unde prodit illa? aut quis eam inducit? ubi extrarius ille tuus opifex est qui ovo suam insculpat formam? …An non luce clarius est, illam ex ovi potentia existere, ut ex eodem ovo pullum universum?" (pp. 68–9). See Figard, op. cit., 203–4 for a discussion of this passage, and Aristotle, Metaphysics V, 9 and 12, and Physics, Book VII, for definitions of active and passive potestas.
19. Ibid. I vi 70–1, 75.

20. Ibid. I vii 97–98; Philiatros sums up the argument so far: the genitor is not cause of the species; the heavens, as higher cause, are.
21. Ibid. II iii 157, 163, iv 167-170, vii pp. 191–3 for the importance of spiritus; I vi 78, II vii 186–7 (Philiatros) for suggestion that heat and spiritus are the divine substance of the soul. For Galen's equivocal position on this point, see above, Part I, esp. pp. 111–2; and de placitis (Kühn V p. 643).
22. Ibid. I viii esp. 101–2; x, esp. 120-3; xi.
23. Pagel, "Harvey revisited" Part II, esp. 8–20; for Schegk, 12, 26–30: p. 27 for quote. My comments on Schegk are based on Pagel's account and directly on Jacobus Schegkius, De plastica seminis facultate, Strasbourg, 1580, esp. fols. A2 ro – B2 vo. Among other examples by which Pagel seeks to place Harvey in an orthodox Peripatetic tradition are Albertus Magnus, Ponce de Santacruz and Cesare Cremonini; the opposition is chiefly represented by Daniel Sennert (11–20), although the 'dualist opponents' Harvey singles out for attack in Exercises 50 and 71 of De generatione (p. 7 e.s and note 76; 20 and note 82) are, according to Pagel, Scaliger and Fernel. For Scaliger, see below. The evident similarities between e.g. Schegk and Fernel suggest a weakness in Pagel's thesis.
24. Joannes Bronzerius, De principio effectivo semini insito, Padua, 1627 (and see Part III Chap. 15, note 53, above); I ix 49- xiii 74; xv–xviii. II iii 131 – iv 136 esp. Much of the text is concerned with answering objections raised, Conciliator-fashion.
25. Piccolomini, op. cit., 1586. The broad scope of the work is suggested in its full title: Anatomicae Praelectiones..Explicantes mirificam corporis humani Fabricam: Et quae animae vires, quibus corporis partibus, tanguam instrumentis, ad suas obeundas actiones, utantur; sicuti tota anima, toto corpore. Book I, Lectio Tertia, 11–14; Quarta, 15–18; Quinta, 19–23 esp. p. 37 (Lectio Nona) for his definition: "Quae sit actio totius animae?… esse vita." He defines plant life as "operatio, plantae conveniens" and notes that this will vary by species: "vita homini conveniens..distat a vita conveniente asino, leoni, bovi..Non enim anima est quod vivit, sed anima est id, quo compositum animatum vivit." So for Piccolomini, as for Fernel, differences between species are the result of differences in their souls – and not, as for Sylvius, merely in their temperaments.
26. Pagel, op. cit., 13–14, citing Sennert's Hypomnemata physica, IV 6–8. Guibelet, op. cit. (note 9 above), xii pp. 180–4.
27. Scaliger, op. cit. (1557), Exercitatio Sexta, ch. 5, "De generatione", fol. 13r–14r. Pagel, op. cit. Part II, pp. 14–19, 20.
28. Scaliger, Ibid.; ch. 8, 15r–v: "Non est igitur spiritus pars corporis, quae formatur… Est igitur in corpore seminis, sicut agens in patiente. Non separatus loco, qui sensu perpiciatur, sed intellectione… Movet enim anima spiritum sine instrumento… Terra nonne descendit sine instrumento? Inter formam enim, et materiam nullam corpus interesse necesse est." This cuts across the arguments of those such as Fernel, that a medium of some sort is necessary to connect soul and body.
29. Ibid., ch. 9: "Est enim canis imperfectus, carens tantum articulationibus…seminis illius anima nihil aliud habet agere, quam vivificare materiam illam, quod

et facit. Fabricare autem non potest, quia extra matricem suam est: quippe extra terra sinum. Sic arboris anima nihil aedificat hyeme ."

30. Ibid. 15v–16v (chs. 9–11). Chapter 6, 14v, for his attack on "Error detestandus de fortuita hominis generatione", which recalls the commentaries on XII Metaphysics surveyed in Part II, above: e.g. Nifo, pp. 203–6.
31. Pagel, op. cit., esp. Part I, pp. 10–11, 13; II, 11, 34, 36. Pagel's attempt to divide the accounts which he discusses into genuine Aristotelian, 'monist' schemes (in which the formative virtue which forms the foetus is an instrument of the parents and the developing foetus is not distinguished into form and matter) and 'dualist' schemes, such as Sennert's (in which the agent is a soul within the seed) is not entirely convincing. It is his attempt to distinguish the two types according to their use of formative virtue or anima which runs into most difficulty: for instance, he is forced to argue that when Schegk, who is cast as a monist, uses "soul", he really means "formative faculty" (p. 28). Another difficulty, as I have already pointed out, is that virtually all the participants in the debate take Aristotle as a major authority. But Pagel is always illuminating, and his careful untangling of Schegk's cryptic doctrine is particularly valuable.
32. See above, pp. 209–12; 235–42; 255–60.

Conclusion: Academic Theories of Generation in the Renaissance

Abstract I expected to be able to identify a category which I could label 'theoretical works on generation' and analyse accordingly. This worked reasonably well for the natural philosophers but not for the medical writers. A very clear lesson is the context in which these works were written, almost all of them designed for university teaching. A second point is the surprising orthodoxy of almost all these writers..I identified a set of 'classic questions' which arose directly out of the attempt to reconcile the opinions of Aristotle and Galen. The very richness of the tradition in which they worked, and which they sought to make whole, produced its own problems for these writers

There are two major themes. The first is continuity between medieval and Renaissance topics, formats, approaches of doctors and natural philosophers. I would also argue that conciliation was more fundamental as a motive than other scholars have suggested. I shall be satisfied if I have managed to make the workings of this undeniably foreign intellectual milieu clearer to those of us who cannot share it, and to provoke a sympathetic understanding of its aims and the extent to which we can judge them to have been achieved.

Victor Cardelinus began his De origine foetus with an observation that might well stand as a warning to the 'conciliatory' theme of this thesis. The fact of generation, he pointed out, was well known, accomplished, night and day, by wise and foolish folk alike;

> yet there is no one who can understand properly the manner in which offspring are generated; and if they do, they cannot persuade the rest; because certainly for a thousand years students of nature have disputed, and will dispute so long as the names of Hippocrates, Aristotle and Galen shall fly around in men's speech. And there is among them such discrepancy, that nothing can bring them together, unless the hidden, unutterable operation of nature, deeper than our thought, can be comprehended.[1]

L. Deer Richardson, *Academic Theories of Generation in the Renaissance*,
History, Philosophy and Theory of the Life Sciences 22,
https://doi.org/10.1007/978-3-319-69336-1

As surveys such as Roger's, Needham's and Adelmann's make clear, the number of generation treatises published in the Renaissance was enormous – quite large enough to justify thoughts of a millenial production schedule. As they also make clear, any attempt to consider these treatises across the board inevitably means that each gets a sentence of praise or blame, or at best a brief characterisation. Throughout this thesis, my aim has been to be intensive rather than extensive: to concentrate on a relatively small number of examples, but to analyse them thoroughly and take their arguments seriously.

To make this plan feasible, I chose, as I explained in the Introduction, to exclude works as varied in their structure, purpose and sources as Raynalde's The Birth of Mankynde and Fabricius's De formatione ovi et pulli.

I expected, when I began, to be able to identify a category which I could label 'theoretical works on generation' and analyse accordingly. This worked reasonably well for the natural philosophers: I was indeed able to identify a small group of texts, either commentaries on Aristotle's Generation of Animals or treatises based on the same themes. In the case of the medical writers, I found a similar group of specialist monographs, all by medical teachers or practitioners, which appeared from their title pages, and the qualifications of their writers, to correspond to these natural philosophical texts. Closer study revealed that the category I was looking for did not exist: there was no identifiable group of medical texts which could be labelled 'on generation theory'. Those I thought I had identified had no recognisable common core of topics and varied wildly in level of sophistication as well as content. A few samples will show the nature of the problem.

The de genitura hominis (1495) of Candidus Decembrius (1399–1477) is a late medieval treatise of the 'women's book' type, based chiefly on Aristotle's Historia Animalium. Charles de Bouelle's Liber de generatione contains nothing about animal generation per se: its subject is generation in the natural philosophical sense, as in Aristotle's De generatione et corruptione. Book VII of Fernel's Physiologia, "De hominis procreatione atque de semine" comes closer to a theoretical treatment, especially in the first half, in which he considers the operations of spiritus and the soul in generation, as we have seen. Later chapters describe the development of the foetus in Hippocratic-Galenic terms. Jacobus Sylvius's de…hominis generatione, published the year after the Medicina, is in contrast entirely practical. Petrus de Peramato's Liber de semine and liber de hominis procreatione are part of his compendium-like Opera medicinalia (1576) and again lean heavily on the Hippocratic and Galenic generation texts in their treatment of questions like, what is seed? where does it come from? what is the contribution of female seed? how do the parts of the foetus develop? They are followed by short practical texts on the regimen of pregnancy and infancy. Terelius's De generatione.. is a disappointingly shallow work, comparing the opinions of the ancients on generation questions (Book I) and the length of gestation in particular (Book II). De generatione and Lucretius are favoured authorities. Style appears to be more important than content, and the tone of the work is 'look what I've read – aren't I clever?' The Concordia ..de humano conceptu.. (1582) is a series of 121 conventional theses, mostly on practical subjects, presented by Andreas Helepyrus for his degree, presumably in medicine. It does at least illustrate the theme of conciliation which we have followed throughout. The de hominis generatione (1597) of Hieronymus Mercurialis, one of the leading Italian professors of his day, appears to be based on his Pisan lecture notes. It illustrates his humanist concern to explain away errors and discrepancies between classical authorities by careful editing and exegesis, and to use such synthesis as the theoretical basis for improved medical practice. And it considers 'classic questions' such as the role of the female and the order of development of the principal parts, as well as the medicolegal question of the length of pregnancy. Costaeus' De Humani Conceptus Formatione, ac Partus Tempore. (1596) deals almost entirely with this last question, for which Hippocrates is the most important authority. In his De natura foetus disputatio (1615) Sisinius puts forward as his aim the attempt to determine which, among ancient opinions on generation, is "more probable, and truer" for a range of questions both theoretical and practical. Cardelinus's de origine foetus is similarly concerned with comparing the opinions of classical authorities, particularly their confusing definitions of innate/vital/natural/genital/animal heat; Gallego's de principiis generationis (1634) considers particularly the question of ensoulment.[2]

The problem is clear: some of these generation texts deal with practical questions of the regimen of pregnancy, exactly similar to those of the 'women's books': those of Candidus, Sylvius, Helepyrus, Costaeus for example. Others – for instance the treatises of Petrus de Peramato, Terelius, Mercurialis, Sisinius, Cardelinus and Gallego – deal with questions similar to those which have been treated in some detail in the rest of this thesis, but do so, as it were, at second hand. The arena in which the classic questions of generation theory were debated is not these derivative compilations labelled 'on generation'; they only codify for students what was thought through elsewhere, in the teaching texts of natural philosophical and medical theoria examined in Parts II and III above.

This helps to explain why the type of text chosen for study is different for the two groups, doctors and philosophers. In each case, I have tried to choose examples which are at the same time typical and most revealing of the theoretical interests of each group. Wanting to explore the context – university teaching – within which medical writers put forward academic theories of generation, I chose the compendia texts because of the insight they gave into the kinds of questions medical writers asked, the answers they expected to get and the uses to which their books might be put. Disputationes, medical and natural philosophical, would have been a possible choice, but their narrow focus and polemical intent give very little indication of the context within which they were written. I have confined myself to mentioning the major debates – Cremonini v Caimo, Scaliger v Cardan, Mundinius v Hofmann – and outlining their terms of reference.

In addition, I chose to concentrate, within these two categories, on texts produced in the lifetime of Jean Fernel and his immediate successors. One reason for this choice was to test my contention (p. 19 above) that "the concept of a radical transformation between the sixteenth and earlier seventeenth centuries is untenable". The time scale also imposed a discipline; and it excluded, in particular, perhaps the chief example of an academic work of generation theory known to modern readers: William Harvey's De generatione animalium. So much has been written on Harvey (and so little, comparatively, on my sources, even including Fernel) that to have included him would have unbalanced the structure of the thesis, as well as adding further to its length. But I hope that what I have had to say will at least cast some light on the theoretical problems and assumptions of medical and natural philosophical writers in the generation preceding Harvey's own.

Limitations of space imposed a further discipline, and in the absence, in most cases, of appropriate secondary references, I have had to make some difficult choices about what to put in and what to leave out. Certain, possibly wearisome, detail was included because it was important for the understanding of an argument, and because I could not assume that a modern reader would have the knowledge of even a second-rate sixteenth century arts graduate. At the same time, I am aware of how much I have had to leave out, or skim over.

If we now stand back from the detail of the previous pages, what general picture do we have? Can we see common themes? I think that we can – and, moreover that they confirm some claims (and fail to substantiate others) which have been made by recent scholars.

A very clear lesson is the importance of the context in which these works were written, and, in particular, the fact that almost all of them were originally designed for university teaching. The importance of set texts, and of the conventions of the lectura, to the style, format and content of Renaissance publications in both medicine and natural philosophy has been shown by a number of recent scholars.[3] A particular feature, which has been studied in some detail by Andrew Cunningham for a slightly later group of medical texts, is the concern which both groups show for the order in which books should be published or taught. The assumption that there is a correct teaching order, which most perfectly reflects the logical order in which texts should be studied, according to the rules of method, is one which we have seen both in the natural philosophers of Part II, especially Accorambonius, Nifo and Cremonini; and in the 'compendia tradition' of Part III, especially as it appears in Fernel. Both traditions go back to ancient authority: to Aristotle's own comments on the relationship between his natural philosophical works, and to Galenic texts such as the De ordine librorum.

The lecture theatre as a source for the printed word is particularly evident in the case of Montanus, whose commentaries were lecture notes, posthumously published; and of Cremonini, many of whose commentaries survive only in manuscript, in the form of student notes.

A second point is the surprising orthodoxy of almost all these writers (even Telesio is working within the Aristotelian tradition) and the small part played by heterodox opinions. Neo-Platonist, Gnostic, atomist, Paracelsian thought have received considerable attention from modern scholars[4]; these figure almost not at all in the academic texts considered above. The classical opinions on generation, the theories of the elements, temperaments, the action of heat and of the soul, come from orthodox sources: Plato, Hippocrates, Aristotle, Galen for the most part. The only deviations from this orthodoxy, if they are such, are the neo-Platonic and 'Stoic' opinions discussed by, for instance, Contarini, Fernel and Scaliger. But none of these authors sees such opinions as in any way contradicting the Aristotelian-Galenic orthodoxy which they also put forward: it is not an either-or situation. R. G. Lewis expresses this area of consensus very well, in her thesis on Rondelet and the circle around him at Montpellier:

> Guillaume Rondelet and the disparate members of his circle of friends and pupils all belonged to the same 'school of thought' in natural philosophy and medicine; so, however, did almost all the academic physicians of Europe at this time. They were all Aristotelians in philosophy and Galenists (and Hippocrateans too) in medicine; that is to say, they had all undergone much the same elementary schooling in the arts, especially in logic, and they had all experienced the effect which the labours of philologists, textual critics and lexicographers had had upon the literature of medical teaching.[5]

Despite the claims of modern scholars and of sixteenth-century writers themselves, the authorities they cited, and the models they chose, suggest that the academic writers considered in this thesis were remarkably conservative. Charles Schmitt and others have emphasised the continued, if not increased, popularity of the works of Aristotle and the commentaries of Averroes in this period; Durling has made the same point for the Galenic texts; Lohr's parallel studies of medieval and Renaissance commentators reinforce the point.[6] This continued and even increased popularity did not just apply to classical authorities; even more surprising, in view of the caustic comments of humanist translators, is the continued popularity of Arabic and medieval texts, compilations and commentaries. Further evidence of this is provided by the use made of Averroes' Paraphrase and Colliget by the natural philosophers in Part II and the parallel influence which, I have suggested, Peter of Abano and later medieval North Italian commentators had on the medical treatises of Part III. The Greek commentators, except for the commentary on De generatione animalium attributed to Philoponus, and the writings of Alexander of Aphrodisias, seem correspondingly less important to both groups. Perhaps this is the effect of the coming to maturity of a new generation of scholars who were able to take the humanist achievement for granted. The generation of writers which immediately followed the publication of Fernel's major works in the 1540s and 1550s were, in effect, the first post-humanist generation; and it is from this same time – mid-century, roughly speaking – that Charles Schmitt and Edward Cranz have dated the appearance of the major new editions of Aristotle and Averroes, and a renewed interest in both the Philosopher and the Commentator.

Averroes and Peter seem to have been important to the development of the discussion of academic theories of generation in the Renaissance in several ways. Each of them was both doctor and natural philosopher, ideally placed to compare the opinions of the major authorities of the two professions: Aristotle for the philosophers, Hippocrates and Galen for the doctors; to identify the major questions and points of controversy which these authorities raised, for the topic of generation; and finally to mediate between them.

This process of comparison, identification of dubia or quaestiones and conciliation was an established tradition, followed not only by Arabic and medieval commentators but by Aristotle and Galen themselves, who had commented on their predecessors in similar manner; the commentators and compendia writers of the Renaissance were practitioners in the same tradition.

What, then, can we say about the relationship between doctors and natural philosophers approaching this topic, generation? We have seen that the theory of generation was a question which interested them both, and which both could claim within their overlapping sphere of knowledge. They shared not only a common educational background and common authorities but also, as the last two sections have shown, a set of common questions to which both groups contributed.

What were these questions, and in what contexts were they discussed? In Parts II and III, I have tried to answer this. I identified a set of 'classic questions' – on the nature of the seed, the role of male and female, the relative importance of the principal parts, the actions of heat and of soul – which arose directly out of the attempt to reconcile the opinions, in particular, of Aristotle and Galen. Behind these questions are other, more fundamental ones, which were of importance to more sophisticated writers: Nifo, Cremonini, Contarini, Fernel, Argenterius for example. These turned on the attempt to explain the miraculous ability of living creatures to reproduce – not a word a Renaissance writer would have used – their kind; using the tools of scientia: logic, observation, authority. How can one explain the apparently divine aspects of animal generation in terms applicable to our knowledge of the natural, including the non-living, world? What is the source of the facultative, self-ordering aspect of the composition of bodies? How do 'celestial' or 'divine' powers operate in the world of the elements? If, as Cremonini claimed, "the study of nature means in the first place the study of man, as it were the end of natural knowledge" (p. 229 above), then doctors and philosophers shared the same aim, and the same challenge: to link an abstract system of natural causes with the observed phenomena, the perceivable results of generation with causes which could be known, and distinguished from one another, only 'by intellection'.

The very richness of the tradition in which they worked, and which they sought to make whole, produced its own problems for these writers; even though they were drawing on only a part of the available spectrum of possible explanations. One problem which at least some of them recognised was the problem of definition, in which a word could have many different meanings for different authors, or for the same author (Galen for example) using different sources. 'Element' could be used to mean at least four things within the body alone: earth, air, fire, water or the system of qualities on which these were based; humours; simple parts; organic or composite parts. Cardelinus notes five different kinds of heat which could be described within the body, and the fever texts added other, praeternatural varieties. Similar problems arose in trying to define anima or soul.

One reason for this confusion of terms was that successive layers of definition, each using the same term, had become confused, as we saw in the case of the elements and temperaments. Another was the narrow base on which this superstructure of definition had been raised. Elements, temperament, innate heat, soul might all be included in a complex explanation of the structure and workings of the living body such as that in Jean Fernel's Physiologia: and logically, for Fernel, each of these levels had a necessary place in his explanation. Each was related, however, to the same basis in sensory experience, and whereas 'leg' or 'heart' or 'fibre' had a clear meaning in terms of this experience, 'soul' or 'heat' or even, in the variety of senses in which it was used, 'blood', did not.

In his physical writings, Aristotle had suggested that the basic actions of alteration, transmutation and concoction arose from the contrary actions of the four elemental qualities, or from the balance or complexio of these actions. Galen similarly made such qualities the basis of the 'natural faculties', the actions which animals share with plants, among them generation. And he suggested that these qualities could be determined by touch, and that their discrimination was the foundation of certain medical knowledge. The reintroduction of the soul into questions of theoretical medicine and natural philosophy, which was a feature of the Renaissance accounts we have studied, was a complication not without its uses. But unlike temperament, soul (or equally celestial heat) is an unknown quantity, which cannot be directly related to the structure of the physical world. It can be understood only by observing its actions within the body, and proved to exist only by arguing backwards from these same bodily actions. And, by focusing attention away from the observable, and onto the 'hidden causes' visible only to the intellect, it advances physiological understanding very little, and complicates it with layers of explanation which Aristotle and Galen were able to do without.

There are two basic themes which run through this thesis, and on which I would like to end. The first is continuity: I believe that the evidence I have presented here argues for continuity between medieval and Renaissance topics, formats, approaches; for continuity between writers of the sixteenth century and those of the first half of the seventeenth, who seem to me to be followers in the same tradition; and for continuity between the authorities, subject matter and concerns of doctors and natural philosophers.

That is not to say that the approach of medicine and natural philosophy was identical, or that their areas of interest did more than overlap. Galen distinguished between the doctor, who asked how, and the philosopher who asked why. This is still a useful distinction for the Renaissance, although Galen would no doubt have also agreed that the 'best doctors' included in Part III – among whom I am certain he would count Fernel – were also philosophers.

I would also argue that the evidence I have presented suggests that the reconciliation of divergent opinions – conciliation – was more fundamental as a motive, and a method of approaching classical authorities, than other scholars have suggested.

The theme of conciliation which I have suggested throughout this thesis does not square with the opinion of Roger, and even of such a distinguished historian as Walter Pagel, that sixteenth century writers on topics such as generation can be divided into two warring camps, usually identified as Aristotelians and Galenists. But as Dr. Lewis pointed out, almost everyone was both; and even on such an apparently clear issue as the existence of female seed there were shades of opinion rather than a clear opposition. This is not to say that there were no controversies, or that academic rivals hesitated to criticise one another. But the most polemic of treatises usually stated as its aim the intention to clear up the controversy once and for all by reconciling and explaining the classical authorities which lay behind the dispute.

More pacific writers – Conciliator, Accorambonius, Caimus, for example – made the reconciliation of authorities a major purpose of their work.

This seems to be closely related to the purpose (teaching) of many of these texts. Formats such as the commentary, guaestiones and dialogue, which allowed for a careful balancing of opinions, served this purpose well. It is easy to see why the comparing of major authorities and balancing of their opinions should be a valued teaching tool. In a sense, conciliation is only an aspect of the tradition of commentary on which arts teaching was based, with its programme of lecture-commentaries and disputations. Teaching also seeks, and sought, to make things clear, to place them in order, to distinguish one from another; as Caimus wrote, its purpose was to "make the implicit explicit, distinguish the confused, spell our the compressed, illuminate the obscure". And all the texts considered in the pages above were, after all, written with teaching in mind.

The theme of concord appears also in the models which these writers chose: models like Averroes and Peter, who bridged the gap between medicine and philosophy, and who provided guidance on methods of approach. And it is, in a sense, almost inevitable that there should be some common ground between two groups of professionals who shared so much, in their education, their sources, their research methods.

In a sense, this thesis is itself a tribute to the methods of its subjects, as I have described them. My aim, like Caimus's, has been to draw out, define, explain, and illuminate. In researching and writing this thesis, there has been a constant challenge, and a continuing fascination, in trying to understand, and make clear to myself and others, patterns of thought and modes of argument which are precise, subtle, and very different from our own. The problems with which these writers struggled are ones for which we still have not found entirely satisfactory answers. Both our answers, and our questions, are different from theirs; that is not really the point.

My interest is not in seeking parallels with modern biology; I shall be satisfied if I have managed to make the workings of this undeniably foreign intellectual milieu clearer to those of us who cannot share it; to catch some of its subtleties; and to provoke a sympathetic understanding of its aims and the extent to which we, as observers and commentators, can judge them to have been achieved.

Appendices

Appendix I: Renaissance Editions of Major Classical Works on Generation

Abstract Appendix I gives first and other major editions for the most important of the texts on generation of Plato, Hippocrates, Galen and Aristotle, together with some information on their previous availability in manuscript.

Work	First printed/major editions	Major French editions
PLATO		
Opera	1484–1485 Florence : Ficino tr.	
	1513 Venice : Aldus (Greek)	
	1546 Basle : Froben	
	1561Basle : Cornarius tr.	
Timaeus	1520 [Rome] : Calcidius' tr.	
	1525 Leonicus tr.	
		1532 Paris (Greek)
		1544 Paris : Ficino tr.
		1551 Paris (French)
		1578 Paris : Chalcidius/Cicero tr.
HIPPOCRATES		
Opera	1525 Rome : 1st ed. in trans.	1546 Paris : Cornarius tr. Wechel
	1526 Venice : Greek 1st ed.	
	1538 Basle : Greek and Cornarius tr.	
	1545 Venice	
De genitura	1525 Basle : Cornarius tr.	1545 Paris : Gorraeus tr.
		1549 Paris : Cornarius tr. Wechel

(continued)

L. Deer Richardson, *Academic Theories of Generation in the Renaissance*, History, Philosophy and Theory of the Life Sciences 22,
https://doi.org/10.1007/978-3-319-69336-1

De natura pueri	1483, 1519, 1534 In Articella	1545 Paris : Gorraeus tr.
	1525 Basle : Cornarius (w. De genitura)	1549 Paris : Cornarius tr. Wechel
		1559 Paris (French) Chrestien tr.
De morbis iv	As works	
ARISTOTLE		
Opera	1497 Venice : Aldus:1st ed. Gk. text	1590 Lyons: ed. Casaubon (Greek/Latin)
	1550–2 Venice : Junta : with Averroes	
De generatione animalium	1476 Venice (Gaza tr.) with Parts and History	1533 Paris : Colines : Gaza tr. with Parts & History
	1526 Greek text with Philoponus' commentary	
Historia animalium	1476	1524, 1533
	1619 Scaliger tr.	
De partibus animalium	1476	1533
GALEN		
Opera	1490 Venice : Pincius (Latin)	[1549]–1551 Lyons : Gesner tr.
	1525 Venice : Aldus (Greek)	
	1541–5 Venice	
De semine	1490 In Opera, 1st ed. (Reggio)	1533 Paris : Colines : Guinterius tr.
De foetuum formatione	1535 Basle : Felicianus tr.	1536 Paris : Guinterius tr. in Opera diversa : Colines
	1536 Basle : Cornarius tr. with de semine, de uteri dissectione	1556 (French) Chrestien tr.
De usu partium	1528	1528 Paris : Reggio tr. : Colines
		1566 Paris (French)
De Hippocratis et Platonis placitis	1534	1534 Paris : Guinterius tr. : Colines
	1535 Basle : Felicianus [In Opera from 1544]	

Appendix II: Renaissance Compendia

Abstract Appendix II lists twenty-six works by twenty-one authors published between 1471 and 1668, which follow the 'compendia' format, with some variations. They are listed in order of their publication (and usually, though not always, of their composition) beginning with the *Canon* and *Conciliator*. The format column in the table shows only the portion of the work which corresponds to the *Canon* I 1 Theses 1-6 (or to Fernel's *Physiologia*).

Author	Short title	Date	Format
Avicenna [980–1037]	Canon	printed from 1479	Bk I Fen i Theses 1–6:
			Definition of medicine
			Elements
			Temperaments
			Humours
			Members (parts)
			Faculties
			[operations]
Peter of Abano [1250–1316]	Conciliator	written c. 1303–1310 pr. 1471 & later eds	II [Quesita 10–63]
			Definition of medicine
			Elements
			Temperaments (complexiones)
			Humours
			Generationem
			Members
			Virtutes (faculties)
			Spirits
			Operationes
Hugo Senensis [d 1448]	Expositio in primam Fen..	1498	As Avicenna
Valla [d 1499]	De inventa medicina	1529	IYSAGOGAE Joanitii medici: De rebus naturalibus
			De quatuor elementis
			De commixtionibus (temps.)
			De compositiones (humours)
			De modis membrorum (parts)
			Virtues or faculties
			Operations
			Spirits
Cornarius [1500–58]	Epigraphe	1529	Table: De rebus naturalibus
			Elementa
			Humores
			Commixtiones
			Membra
			Virtutes
			Operationes
			Spiritus (3)

(continued)

Fernel [1497–1558]	De naturali parte medicina	1542	Parts
			Elements
			Temperaments
			Spirits & Innate Heat
			Faculties of the Soul
			Humours
			Procreation of Man
Forliviensis [d 1413]	Expositio	1547	As Avicenna: incl. (72r) Capitulum 7 doctrina sextae quod est de operationibus
Fuchs [1501–1566]	Institutionum medicae	1555 enlarged 1618 and improved by Stupanus	Order as Conciliator
Fernel	Medicina	1554	As 1542
Cornarius	Medicina	1556	Division of medicine into 5 parts
			Prima into 3 divisions Res naturales: E, T, H, P, F/ Actiones/Spirits
Valles [1524–92]	Controversarium medicarum ac philosophicarum	1556	Elements temperaments humours parts: including questions on generation Faculties
Montanus [1498–1552]	Methodus medicinae universalis	1558 (in Opuscula)	All broadly follow order of Avicenna
	Commentaria	1558	
	Medicina universa		
	113–120 :elements		
	121–158 :temperaments		
	159–194 :humours		
	195 members/parts		
			As Avicenna
Oddis [1478–1558]	In primam primi Fen	1575	Elements
Peramato	Opera medicinalia	1576	Humors
			Temperaments
			Faculties
			Liber de Semine
			Liber de hominis procreatione
			Entirely in the form of tables
Wecker	Medicinae Utriusque Syntaxes	1583	Bk I part I: PHYSIOLOGIKE
			De Elementis

(continued)

			De Semine
			De Humoribus
			De Urina
			Similar & instrumental parts
			Spirits(De spiritibus)
			Temperaments (De temporum diff.)
			Animal faculties
			De actionibus seu functionibus
Bertinus [fl 1586]	Medicina. Methodice	1587	I De Elementis
			II De Temperamentis
			III De Humoribus
			IV De Semine et Sanguine Menstruo...
			Bks I–VII correspond with books of Physiologia
Costaeus [d 1603]	Disquisitionum Physiologicarum	1589	Based on, and follows, Avicenna I i, 1–6
			Generation in members
Riolan [15387–1605]	Commentarius ad Fernelii Librum de.	Originally published 1575–8	
	Opera	1610	Follow Fernel, omitting Parts
	Praelectiones in libros Fernelii physiologicas	1601	Physiologia: E, T, S & C I, H, P, Fac. An, Functionibus
	Universae medicinae compendium Opera:	1610	As Avicenna
Pouce de Santa Cruz [c 1570–1650]	In Avicennae primam primi..	1622	As Avicenna
Sanctorius	Commentaria in primam Fen..	1626	Naturals: Elementa
Campanella [1568–1639]	Medicinalium..libri septem	1635	De mentis.
			De spiritus
			De humorum.
			De solidarum
			partium
			De sanitatis subjecto:
Blasius	Medicina universa	1665	Solids
			Humors
			Spirits
			Physiologia:
	Institutionum medicarum compendium	(1666)	De corporis humani.. Principiis

(continued)

		1667	De partibus solidis
			De humoribus
			De spiritibus
			De forma corporis humani
			De sanitate:
Schenckius	Synopsis	1668–71	The nature of health
			Solids
			Humors
			Spirits & native heat
			De causis sanitatis
			Soul
			Temperaments
			Occult qualities
			Dissimilar parts [of the unity of the body]
			Six non-naturals

Appendix III : A Note on Parts; the Anatomical Assumptions of Generation Theorists

Abstract Appendix III briefly considers 'parts', since anatomy is one aspect of Renaissance medicine which has received considerable attention from modern scholars, and also because anatomies are among the 'practical' works on generation which I have excluded.

Jean Fernel included parts, or members, within the natural part of medicine (or physiology). In both 1542 and 1554, he placed it first, not fifth, as in the Canon. And he took pains to explain his reasons for this change, and the importance of such knowledge for the physician.[7]

Why then, did Riolan omit to comment on this book, leaving his son to complete the father's work with his Anatome?[8] And why have I placed 'parts' in an appendix?

Parts, for Fernel, meant two things: theoretically, the simple and composite parts of which the body was said to be composed – a definition going back to Aristotle; and practically, the details of anatomy. Parts in the former sense, in particular the simple or homogeneous parts, are considered in the chapter on elements and temperaments (Part III, Chap. 14) above. Parts in the sense of anatomy were described both in compendia like the ones listed in Appendix II, and in textbooks of anatomy, one of the two classes of practical treatise which I identified in the Introduction. These might be editions of medieval treatises arranged from head to heel like Mundinus's Anatomia; straightforward dissection manuals in the order most to least perishable, like da Carpi's Isagoge; opulently produced and illustrated treatises, like those of Vesalius and Colombo, arranged – like Book I of the Physiologia – according to the system of simple parts; or, like Piccolomini's Praelectiones, grouped by faculty. They might be separate treatises on the organs serving generation; in the 'women's books', too, a description of the figure and location of the genital parts often preceded the description of their use.[9]

I pointed out in the Introduction that much of the past history of generation theory has been written by anatomists, who emphasised anatomical approaches and discoveries, such as Fallopio's discovery of the uterine tubes. More generally, the history of anatomy and its teaching has been a favourite theme of recent historians of medicine, with Vesalius and Harvey as focal points.[10] Much of this work is excellent; but whatever the impact of anatomical discoveries may have been on the understanding of the cardio-vascular system, they were, I believe, of minimal importance in clarifying understanding of the theory of generation in the period covered by this thesis.

The basic structures of the 'parts serving generation' had been known, and described, from the time of Herophilus. The descriptions offered by sixteenth-century anatomists of the stature of Fallopio, Eustachio, Colombo, Coiter and Platter were, in the words of Jill Lewis, "simply adding to a corpus of information going right back to Galen, and differing from it scarcely at all in fundamentals".[11] Moreover, they would have been recognisable to, and indeed differed remarkably little from, the sort of description, in terms of gross anatomy, which might be given by a competent nursing student today.

The sixteenth-century doctor and the twentieth-century student would, of course, disagree in their interpretation of the anatomical 'facts' before them. And as Wightman has pointed out page 15 and note 23, (see above) this is what counts. A straightforward, 'value-free' anatomical description is impossible; what matters is the interpretation which the anatomist puts on the mass of tubes and tissues before him. The interpretation which Renaissance anatomists placed on what they saw came from theoria – that is, from the classical authorities of Part I, and their interpretation by doctors and by natural philosophers.

Thus when da Carpi gives a straightforward description of the vessels supplying blood to the testes, he follows it immediately with the observation – which is not an anatomical one – that the right testis produces males because the seed produced there is hotter than in the left. Similarly, the early Renaissance anatomical treatises discussed by Lind often show a clear debt to Hippocrates for the details of their description of the development of the foetus.[12] The anatomical drawings of Leonardo combine carefully-observed detail with sketches which illustrate the opinions of Hippocrates, Galen and Avicenna. A famous drawing in the Royal Collection at Windsor shows a passage from the spinal cord into the penis, recalling Platonic and Hippocratic theories of the origin of the seed; another tube joins the uterus and the breasts, illustrating the theory that menstrual blood is converted to milk in pregnancy. Other sketches show the arrangement of the vessels supplying the womb, again based not on observation but on theory.[13]

Renaissance gross anatomy is not, of course, the same as modern anatomy; but for the purposes of understanding the subject matter of this thesis only a few differences are relevant. The most important differences relate to the arrangement of the arteries and veins supplying the testes and ovaries, and the question of what happens to the female seed after it leaves the ovary. Both are points of theoria, not anatomy.

On the first point: the arteries and veins supplying the (male and female) testes were often considered as preparatory seminal vessels. This seems to have been in an effort to reconcile Aristotle's contention that the seed was not elaborated in the testes, with Galen's view that these organs were indeed important to generation.[14] Anatomical illustrations, as late as the end of the sixteenth century, show the spermatic vein and artery joining to form vasa praeparantia before they enter the testis (see Figs. 1 and 2 below). Similarly, as the illustrations from Guibelet also show, in the female these vessels were shown as dividing again, with one branch going to the testis and one to the uterus. Very similar illustrations appear in Laurentius' Historia Anatomica. Both works postdate Fallopio's discovery, but the plates appear to be based on an earlier source; they are similar to the illustrations in Vesalius's Tabulae Sex and De Fabrica, but show the details more clearly.[15]

There was some question as to the route followed by the female seed when it left the testis, which again Fallopio seems not to have cleared up entirely. Did the seed enter the uterus at the 'horns' (Fallopian tubes) or at the vagina? Both the Laurentius and Guibelet plates show insertion to the body of the uterus. Once again, the choice was based on theoretical considerations, not anatomical ones. Female seed which

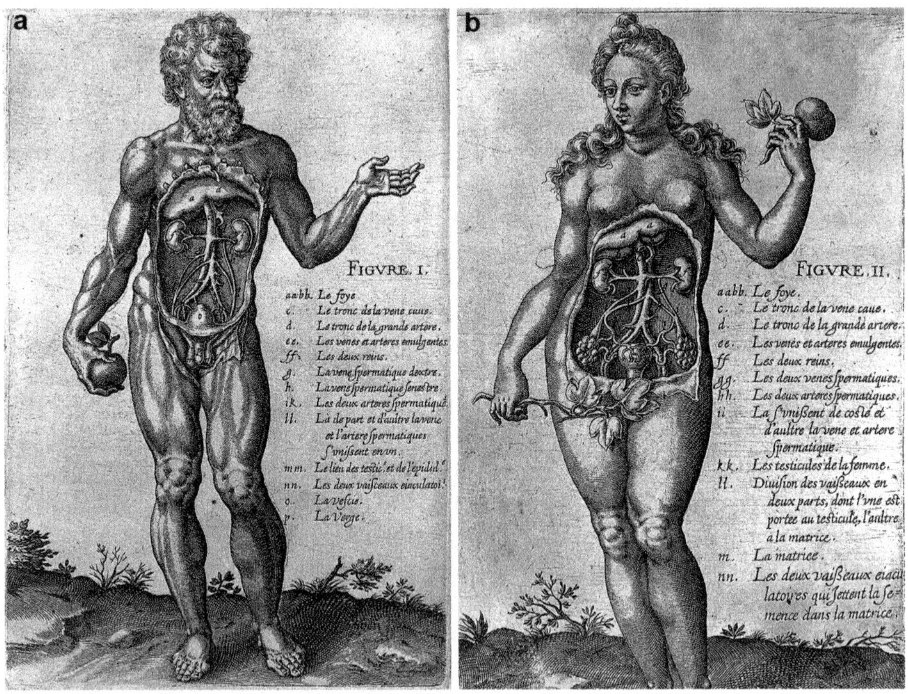

Fig. 1 Male and female anatomy of generation from Trois Discours Philosophiques, 1603 (Courtesy Wellcome Collection)

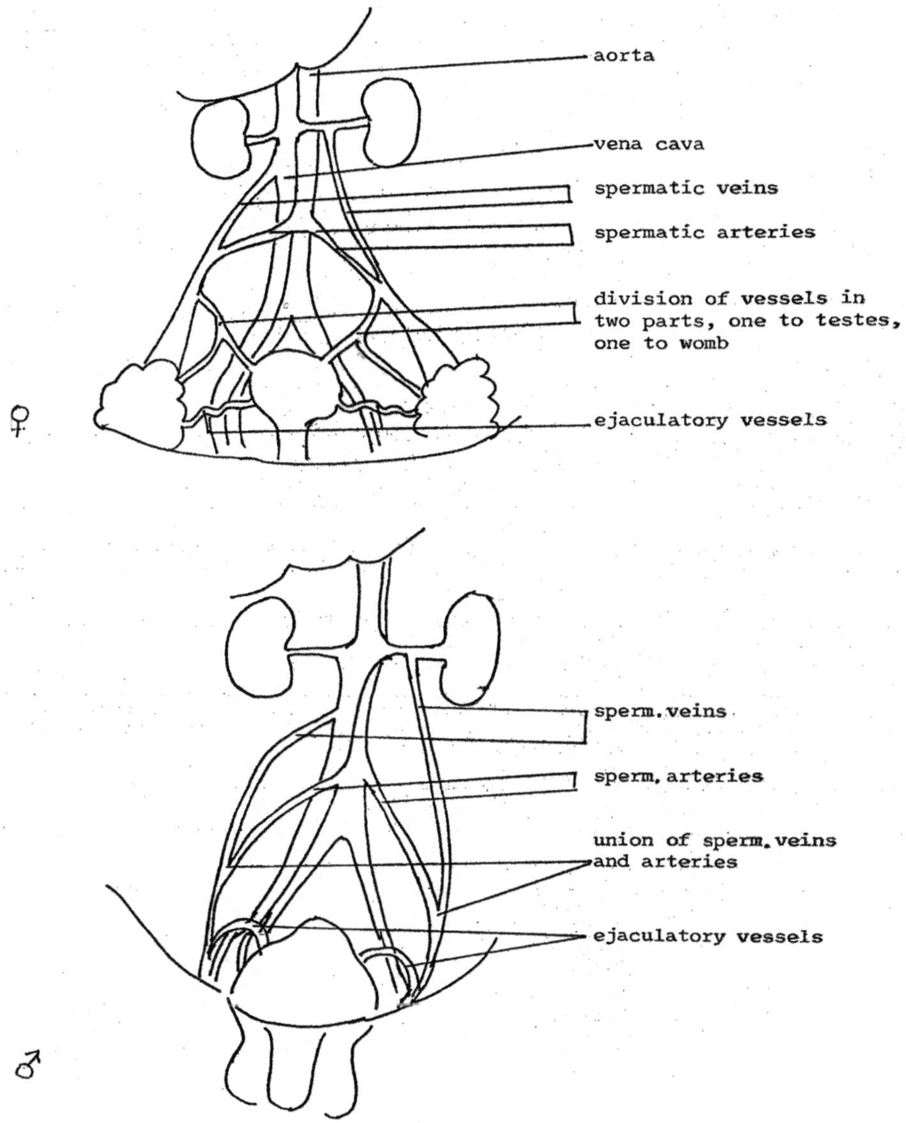

Fig. 2 Sketch of anatomical figures, Guibelet (1603)

entered the body of the uterus could be assumed to be a constitutive part of the foetus, as Hippocrates and Galen argued; seed which entered at the neck was likely to be identified with vaginal moisture, as in the arguments of Aristotle and Averroes. The latter arrangement may also have been based on descriptions derived from Herophilus, and preserved in Galen, which placed the exit point for female seed in the neck of the bladder – by analogy with the male arrangement.[16] Alessandro Benedetti argued for both: a prolific seed which formed a mass within the uterus (and which he claimed to have seen in a dissected deer, half a century before Harvey's De generatione) and a non-prolific seed which corresponded to Aristotle's description.[17]

The question concerning the number of chambers within the uterus – one, two, three or seven – does not concern us directly. It is based, not just on descriptions taken from animal anatomy, but also on the theories of sex-determination found in, for example, Galen, De semine, which suggest that males are formed in the right, hotter, side of the womb, females in the left.

The object of this survey is not to criticise Renaissance anatomists but simply to point out the different assumptions which would otherwise make their descriptions confusing to a modern reader. I have found the figures in Guibelet useful in sorting out my own puzzlement, and offer them as a guide to Renaissance understanding of the detailed arrangement of the parts serving generation.

Bibliography[18]

Primary

ACCORAMBONIUS, Felix. Felicis Accorambonii.. Interpretatio obscuriorum locorum et sententiarum omnium operum Aristotelis, & praecipuorum dubiorum. quae in singulis eius libris moveri solent, …declaratio..& omnium ferme controversiarum, quae versantur inter Platonicos, Galenum, & Aristotelem examinatio. His additur.. difficilium Theophrasti in libris de Plantis explanatio; & in librum Galeni de Temperamentis Annotationes … Rome, 1590.

ACTUARIUS, Joannes. Actuarii Joannis filii Zachariae opera. De actionibus et spiritus animalis affectibus eiusque nutritione Lib. II. De urinis Lib VII. Methodi medendi Lib. VI. Paris, 1556.

ALBERTUS, F. Aloysius. De Nutritione, Augmento, et Generatione, Disputationes. In quibus Aristoteles defenditur adversus Galenum. F. Aloysii Alberti Patavini, Ord. Eremit. S. Augustini Philosophi, & in Patrio Gymnasio Sacrae Scripturae interpretis. Venice, 1627.

ALBERTUS, Magnus. Divi Alberti Magni De Animalibus Libri vigintisex Novissime impressi. [emendati per M. A. Zimaram]. Venice, 1519.

Pseudo-ALBERTUS, Magnus. Magni Alberti secreta mulierum et virorum: philosophi decantatissimi: nuper juxta veritatem veterum exemplationum correcta et emendata. [s.l., s.n.] 1542.

ALEXANDRINUS, Julius. Julii Alexandrini Serenissimi Romanorum Regis Physici, De Medicina et Medico, Dialogus, Libris Quinque Distinctus. Zurich, 1557.

ALTOMARI, Donatus Antonius ab. Donati Antonii ab Altomari Medici atque Philosophi Neapolitani. Omnia, quae hucusque in lucem prodierunt, Opera, nunc primum in unum collecta, & ab eodem Auctore diligentissime recognita & aucta … Lyons, 1565.

ARGENTERIUS, Joannes. Joannis Argenterii Medici de somno et vigilia Libri Duo, in quibus continentur duae tractationes de calido nativo, et de spiritibus. Omnia non solum mediciis, sed etiam universis bonarum artium studiosis utilia, & iucunda, atque aliter explicata, quam hactenus fuerint ab alii tradita. Florence, 1556.

[ARGENTERIUS] Joannis Argenterii Castellonovensis Pedemontii, Philosophi ac medici acutissimi simul et gravissimi. In Florentissimis Italiae Gymnasiis, Neapolitano, Pisano, Taurinensi, &c., quondam Medicinae cum Theoricae, tum Practicae Professoris primarii, ac celeberrimi.

OPERA NUNQUAM EXCUSA. Iamdiu desiderata, ac e tenebris in lucem prodita. In duas partes distincta. Quarum prior commentario in Hippocratis Aphorismam primam, secundam & quartam sectiones plus sex & triginta annorum spatio elaboratos. Altera vero de Febribus tractatum singularem: & primi libri ad Glauconem praeclaras explanationes: Item de Calidi significationibus, ac Calido Nativo doctissimum libellum complectitur.

In quibus potissimae, ac difficilimae totius Medicinae controversiae, subtiliter, & exacte dijudicantur, ipsaque sanitatis Tuendae, ac Restituendae ratio planissime traditur. Venice, 1556.

L. Deer Richardson, *Academic Theories of Generation in the Renaissance*, History, Philosophy and Theory of the Life Sciences 22,
https://doi.org/10.1007/978-3-319-69336-1

— Joannis Argenterii Pedemontani in Artem Medicinalem Galeni, Commentarii Tres, Non solum mediciniae professoribus utiles & necessarii: sed etiam Philosophis, & universis, qui rerum scientia delectantur, summopere iocundi. Mondovi, 1566.

ARISTOTLE. De natura animalium, De partibus animalium, De generatione animalium. [Translated by Theodorus Gaza], Venice, 1476 (Later editions in 1492, 1498, 1533).

— Aristotelis quinque de Animalium generatione libri ex inter-pretatione Theodori Gazae cum Philoponi: seu Joanis Grammatici Commentariis per Nicolaum Petrum Corcyraeum Accuratissime e graeco in latinum conversis. Venice, 1526.

— In hoc volumine haec continentur. Aristoteles de historia animalium libri ix. De partibus animalium ... libri iiii. De generatione animalium libri iv. Theodoro Gaza interprete. De communi animalium gressu liber 1. De communi animalium motu liber 1, Petro Alcyonio interprete. Paris, 1533.

— [Opera omnia] Sextum Volumen, Aristotelis Stagiritae Libri Omnes, Ad Animalium Cognitionem Attinentes, cum Averrois Cordubensis variis in eosdem commentariis.. Venice (Juntas), 1550.

— **'Αριστοτελους ...τα σωζομενα** Operum Aristotelis...nova editio, Graece & Latine... adscriptus ad oram libri... emendationibus.. Latinae interpretationes adjecta sunt..Accesserunt... fragmenta quaedam, etc. 2 vols, [ed. I. Casaubon] Lyons, 1590.

— **'Αριστοτελους περι Ζοον ἱστοριας** Aristotelis Historia de Animalibus Julio Caesare Scaligero Interprete, cum eiusdem Commentariis. Toulouse, 1619.

— The Works of Aristotle translated into English under the Editorship of W. D. Ross. Oxford, 11 vols, 1908–1931.

— A. L. Peck (editor and translator), Aristotle: Generation of Animals, Loeb Classical Library, 1953.

— Aristoteles Latinus XVII, 2 v, De generatione animalium (Translatio Guillelmi), ed. H. J. Drossaart, Leiden 1966. Aristoteles Latinus: Codices. Supplementa Altera, ed. L. Minio-Paluello, Bruges and Paris, 1961.

— Balme, D. M. (editor and translator), Aristotle's De partibus animalium I and De generatione animalium I (with passages from II 1-3), Oxford, 1973.

— **ΑΡΙΣΤΟΤΕΛΟΥΕ ΤΑ ΜΕΤΑ ΦΥΣΙΚΑ** Aristotle's Metaphysics. A Revised Text with Introduction and Commentary by W. D. Ross. Oxford, 1975 (first edition 1924).

Pseudo-Aristotle. The Problems of Aristotle, with other Philosophers and Physicians. Wherein are contained, Divers Questions with their Answers, touching the Estate of Mans Body, with Marcus Antonius Zimaras Sanctipertias his Problemes. Alexander Aphrodiseus His Problems. London, 1670.

— Secreta Secretorum [Facsimile of the edition of] London 1528. Amsterdam and New York, 1970.

ARISTOTLE (Commentaries) Commentaria in Aristotelem Graeca, vol. 14, pt. 3, Joannis Philoponi (Michaelis Ephesii) in Libros De Generatione Animalium Commentaria, edited by Michael Hayduck, Berlin, 1903; vol. 22, i, Michaelis Ephesii in Parva Naturalia Commentaria, ed. Paul Wendland, Berlin, 1903; 22, 2, Michaelis Ephesii in Libros De Partibus Animalium, De Animalium Motione, De Animalium Incessu Commentaria, ed. Michael Hayduck, Berlin, 1904.

AUBERT, Jacques. Progymnasmata in Joannis Fernelii librum de abditis rerum et medicamentorum causis: quibus adduntur quorundam gravissimorum morborum curationes. Basle [1579].

AVERROES. Collectaneorum de re medica Averrhoi philosophi, post Aristotelem atque Galenum facile doctissimi, Sectiones tres,

i De Sanitatis functionibus, ex Aristot. & Galeno.

ii De Sanitate tuenda, ex Galeno

iii De curandis morbis.

A Joanne Bruyerino Campegio, prudentissimi, literatissimique Cardinalis Turnonii medico, nunc primum Latinate donate. Lyons, 1537.

— Colliget Averrois Totam medicinam ingentibus voluminibus ab aliis traditam mira quadam brevitate & ordine sic ad amusim complectens, ut jure ab omnibus liber iste medicorum stimulus cognominetur...Accesserunt postremo M. Anto. Zimarre dubia & solutiones in supradictam Averrois colliget, novum corollarium. Venice, 1542.

AVICENNA. Liber Canonis. 1964 Georg Olms Verlagsbuchhandlung, Hildesheim. (Facsimile reprint of 1507 ed. Venice).

BERENGARIO DA CARPI, Jacopo. A short Introduction to Anatomy (Isagoge Breves) translated with an Introduction and Historical Notes by L. R. Lind and with Anatomical Notes by Paul G. Roofe. Chicago, 1959.

BERTACCHIUS, Dominicus. De Spiritibus libri IV. Venice, 1584.

BERTINUS, Georgius. Georgii Bertini Campani Medicina Libris Viginti Methodice absoluta. In qua mutuus Graecorum & Arabum CONSENSUS, legitima veteris Medicinae adversus Paracelsistas DEFENSIO, vera Animadversionum Argenterii in Hippocratem & Galenum CONFUTATIO, dilucida Controversiarum & Locorum difficilium EXPLICATIO, atque accurata novarum hoc seculo Rerum OBSERVATIO continentur. Basle, 1587.

BLASIUS, Gerardus. Medicina universa; hygiene & therapeutices fundamenta methodo nova brevissime exhibens. Amsterdam, 1665.

— Institutionum medicarum compendium, disputationibus xii ... Amsterdam, 1667.

BOUELLES, Charles de (Carolus Bovillus Samarobrinus). Que hoc volumine continentur. Liber de intellectu. Liber de sensu. Liber de nichilo. Ars oppositorum. Liber de generatione. Liber de sapiente. Liber de duodecim numeris. Epistoles complures. [Paris, 1510].

BRISIANUS, Hieronymus. Physiologiae Libri ii, quibus naturae miracula...explicantur. Venice, 1596.

BRONZERIUS, Joannes. De innato calido et naturali spiritu disputatio Io. Hieronymi Bronzerii. In qua pro rei veritate Galeni doctrina passim explicatur, defenditur. Padua, 1626.

— De principio effectivo semini insito disputatio Io. Hieronymi Bronzerii. In qua complura attinentia ad seminis Corpulentiam, Spiritus, Calidum innatum, Animationem, nec non ad Animae materialis naturam & facultates Discutiuntur & enodantur, passimque Galeni roboratur Physiologia. Padua, 1627.

CAESALPINUS, Andreas. Andreae Caesalpini Aretini, Quaestionum Peripateticarum Libri V ... Daemonum investigatio Peripatetica ...Quaestionum Medicarum Libri II. De Medicamentis facultatibus Libri II... Venice, 1593.

— Andrae Caesalpini Artis Medicae Liber VII. De Morbis Ventris [bound with] Andreae Caesalpini Artis Medicae Liber Octavus De Genitalium Morbis. Rome, 1603.

CAIMUS, Pompeius. De calido innato Libri Tres, In quibus non solum eius natura explicatur, sed solida etiam Medicorum in hoc argumento doctrina ostenditur, et Galenica praecipue a Neotericorum objectionibus vindicatur. Venice, 1626.

CAMPANELLA, Thomas. Medicinalium, Juxta Propria Principia libri septem. Opus non solum medicis, sed omnibus naturae et private valetudinis studiosis utilissimum. Lyons, 1635.

CANDIDUS DECEMBRIUS . De genitura hominis. Rome, [c 1495].

CARDANUS, Hieronymus. Libelli quinque... I. De supplemento almanach. II. De restitutione temporum et motuum coelestium. III. De judiciis geniturarum. IV. De revolutionibus. V. De exemplis centum geniturarum... Ejusdem, antea non edita, Aphorismorum astronomicorum segmenta VII. Nuremberg, 1547.

— De subtilitate libri xxi. Nuremberg, 1550.

— Hieronymi Cardani Medici Mediolanensis. Contradicentium medicorum libri duo, quorum primus centum et octo, alter vero totidem disputationes continet.. [with other works]. Paris, 1564.

— Les livres intitulez de la subtilité, et subtiles inventions, ensemble les causes occultes, et raisons d'icelles. Traduits de latin en françoys par Richard le Blanc. Nouvellement revuez. Paris, 1578.

CARDELINUS, Victor. De origine foetus Libri duo. In quibus praeter caetera luculentis Graecorum auctoritatibus calidum à calore dirimitur. Ac expenditur, quid utrunque apud eosdem significet cum adjectione innati, vitalis, naturalis, genitalis, & animalis. Quidve sit archigonum. Victoris Cardelini Bassanensis Art. ac. Leq. Doct. Vincenza, 1628.

CHAMPIER, Symphorien. Ars parva Galeni ...Subjunguntur Paradoxa Domini Simphoriani Champerii: ...in Artem parvam Galeni. In quibus ... Que a Trusiano. Genitili. Jacobo

Forliviensi. Sermoneta. & Ugone Senensi: omnibusque neotericis scripta sunt ad medicos breviter: clareque narrantur...Itemque additiones Haly Rodoam admodumque acute : ac docte. Lyons, 1516.
— Simphoriana Champerii Lugdunensis: in libros demonstrationum Galeni Cathegorie medicinales. In quibus..que Galenus in Demonstrativis sermonibus: et Aristoteles in Cathegoriis & naturalium libris scripserunt: breviter clareque et placido stilo pertractantur atque declarantur sententiae. Lyons, 1516.
— Symphonia Platonis cum Aristotele: & Galeni cum Hippocrate, Paris, 1516.
— Speculum Galeni. Epitome Galeni sive galenus abbreviatus vel incisus aut intersectus quecumque in speculo domini Simphoriani Champerii... [includes: De elementis Galeni epitoma; De generatione animalium epitoma...; De gyneciis liber hoc est de passionibus mulierum...] Lyons, 1517.
CICERO. The Nature of the Gods, translated by Horace C.P. McGregor. Penguin Classics, Harmondsworth, 1972.
Coiter, Volcher. Externarum et internarum principalium humani corporis partium Tabulae et Exercitationes. Nürnberg, 1573.
COLUMBUS, Realdus. De Re Anatomica libri XV. Venice, 1559.
CONTARINI, Gaspar. Gasperis Contareni Cardinalis Amplissimus philosophi sua aetate praestantissimi de Elementis et eorum mixtionibus libri quinque ...nunc primum in lucem aediti. Paris, 1548.
CORNARIUS, Janus. Universa rei medicae EPIGRAPHE seu Enumeratio, compendio tractata, Jano Cornario Zuiccaviensi autore. Basle, 1529.
— Iani Cornarii/medici physici/Medicina, sive Medicus, Liber/unus/eiusdem orationes duae: altera Hippocrates, sive Doctor verus: altera De rectis medicinae studiis amplectendis. Basle, 1556.
COSTAEUS, Joannes. Disquistionum Physiologicarum Joannis Costaei Laudensis, In primam primi Canonis Avic. sect. Libri Sex. Bologna, 1589.
— De Humani Conceptus Formatione, ac Partus Tempore. Joannis Costaei Laudensis Medici Clarissimi, Olimque in Bononiensi Gymnasio Medicinae supraordinario loco professoris. Pavia, 1604 (first edition, Bologna, 1596).
CREMONINI, Cesare. Excellentissimi Caesaris Cremonini Centensis Lecturae Exordium Habitum Patavii. VI. Kalend. Februar. MDXCI. Quo is primum tempore philosophiae interpres ordinarius eò est Profectus. Ferrara, 1591.
— Caesaris Cremonini Centensis In Schola Patavina Philosophi Ordinarii Explanatio Prooemii Librorum Aristotelis De Physico Auditu. Cum Introductione ad Naturalem Arist. Philosophiam.. Descriptionemque universae naturalis Aristoteliae philosophiae.. Padua, 1596.
— Caesaris Cremonini Centensis In Schola Patavina Philosophi primae Sedis: De Formis quatuor Corporum Simplicium, quae vocantur Elementa, Disputatio. Venice, 1605.
— Caesaris Cremonini Centensis Disputatio de coelo: adjecta est apologia dictorum Aristotelis de via lactea, de facie in orbe lunae. Venice, 1613.
— Apologia dictorum Aristotelis de quinta coeli substantia adversus Xenarcum, Joannem Grammaticum et alios. Caesaris Cremonini in schola Patavina Philosophi primae sedis. Venice, 1616.
— Apologia dictorum Aristotelis de calido innato. Adversus Galenum. Caesaris Cremonini Centensis in Schola Patavina Phylosophi Primae Sedis. Venice, 1626.
— Caesaris Cremonini Centensis In Schola Patavina Philosophi Primae Sedis, Tractatus de Semine, pro Aristotele contra Galenum. (Bound with de calido innato). Leyden, 1634.
— Apologia dictorum Aristotelis De origine, et principatu Membrorum. Adversus Galenum. Caesaris Cremonini Centensis in Schola Patavina Philosophi primae Sedis. Venice, 1627.
— Cesare Cremonini nel Primo delle sue Meteori a diffessa per Aristotile. Contro li Astrologhi Guidiciarii. in Raccolta medica et astrologica, divisa in doi discorsi L'uno per Hippocrate contro Galeno, dell'Abuso commune di cavar sangue col salasso nelle Febri L'altro per Hippocrate, & Aristotele contro li Astrologhi Giudiciarii, cosi in generale come per uso di Medecina.. Lootri Nacattei Compilatore, & conversore. Venice, 1645.

— Expositio super caput tertium libri secundi de generatione animalium (undated manuscript treatise). Wellcome Microfilms: M.S. 210a = Padua, Museo Civico, M.S. CM494 (xvii) Item 2 M.S. 212a = Venice, Biblioteca Marciana, M.S. Lat. VI, 179 [= 2825] ff 75r-142v.
DE GARBO, Thomas. Summa Thome de Garbo. Thome de Garbo Florentini/doctoris praeclarissimi/Summa Medicinalis: cum duobus eiusdem Tractatibus: unus/De restauratione humidi radicalis: alter/ De reductione medicinarum ad actum: Lyons, 1529.
DELPHINIUS, Julius. In III Galeni artis medicinalis libros explanatio. Eiusdem de ratione medicamentorum praescribendorum liber. [Book III only] Venice, 1557.
— Julii Delphini Mantuani Medici Clarissimi Quaestiones Medicinales, nunc primum in lucem editae. Venice, 1559.
DU GARDIN, Louis. De Animatione Foetus Quaestio: In qua ostenditur, quod Anima Rationalis, ante Organizationem non infundatur. Douai, 1623.
— Anima rationalis restituta in integrum: sive altera refutatio opinionis, quae sibi persuadent, Anima Rationalem, ante omnem Organizationem, infundi in Semen. Douai, 1629.
FALLOPPIUS, Gabrielus. Opera omnia, in unum congesta, et in medicinae studiosorum gratiam excusa; Frankfurt, 1550.
FERNEL, Joannes. Joannis Fernelii Ambianitis de proportionibus Libri duo. Prior, qui de simplici proportione est, & magnitudinum & numerorum tum simplicium tum fractorum rationes edocet. Posterior, ipsas proportiones comparat: earumque rationes colligit. Paris, 1528.
— Joannis Fernelii Ambianatis, De naturali parte medicinae Libri septem, Paris, 1542.
— De Abditis Rerum Causis Libri Duo, Venice, 1550.
— Jo. Fernelii Ambiani, Medicina, Paris, 1554.
FIENUS, Thomas. De Formatrice foetus liber in quo ostenditur animam rationalem infundi tertia die. Antwerp, 1620.
— De formatrice foetus liber secundus in quo prioris doctrina plenius examinatur et defenditur. Louvain, 1624.
— Thomae Fieni Pro sua de animatione foetus tertia die Opinione Apologia adversus Ant. Ponce Sanctacruz. Louvain, 1629.
FRACASTORIUS, Hieronymus. Hieronymi Fracastorii Veronensis, Opera Omnia. Venice, 1555.
FRANZOSIUS, H. Hieronymi Franzosii Veronensis. Philosophi et Medici. Tractatus Apologeticus De Semine Pro Aristotele adversus Galenum. Verona, 1645.
FUCHS, Leonhard. Institutionum medicinae ad Hippocratis, Galeni, aliorumque veterum scripta recte intelligenda...libri quinque.. Lyons, 1555.
— Institutionum medicinae, seu medendi methodi... E. Stupani.. repurgati..plurimisque in locis auctiores redditi. Basle, 1618.
GALEN. Exhortatio ad bonas arteis ... de optimo docendi genere, & qualem oporteat esse medicum, Basle, 1526.
— Cl. Galeni Pergameni Libri Tres. Primus, de facultatum naturalium substantia. Secundus, quod animi mores, corporis temperamentum sequuntur. Tertius, de propriorum animi cuiusque affectuum agnitione & remedio. Guinterio Ioanne Andernaco interprete. Paris, 1528.
— Claudii Galeni De Elementis: Libri Duo, Victore Trincavelio Interprete. Adiecimus in calce Hippocratis librum de Elementis, unà cum commentario in eundem Jacobi Sylvii medici. Lyons, 1550.
— Claudii Galeni Pergameni de Temperamentis Libri III. De inequali intemperie Liber unus. Thoma Linacro Anglo Interprete. Cum Isagoge in eosdem libros, & scholiis marginalibus longè doctissimus, per Jacobum Sylvium. 1549.
— Claudii Galeni Pergameni de Naturalibus Facultatibus Libri Tres, Thoma Linacro Anglo interprete..Additusque est de naturalium facultatum substantia liber, & an sanguis in arteriis natura contineatur, Victore Trincavelio interprete..Lyons, 1550.
— Galeni Operum Omnium Sectio Prima [etc.] Illustriores quam unquam antea prodeunt in lucem omnes hi Galeni libri. Accesserunt enim his ex veterum Graecorum exemplarium collatione adnotationes luculentissimae singulis locis in margine appositae, Augustino Ricco Medico Lucensi authore. Venice, 1541–1545.

— De la formation des enfants au ventre de la mère. Traduit du grec en français par G. Chrestian, Docteur en médecine. Paris, 1556.
— Oeuvres anatomiques, physiologiques et medicales, traduites (...) par Ch. Darmeberg. Paris, 1854–1856.
— [Quomodo animetur fetus ad Gaurum] Die neuplatonische fälsliche dem Galen zugschriebene Schrift πρὸς ταῦρος φερί του φῶς εμψυχοῦται τὰ ἐμβρυα aus der pariser Handschrift zum ersten Male herausgegeben, von Dr. Karl Kalbfleisch.. Berlin, Verlag der kön. Akademie, 1895.
— Galen on the Natural Faculties, with an English translation by Arthur John Brock, M.D. London, 1928. (Loeb Classical Library)
— De uteri dissectione; translation by Charles Mayo Goss, "On the Anatomy of the Uterus", Anatomica Record, v, 144, No. 2, October 1962.
— Claudii Galeni Opera Omnia. Editionem curavit C. G. Kühn 19 vols. Hildesheim, 1964–1965 (Facsimile of Leipzig, 1821–1830).
— On the Usefulness of the Parts of the Body περὶ χρειας μορίων De usu partium. Translated from the Greek with an Introduction and Commentary by Margaret Tallmadge May. Ithaca, 1968.
GALLEGO DE LA SERNA, Joannes. Joannis Gallego de la Serna Malachensis, Catholicorum Philippi III & IV Hispaniarum Regum Archiatri; Necnon Christianissime Gallorum Reginae Primarii quondam Medici, Opera Physica, Medica, Ethica, quinque Tractatibus comprehensa: quorum
I Agit de Principiis generationis omnium viventium
II De Conservatione infantis in utero..
III De puerorum aliendi ratione, & sanitate tuenda..
IV De communi puerorum educandi ratione..
V De optimi Regis educandi ratione.
Lyons, 1634.
GORRAEUS, Joannes. Jo. Gorraei Parisiensis, Definitionum Medicarum Libri XXIIII. literis Graecis distincti. Paris, 1564.
GUARINONIUS, Christophorus. Christophori Guarinonii Fontani Veronensis, Philosophiae et Medicinae Doctoris, Rodolphii. II Imperatoris maximi et gloriosissimi a consiliis valetudinis, Commentaria in Primum Librum Aristotelis, de Historia animalium. In quibus Animantium differentiae conspicuae fiunt, & sedulo ad certam divisionis normam ordinatur. Frankfurt, 1601.
— Christophori Guarinonii Fontana Veronensis, Philosophiae et Medicinae Doctoris, apud Rudolphum II Imperatorem supremi a Consilii valetudinis, De Natura Humana Sermones Quatuor Veronae in consessu literatorum habiti. Frankfurt, 1601.
GUIBELET, Jourdain. Trois Discours Philosophiques.
Le I. de la comparaison de l'Homme avec le Monde
Le II du Principe de la generation de l'Homme
Le III. de l'Humeur melancholique.
Mis de nouveau en lumiere par Jourdain Guibelet. Evreux, 1603.
HARVEY, William. Exercitationes de generatione animalium. Amsterdam, 1651.
— The Works of William Harvey, M.D. trans. Robert Willis, London, 1847.
HELEPYRUS, Andreas, Respondente. Concordia Medicorum et Physicorum de Humano Conceptu, atque Foetus Corporatura, Incremento, Animatione, Mora in Utero ac nativitate... CXXI thesibus publicè Novemb. disputandis compraehensa. Praeside D. Joanne Lonaeo Boscio Medicinae ac Mathematices Professore. ordinario, atque M. Andrea Helepyro Monachiensi, respondere pro viribus parato. Ingolstadt, 1582.
HIPPOCRATES. Oeuvres complètes d'Hippocrate, traduction nouvelle avec le texte grec en regard, collationné sur les manuscrits et toutes les éditions; accompagnée d'une introduction, de commentaires medicaux, de variantes et de notes philologiques; par E. Littré. Paris, 10 vols., 1839–1861.
— Hippocratic Writings. Edited with an introduction by G.E.R. Lloyd. Translated by J. Chadwick and W. N. Mann, I. M. Lonie, E. T. Withington. Penguin Classics, Harmondsworth, 1978.

HOFMANN, Caspar. De Generatione Hominis Libri Quatuor, Contra Mundinum Mundinium. Adjecimus sententiam eiusdem de Formarum Origine secundum Aristotelem. Frankfurt, 1579, 1629.

HORST, Gregor. Conciliator Enucleatus, seu Differentiarum Philosophicarum et Medicarum Petri Apponensis Compendium Opera Gregori Horsti, D. Profess. Med. Primar. in Acad. Giessena & Archiatri Hassiaci elaboratum. Editio nova Giessen, 1621.

HUGO SENENSIS . Expositio Ugonis Senensis super primo Canonis Avicenne cum quaestionibus eiusdem. [Venice, 1498].

JACOBUS FOROLIVIENSIS . Jacobi forliviensis verissimi medicine interpretis Egregia & subtilis expositio cum questionibus supra aureum capitulum de generatione embrionis feliciter incipit. Pavia, 1479.

— Jacobi de Forlivio/Scriptium/Super Avicenna Canonis Lib III Fen XXI Tractatus 1 Capitulum II/ Venice, 1489.

— & others. Jacobi forliviensis verissimi medicine interpretis egregia & subtilis expositio cum questionibus supra aureum capitulum de generatione embrionis. Venice, 1502.

— Jacobi Foroliviensis Medici Singularis Expositio et Quaestiones in Primum Canonem Avicennae. Adjecta Jacobi De Partibus in VII et VIII Cap. Doct ii Fen iii expositione, ac Ugonis quaestione, de malitia complexionis diversae. Venice, 1547.

LAURENTIUS, Andreas. Historia Anatomica Humani Corporis et Singularum eius Partium multis controversiis et observationibus novis illustrata. Frankfurt [1599].

LEMNIUS, Levinus. De Miraculis Occultis Naturae, Libri III. Item, de vita cum animi et corporis, incolumitate recte instituenda, liber unus. Illi quidem iam postremum emendati, & aliquot capitibus aucti: hic vero nunquam antehac editus. Antwerp, 1574.

LEONICENUS, Nicolaus. De virtute formativa. Venice, 1506.

— In In libros Galeni e graeca in latinam linguam a se translatos prefatio communis. Eiusdem in Artem medicinalem Galeni..praefatio. Galeni Ars medicinalis Nicolao Leoniceno interprete.. Eiusdem ad Franciscus Castellum.. in opus de tribus doctrinis ordinatis secundum Galeni sententiam prefatio. Eiusdem de tribus doctrinis ordinatis secundum Galeni sententiam opus. [With Laurentianus' translation of De differentiis febrium.] Venice: J. Pontius de Leucho, 1508. [With Leonicenus' translation of Galen on Hippocrates' Aphorisms] Ferrara, 1509.

LEONICENUS N./Monardus, J. Liber absolutissimus, quem hoc est Artem medicinalem, inscripsit. Latinate donatus partim a Nicolao Leoniceno Vicentino: partim a Joanne Monardo... Idem liber artis medicinalis.. Graece additus... Commentarii perbreves et familiares in eundem librum... in lucem editi per Joannem Agricolam Ammonium. Basle, 1541.

Licetus, Fortunius. De ortu animae humanae libri tres. Fortunii Liceti Genuensis, Philosophi, et Medici, in Pisana Academia Professoris. Genoa, 1602.

— Fortunii Liceti Genuensis Phil. Medici.. De perfecta Constitutione hominis in utero liber unus. Padua, 1616.

LICETUS, Fortunius. Fortunii Liceti Genuensis in Patavino Lyceo Physiologi Ordinarii. De Anima subjecto corpori nil tribuente, Deque Seminis Vita, & Efficientia primaria in formatione Foetus Liber Unus: in quo respondetur oppositionibus Aut. Ponce Santacrucii.. Padua, 1631.

LUCRETIUS. In Carum Lucretium poeta Commentarii a Joanne Baptista Pio editu: [Rome, 1514].

— Titi Lucretii Cari De Rerum Natura Libri Sex. [trans. Dionysius Lambinus] Paris, 1563.

MERCURIALIS, Hieronymus. Praelectiones Pisanae Hieronymi Mercurialis Foroliviensis Medicinam Pisis Loco Supra-ordinario Vocato Profitensis In Epidemicas Hippocratis Historias, Non minus ad Theoricam, atque Practicam Medicinam utiles, quam ob eruditionem iucundae. Nec non Tractatus Primo De Hominis Generatione Secundo, De Balneis Pisanis, Tertio, De Vino & Aqua...Venice, 1597.

MICHAEL OF EPHESUS (JOANNIS PHILOPONI) Joannis Philoponi (Michaelis Ephesi) in libros de generatione animalium commentarium edidit Michael Hayduck. In: Commentaria in Aristotelem Graeca edita consilio et auctoritate Academiae Litterarum Regiae Borussicae, XIV, pars III. Berlin.

MONTANUS, Joannes Baptista. In artem parvam Galeni explanationes. A Valentino Lublino Polono editae. Venice, 1554.
— Methodus Joannis Baptistae Montani De Elementis, Iatriae Candidatis apprimè utilis ac necessaria.. Vienna, 1554.
— Johannis Baptistae Montani ...In Primi Lib. Canonis Avicennae Primam Fen, Profundissima Commentaria. Venice, 1558.
— Opuscula varia ac praeclara: in quibus tota fere medicina methodice explicatur... Basle, 1558.
— Medicina universa Johannis Baptistae Montani Veronensis, Philosophi, et Medici, summi; Ex Lectionibus eius, caeterisque opusculis, tum impressis, tum scriptis collecta, & in tres tomos nunc primum decenti ordine digesta, studio & opera Martini Weindrichii Vratislaviensis... Frankfurt, 1587.
MUNDINIUS, Mundinus. Mundini Mundinii Vincentini Disputatio in tres partes divisa, In qua ea, quae de semine sunt controversa inter peripateticos, & Veteres Medicos, & doctissimos quosdam Neotericos accuratissime discutiuntur. Treviso, 1609.
— De genitura pro Galenicis adversus Peripateticos et nostrae aetatis Philosophos, ac Medicos, Disputatio. In qua nova praesertim dogmata spectantia ad foetuum generationem, similitudines, morbos haereditarios, notas corporis, facultatem formatricem, calidi nativi, animarumque corruptibilium originem refelluntur. Nec non Animatio Seminis adversus Recentiores omnes constanter Defenditur. Venice, 1622.
MUNDINUS, Vincentius (fl. 1586) Expositio in Galeni liber artis medicinalis. Bologna, 1586.
NIPHUS, Augustinus. Philosophorum hac nostra tempestate monarche Augustini Niphi Suessani In duodecimum Metaphysices Aristotelis & Averrois volumen ad etatis nostre decus Antonium Justinianum patritium Venetum: Commentarii in lucem castigatissimi nuperrime prodeuntes. At end: Venetiis impensa heredum quondam Domini Octaviani Scoti Modoetiensis: ac sociorum. 12 Februari 1518.
— Augustini Niphi Medicis. Philosophi Suessani Expositiones in omnes Aristotelis Libros. De Historia Animalium Lib IX De Partibus animalium, & earum Causis Lib IIII. Ac de Generatione animalium Lib V. [with Index to text & commentary] Venetiis. Apud Hieronymum Soctum. 1546.
— Augustini Niphi Medicis Philosophi Suessani Expositiones, in Aristotelis Libros Metaphysices. Venice, 1558.
— Augustini Niphi Medices Philosophi Suessani Expositio Subtilissima nec non et Collectanea Commentariaque in Tres Libros Aristotelis De Anima nuperrime accuratissima diligentia recognita. Venice, 1559.
— Augustini Niphi sua Tempestate Philosophi Omnium Celeberrimi Opuscula Moralia et Politica: cum Gabrielis Naudaei de eodem Auctore Judicio. Paris, 1645.
NOLANUS, Ambrosius. Ambrosii Nolani Philosophi ac medici praestantissimi, Castigationes adversus Averroem, recens impressae, Cum eiusdem Archetypo diligenter recognitae, cunctisque mendis expurgate.... Venice, 1532.
ODDIS, Oddo de. Expositio in librum Artis medicinalis Galeni. Nunc primum in lucem edita et castigata laboribus, et vigilius Marci Oddi, medici eiusdem filii. Venice, 1574.
— Oddi de Oddis Patavini Physici ac Medici celeberrimi et in Florentissima Patavina Academia Publici Olim ac Ordinarii Professoris. In Primam totam Fen primi Libri Canonis Avicennae dilucidissima et expectatissima Expositio. Venice, 1575.
PAPARELLA, Sebastian. De calido libri tres. Perugia, 1573.
PARÉ, Ambroise. Les Oeuvres d'Ambroise Paré, Conseiller, et Premier Chirurgien du Roy.. Quatriesme Edition. Paris, 1585.
PERAMATO, Petrus de. Liber de Facultatibus nostrum corpus dispensantibus... De semine tractatus... De hominis procreatione a conceptu in partum... pueri & puerperae regimen ... Petro de Peramato Excellentissimi Ducis de Medina Sydonia Protomedico Autore. San Lucar a Barrameda, 1576.
— Petri de Peramato Illustrissimi ac excellentissimi Ducis de Medina Sydonia Proto-medici, Opera Medicinalia. De Elementis. De Humoribus. De Temperamentis. Alia insuper utilissima

traduntur opera, quae novo titulo & libri principio comprehenduntur. San Lucar de Barramede, 1576, 1596.
PETRUS de Abano. Expositio problematum Aristotelis. Venice, 1482.
— Conciliator differentiarum philosophorum & praecipue medicorum.. Venice, 1483.
— Petri Aponensis medici ac philosophi celeberrimi Liber Conciliator differentiarum philosophorum praecipueque medicorum appellatus: noviter castigatissime impressus: Venice, 1521.
PICCOLOMINI, Archangelo. Anatomicae Praelections Archangeli Piccolhomini Ferrariensis civisque Romani. Explicantes mirificam corporis humani Fabricam: Et quae animae vires, quibus corporis partibus, tanquam instrumentis, ad suas obeundas actiones, utantur; sicuti tota anima, toto corpore. Rome, 1586.
PLATO. Inc. Prohemium Marsilii ficini floretini in libros Platonis ad Laurentium medicem Virum Magnanimum.. [The works of Plato, translated by Marsiglio Ficino]. Florence [1484, 1485].
— Chalcidii Viri Clarissimi Luculenta Timaei Platonis traductio, & eiusdem explanatio. [Rome,] 1520.
— Platonis Philosophorum omnium seculorum longe principis Timaeus, vel de Natura, Marsilio Ficino Florentino interprete. Paris, 1544.
— Timaeus Platonis, sive de Universitate, Interprete M. Tullio Cicerone, & Chalcidio, una cum eius docta explanatione. Paris, 1578.
— Timaeus a Calcidio translatus commentarioque instructus. In Societatem Operis conjunctio P. J. Jensen. Edidit J. H. Waszink. (Plato Latinus, edidit Raymundus Klibansky, vol. 4) London, 1962.
POLYBUS. Polybi Illustris Medici, Hippocratis quondam & discipuli & successoris, Opuscula aliquot nunc primum e Graeco in Latinum conversa, nempè De tuenda valetudine, sive de ratione victus sanorum, Liber I. De seminis humana natura, Liber I. De morbis, sive affectibus corporis, Libri II. Albano Torino Vitodurano interprete.. Basle, 1544.
PONCE DE SANTA CRUZ, Antonio. In Avicennae Primam Primi, [ad Philippum IIII Hispaniarum, Indiarumque Regem potentissimum] Madrid, 1622.
RIOLAN, Jean, The Elder. Joannis Riolani Praelectiones in Libros Fernelii Physiologicos, et de Abditis rerum caussis: Cum brevibus Scholiis eiusdem Auctoris. Accesserunt opuscula quaedam Philosophica. Paris, 1601.
— Joannis Riolani Ambiani, Medici Parisiensis, viri clarissimi Opera omnia.. quibus universam medicinam fideliter & accuratè descripsit, atque illustravit. Paris, 1610.
ROGERIUS, Joannes. Joannis Nicolai Rogerii Salernitani. Philosophi ac Medici Ordinariam Medicinam Publice Profitensis. In Gimnasio Salernitano, Quaesitum difficillimum accurate explicatum De sede Animae, membrorumque Principatu ex Galeni Hippocratis, Platonisque placitis adversus Philosophos nunc primum in lucem editum. Naples, 1574.
Rösslin (or ROESSLIN), Eucharius. Der Swangeren Frawen und Hebammen Rosengarten..., Frankfurt, 1513 (and later editions).
— De partu hominis..., Frankfurt, 1532.
RUEFF, Jakob or Jacobus. Ein schön lustig Trostbuchle, Zurich, 1544.
— De conceptu et generatione hominis... libri sex, Zurich, 1554.
RUMBAUM, Christopher. De Partibus Corporibus Humani Exercitationes Quaedam: Quibus generatio, substantia, usus, sanitas, morbus & curatio illarum exponitur. Authore Christophoro Rumbaum Iaurano Silesio. Basle, 1586.
SANCTORIUS, Sanctorius. Commentaria in Primam Fen Primi Libri Canonis Avicennae. Venice, 1626.
SCALIGER, J. C. Julii Ceasaris Scaligeri Exotericarum Exercitationum Liber Quintus Decimus, De Subtilitate, ad Hieronymum Cardanum. Paris, 1557.
— Julii Caesaris Scaligeri Epistolae Aliquot nunc primum Vulgatae. Accedunt praeterea alia quaedam eiusdem opuscula, & fragmenta praefationis in Aristot. Historiam de Animalibus. Toulouse, 1620.

SCHEGKIUS, Jacobus. Jac. Schegkii Schorndorffensis Medici et Philosophi Eximii, De Plastica Seminis Facultate Libri tres. Eiusdem De calido et humido nativus Liber unus. De primo Sanguificationis Instrumento Liber unus. Strasbourg, 1580.

SCHENCK (or Schenckius). Johann Theodor. Synopsis Institutionum Medicinae disputatoriae, Jena, 1668–1671.

SCHROTER, Joannes Fridericus. De omnibus in universum totius corporis humani humoribus. Liber, [conscriptus à Joanne Friderico Schrotero Genesi, Philosophiae & Medicinae Doctore]. Patavii, 1582.

SISINIUS, Joannes. De Natura Foetus Disputatio. Auctore Joanne Sisinio Amabili. Rome, 1615.

SORANUS. Soranus's Gynaecology, translated and with an introduction by Owsei Temkin, Baltimore, 1956.

SYLVIUS, Jacobus. De mensibus mulierum, et hominis generatione, Jacobi Silvii medicae rei apud Parrhisios interpretis Regii, Commentarius. Paris, 1555.

TELESIO, Bernardino. Bernardini Telesii Consentini De Rerum Natura juxta propria principia. Libri IX. Naples, 1587.

— Quod animal universum ab unica Animae substantia gubernatur, adversus Galenum liber unicus. In Varii de naturalibus rebus libelli, ab A. Persio editi. Venice, 1590.

TERELIUS, Dominicus. De Generatione et Partu Hominis, Libri Duo. Lyons, 1578.

THEOPHRASTUS (trans. FURLANUS) Theophrasti Eresii, Peripateticorum post Aristotelem principis, Pleraque antehac Latine nunquam, nunc Graece et Latine simul edita. Interpretibus, Daniele Furlano Cretensis, Adriano Turnebo Accesserunt, Liber de innato spiritu, Aristoteli attributus, & Danielis Furlani Uberes ad omnia Commentaria. Hanau, 1605.

THRIVERIUS BRACHELIUS, Hieremias. In τεχνην Galeni clarissimi commentaria. Lyons, 1547 [col. 1548].

— Universae Medicinae Brevissima Absolutissimaque Methodus, auctore Hieremia Thrivero Medico clarissimo, ac quondam in celeberrima Academia Louaniensi professore publico & primario. Leyden, 1592.

UFFENBACH, Peter. Theses medicae de semine et sanguine menstruo. Quas propositas ab Israele Spachio D. Medico Respondendo tuebitur M. Petrus Uffenbachius, Francofurtanus ad diem 8. Julii privati exercitii gratiâ. Argentorati [Strasbourg] 1592.

VALLA, Georgius. Georgii Valla Placetini viri clarissimi. De expetendis, et fugiendis rebus opus... Venice, 1501.

— De Inventa Medicina, & in quot partes distributa sit. Strasbourg, [1529?].

VALLES, Franciscus. Controversiarum medicarum et philosophicarum libri decem. Alcala, 1556.

VESALIUS, Andreas. De Humani corporis fabrica Libri septem. Basle, 1543.

VINCENTIUS Bellovacensis. Speculum Naturale. Strasbourg [not after 1478].

WECKER, Joannes Jacobus. Medicinae Utriusque Syntaxes, Ex Graecorum, Latinorum, Arabumque Thesauris per Joan. Jacob. Weckerum, Bas. Reipublicae Colmariensis Physicum, singulari fide, methodo atque industria collectae & concinnatae. Lyons, 1583.

ZABARELLA, Jacobus. Jacobi Zabarellae Patavini. De Rebus Naturalibus Libri XXX. Quibus quaestiones, quae ab Aristotelis interpretibus hodie tractari solent, accuratè discutiuntur. Cologne, 1590.

Secondary

ADELMANN, Howard B. (editor and translator) The Embryological Treatises of Hieronymus Fabricius of Aquapendente. The Formation of the Egg and of the Chick [De Formatione Ovi et Pulli], The Formed Fetus [De Formato Foetu]. A Facsimile Edition, with an introduction, a translation, and a commentary by Howard B. Adelmann... Ithaca, 1942.

— Marcello Malpighi and the Evolution of Embryology (five volumes). Ithaca and London, 1966.

ANTONIOLI, Roland. Rabelais et la Médecine. Études Rabelaisiennes Tome XII. Genève, Librairie Droz, 1976.

ARISTOTELISMO padovano e filosofia aristotelica, vol IX of the Atti del XII congresso internazionale di filosofia. Venice, 1958, 1960.

BALSS, Heinrich. Aristoteles biologische Schriften. Munich, 1943.

BARRAUD, G. L'Humanisme et la médecine au XVIe siècle. Paris, 1942.

BLOCH, Bruno. Die geschichtlichen Grundlagen der Emryologie bis auf Harvey, Halle, 1904. (reprinted from Abh. der Kaiserl. Leop.-Carol.Deutschen Akademie der Naturforscher, Nova Acta, Band LXXXII, Nr. 3, pp. 217–334.)

BOLGAR, R. R. The Classical Heritage and its Beneficiaries, Cambridge, 1954.

BROCKLISS, L. W. B. "Medical teaching at the University of Paris, 1600–1720", Annals of Science, 35(3), May 1978, pp. 221–251.

BULLOUGH, Vern, "Medieval Bologna and the Development of Medical Education", Bull. Hist. Med., 1958, 32:201–215.

BYLEBYL, Jerome J. "The School of Padua: humanistic medicine in the sixteenth century", pp. 335–370 in Charles Webster (ed.) Health, medicine and mortality in the sixteenth century, Cambridge, 1979.

CHURCHILL, Frederick B. "The History of Embryology as Intellectual History", Journal of the History of Biology, iii (1970) 155–168.

CLAGETT, Marshall. Giovanni Marliani and Late Medieval Physics, New York, 1941.

— The Science of Mechanics in the Middle Ages, Madison, Wisconsin, 1959.

CLARK, Stephen R. L. Aristotle's Man: Speculations upon Aristotelian Anthropology, Oxford, 1975.

COLE, Francis Joseph. Early Theories of Sexual Generation. Oxford, 1930.

COLEMAN, William and LIMOGES, Camille, editors. Studies in the History of Biology, volume 1. Baltimore and London, 1977.

COPENHAVER, Brian P. Symphorien Champier and the Reception of the Occultist Tradition in Renaissance France, The Hague, 1978.

CORLIEU, A. L'Ancienne Faculté de Médecine de Paris. Paris, 1877.

CORNFORD, Francis MacDonald. Plato's Cosmology. The Timaeus of Plato translated with a running commentary. London & New York, 1971 (first edition 1937).

CUNNINGHAM, Andrew. "Aspects of the history of medical education in Britain in the seventeenth and early eighteenth centuries", London, 1974 (unpublished Ph.D. thesis).

— "The Kinds of Anatomy", Medical History, 1975, vol. 19(1), pp. 1–19.

DARMON, Pierre. Le Mythe de la procréation à l'âge baroque. Paris, 1977.

Cranz, F. E. "Editions of the Latin Aristotle Accompanied by the Commentaries of Averroes" in Philosophy and Humanism: 1976, 116–28.

DEBUS, Allen G. ed. Science, Medicine and Society in the Renaissance: Essays to Honor Walter Pagel, London, 1972, vols I and II.

— Medicine in Seventeenth-century England: A Symposium held at UCLA in Honor of C. D. O'Malley. Berkeley, Los Angeles and London, 1974.

DEL TORRE, Maria Assunta. Studi su Cesare Cremonini: Cosmologia e Logica nel Tardo Aristotelismo Padovano. Padua (Editrice Antenore) 1968.

DIELS, Harman and Kranz, Walther Die Fragmente der Vorsokratiker 3 vols. Zurich/Berlin, 1964.

DURLING, Richard J. "A Chronological Census of Renaissance Editions and Translations of Galen" Journal of the Warburg and Courtauld Institutes Vol. XXIV

Nos. 3–4, 1961, pp. 230–305.

FERRARI, Sante. I tempi, la vita, le dottrine di Pietro d'Abano. Genoa, 1900.

— Per la biographia e per gli scritti di Pietro d'Abano. Rome, 1918.

FIGARD, L. Un medecin philosophe au XVIe siecle. Etude sur la psychologie de Jean Fernel. Paris, 1903.

FREEMAN, Kathleen. Ancilla to the Pre-Socratic Philosophers: A Complete Translation of the Fragments in Diels.. Oxford, 1962.

GARIN, Eugenio. Storia della filosofia Italiana. 3 vols. Turin, 1966.
GASKING, Elizabeth. Investigations into Generation, 1651–1828. London, 1967.
GERMAIN, A. La Renaissance à Montpellier. Étude Historique d'après les documents originaux. Montpellier, 1871.
GILBERT, Neal Ward. Renaissance Concepts of Method. Columbia U.P., New York, 1960.
GILSON, Étienne. Études sur le rôle de la Pensée médievale dans la formation du système cartesien. Paris, 1930.
GOODFIELD, June. The growth of scientific physiology. Physiological method and the mechanist-vitalist controversy, illustrated by the problems of respiration and animal heat. London, 1960.
HALL, Thomas S. Ideas of Life and Matter: Studies in the History of General Physiology 600 B.C. - 1900 A.D. Chicago, 1969.
HASKINS, C. H. Studies in the History of Medieval Science. 2nd ed. New York, 1926.
HERRLINGER, Robert and FEINER, Edith. "Why did Vesalius not discover the Fallopian tubes?" Med. Hist. 1964, 8: 335–41.
HEWSON, M. Anthony. Giles of Rome and the Medieval Theory of Conception: A study of the De formatione corporis humani in utero. London, 1975.
HUXLEY, J. and De Beer, G. R. Elements of Experimental Embryology. Cambridge, 1934.
IRSAY, Stephen d' Histoire des universités francaises et étrangères des origines à nos jours. 2 vols. Paris, 1933–1935.
JACOB, François. La logique du vivant: Une histoire de l'hérédité. Paris, 1970.
JAEGER, Werner. Diokles von Karystos: Die Griechische Medizin und die Schule des Aristoteles. Berlin, 1938.
JARDINE, N. "Galileo's Road to Truth and the Demonstrative Regressus." Studies in History and Philosophy of Science. Vol. 7, no. 4, 1976, pp. 277–318.
JAYAWARDENE, S. A. "Western Scientific Manuscripts before 1600" Annals of Science, 1978, 35(2): 143–172.
KIRK, G. S. and RAVEN, J. E. The Pre-Socratic Philosophers: A Critical History with a Selection of Texts, Cambridge, 1957.
KNOWLES, D. The evolution of medieval thought. London, 1962.
KRISTELLER, P. O. "Paduan Averroism and Alexandrism in the Light of Recent Studies" in Aristotelismo Padovano e Filosofia Aristotelica, 1960, 147–155.
— Renaissance thought, the classic, scholastic and humanist strains. New York, 1961.
— La Tradizione Aristotelica nel Rinascimento, Padua, 1962.
— "Renaissance Aristotelianism", Greek, Roman and Byzantine Studies, 1965, 6: 157–174.
LEHOUX, Françoise. Le cadre de vie des médecins Parisiens aux XVIe et XVIIe siècles. Paris, 1976.
LESKY, Erna. "Die Zeugungs und Vererbungs-lehren der Antike und ihr Nachwirken." Abh. Akad. d. Wiss. u.d. Lit., Mainz, geistes-u. sozialwiss. Kl. Jhrg, 1950. 1227–1425.
LEWIS, R. G. "Guillaume Rondelet and his circle." Oxford, 1967 (unpublished D.Phil. thesis.)
LIND, L. R. Studies in Pre-Vesalian Anatomy: Biography, Translations, Documents, Philadelphia, 1975.
LLOYD, G. E. R. Aristotle, the Growth and Structure of his Thought, Cambridge, 1968.
LOHR, C. H. "Medieval Latin Aristotle Commentaries" Traditio 23 (1967) 313–413; 24 (1968) 149–245; 26 (1970) 135–216; 27 (1971) 251–351; 28 (1972) 281–396; 29 (1973) 93–197.
— "Aristotle in the West: Some recent Books" 25 (1969) 417–31.
— "Renaissance Latin Aristotle Commentaries" Studies in the Renaissance, xxi (1974) 228–289; Renaissance Quarterly, xxviii (1975) pp 689–741; Ibid., xxix (1976) pp. 714–745; Ibid., xxx (1977) pp. 681–741; remaining sections forthcoming.
LONIE, I. M. "A Structural Pattern in Greek Dietetics and the Early History of Greek Medicine." Medical History, 1977, vol. 21, pp. 235–260.
— "Fever pathology in the sixteenth century: orthodoxy and heterodoxy." Paper presented at Wellcome Symposium on Theories of Fever from Antiquity to the Enlightenment, Friday 13 June 1980; to be published in Medical History.

MABILLEAU, L. Étude historique sur la philosophie de la Renaissance en Italie: Cesare Cremonini. Paris, 1881.
MACLEAN, Ian. Woman triumphant: feminism in French literature, 1610–1652. Oxford, 1977.
McVAUGH, Michael R. "Quantified Medical Theory and Practice at Fourteenth Century Montpellier." Bull. Hist. Med. 1969, 43 (5), pp. 397–413.
— "The 'Humidum Radicale' in Thirteenth Century Medicine", 1974, Traditio, 30: 259–283.
— "An early Discussion of Medicinal Degrees at Montpellier by Henry of Winchester". Bulletin of the History of Medicine 1975, 49(1), pp. 57–71.
MAHONEY, E. P. "Nicoletto Vernia and Agostino Nifo on Alexander of Aphrodisias: an unnoticed dispute", Rivista critica di storia della filosofia, 1968, 23: 268–296.
— "Pier Nicola Castellani and Agostino Nifo on Averroes' doctrine of the Agent Intellect", Ibid. 1970, 25: 387–409.
— "Agostino Nifo's early views on immortality", Journal of the History of Philosophy, 1970, 8: 451–460.
— "A note on Agostino Nifo", Philological Quarterly, 1971, 1: 125–132.
— "Antonio Trombetta and Agostino Nifo on Averroes and intelligible species. A philosophical dispute at the University of Padua." in Storia e cultura al Santo fra il XIII a il XX secolo, Vicenza, 1976, pp. 289–301.
— ed. Philosophy and Humanism: Renaissance Essays in Honor of Paul Oskar Kristeller, Leiden, 1976.
MARSHALL, F. H. A. Physiology of Reproduction. London, 1922.
MEYER, Arthur William. Essays on the History of Embryology. Reprinted with a few changes and omissions, and with references, from California and Western Medicine, Volumes 35–37, 1931–2.
— The Rise of Embryology. Stanford, California and London [1939].
MULTHAUF, Robert P. "J. B. van Helmont's reformation of the Galenic doctrine of digestion" Bull. Hist. Med. 1955, 29: 154–163.
MURDOCH, J. E. and SYLLA, E. eds. The cultural context of medieval learning. (Proceedings of the First International Colloquium on Philosophy, Science and Theology in the Middle Ages, September, 1973) Dordrecht & Boston, 1975.
NAPOLI, Giovanni di. L'immortalità dell'anima nel Rinascimento. Turin, 1963.
— Studi sul Rinascimento. Naples, 1973.
NARDI, Bruno. Saggi sull'Aristotelismo Padovano dal Secolo xiv al xvi. Universita degli Studi di Padova Centro Aristotelico Studi sulla Tradizione Aristotelica nel Veneto, vol. I. Florence (G. C. Sanzoni), 1958.
NAUDE, Gabriel and PATIN, Guy. Naudeana et Patiniana, ou singularitez remarquables, prises des conversations de Mess. Naude et Patin. Paris, 1701.
NEEDHAM, Joseph. A History of Embryology. 2nd ed. Cambridge, 1959.
NIEBYL, P. H. "Old Age, Fever and the Lamp Metaphor" Journal of the History of Medicine 1971, 26: 351–368.
OGLE, W. Aristotle on the Parts of Animals. London, 1882.
O'hara, May, Jane (Andrea Janette) "A study of English dietary advice in the second half of the sixteenth century." Ph.D. Thesis, University of London, 1969.
— Elizabethan Dyetary of Health. Lawrence, Kansas, 1977.
O'MALLEY, C. D. Andreas Vesalius of Brussels, 1514–1564. Berkeley, California, 1965.
PAGEL, Walter. Review of: Sir C. Sherrington: The endeavor of Jean Fernel. Bull. Hist. Med. 1946, 20, 587–9.
— The reaction to Aristotle in seventeenth-century biological thought. Campanella, Van Helmont, Glanvill, Charleton, Harvey, Glisson, Descartes. In: Science, Medicine and History. Essays on the evolution of scientific thought and medical history written in honour of Charles Singer. Oxford, 1953, I, 489–509.
— Paracelsus. An introduction to philosophical medicine in the era of the Renaissance. Basle and New York, 1958.

— "William Harvey Revisited" Hist. Sci. 1969, 8: 1–31 and 1970, 9: 1–41.
— and WINDER, Marianne. "The higher elements and prime matter in Renaissance Naturalism and in Paracelsus." Ambix 1974, 21: 93–127.
— "Die Konjunktion der himmlischen und irdischen Elemente in der Renaissancephilosophie und im echten Paracelsus" in Paracelsus Werk und Wirkung, Festgabe fur Kurt Goldammer zum 60. Geburtstag, Wien, 1975., 187–204.
PARK, Katherine S. "The Imagination in Renaissance Psychology", unpublished M.Phil. thesis, London, 1974.
PECK, Arthur L. "The connate pneuma, an essential factor in Aristotle's solutions to the problems of reproduction and sensation." in Science, Medicine and History (1953) v 1, 111–121.
PINTARD, René. Le libertinage erudit dans la première moitié du XVIIe siècle. Paris, 1943.
POPPI, Antonino. Introduzione all'Aristotelismo Padovano. Padua, 1960.
PREUS, Anthony. "Science and Biology in Aristotle's Generation of Animals." Journal of the History of Biology iii (1970) 1–52.
— "Galen's Criticism of Aristotle's Conception Theory." Journal of the History of Biology, vol. 10, no. 1, 1977, pp. 65–85.
RANDALL, John Henry. The School of Padua and the Emergence of Modern Science, Padua, 1961.
— The Career of Philosophy from the Middle Ages to the Enlightenment, New York, 1962.
— "Paduan Aristotelianism Reconsidered" in Philosophy and Humanism, pp. 273–282.
RENAN, Ernest. Averroès et l'Averroïsme, essai historique. Paris, 1866 (3rd edition)
ROGER, Jacques. Les Science de la Vie dans la Pensée Française du XVIIIe siècle: La Génération des Animaux de Descartes à l'Encyclopédie. Paris, 1963.
ROTHSCHUH, K. E. "Das System der Physiologie von Jean Fernel (1542) und seine Wurzeln", Verh. XIX int. Kongr. Geschichte der Medizin, Basel, 1964, pp. 529–536.
— "Technomorphes Lebensmodell contra Virtus-Modell (Descartes gegen Fernel) Sudhoffs Archiv, 1970, 54/4: 337–354.
RUSSELL, E. S. Form and Function: A Contribution to the History of Animal Morphology. London, 1916.
SARTON, George. Galen of Pergamon. Lawrence, Kansas, 1954.
— Introduction to the History of Science. Baltimore Volume I, 1927; II i and ii 1931; III i and ii 1947.
SCHMID, Magnus. "Die Lehre von den Homoiomerien in der Physiologie Fernels" Sudhoffs Archiv 1957, 41: 317–335.
SCHMITT, Charles B. "Towards a Reassessment of Renaissance Aristotelianism" History of Science 1973, vol. 11, pp. 159–193.
— "Paduan Averroism studied through the Venetian editions of Aristotle-Averroes (with particular reference to the Giunta edition of 1550–2)", forthcoming.
SCREECH, M. A. The Rabelaisian Marriage. London, 1958.
SHERRINGTON, Charles. The endeavour of Jean Fernel, with a list of the editions of his writings. Cambridge, 1946.
SHUTE, Richard. On the history of the process by which the Aristotelian writings arrived at their present form. Oxford, 1888 [reprint 1976].
SIEGEL, Rudolph E. Galen's System of Physiology and Medicine. Basel and New York, 1968.
SINGER, Charles. "Some Vesalian problems" B.H.M. 1945, 17: 426–428.
SIRAISI, Nancy G. Arts and Sciences at Padua: The Studium of Padua before 1350. Toronto, 1973.
— "The Expositio Problematum Aristotelis of Peter of Abano." Isis, 1970, 61: 321–339.
— & Demaitre, Luke, eds. "Science, Medicine and the University: 1200–1550. Essays in Honor of Pearl Kibre." Part I, Manuscripta vol. 20 no. 2, 1976.
SOLMSEN, Friedrich. "The Vital Heat, the Inborn Pneuma and the Aether." Journal of Hellenic Studies, 1957, 77: 119–123.
TAYLOR, A. E. A Commentary on Plato's Timaeus. Oxford, 1972 (reprint of first edition, 1928).
TEMKIN, Owsei. "On Galen's pneumatology." Gesnerus 1951: 180–9.

— Galenism: Rise and decline of a medical philosophy, Ithaca, New York, 1973.
THORNDIKE, Lynn. A History of Magic and Experimental Science Volume II New York, 1923, Volumes V and VI, 1941.
— "Translations of Works of Galen from the Greek by Peter of Abano" Isis, 1942, 33: 649–653.
— "Manuscripts of the writings of Peter of Abano" Bull. Hist. Med. 1944, 15: 201–219.
— "Peter of Abano and another commentary on the problems of Aristotle" Bull. Hist. Med., 1955, 29: 517–23.
— 'De complexionibus', Isis, 1958, 49: 398–408.
— and Kibre, Pearl. A Catalogue of incipits of mediaeval scientific writings in Latin, Cambridge, Massachusetts, 1937 (revised edition London, 1963). Supplements in Speculum, 1939, 14: 93–105; 1942, 17: 342–366; 1951, 26: 673–695; 1965, 40: 116–122; and 1968, 43: 78–114.
TROILO, Erminio. Averroismo e Aristotelismo Padovano. Padua, 1939. (Opuscoli Accademia, editi a cura della Facoltà di Lettere e Filosofia della R. Università di Padova, Number 7).
UNDERWOOD, E. A., ed. Science, Medicine and History. Essays on the evolution of scientific thought and medical practice written in honour of Charles Singer, Oxford, 1953. 2 vols.
VERBEKE, G. L'Evolution de la doctrine du Pneuma du Stoicisme à S. Augustin. Paris, 1945.
VLASTOS, Gregory. Plato's Universe, Oxford, 1975.
walker, D. P. Spiritual and Demonic Magic from Ficino to Campanella. London, 1958.
— "The Astral Body in Renaissance Medicine", Journal of the Warburg and Courtauld Institutes, 1958. 21: 119–133.
— "Francis Bacon and Spiritus" in Science, Medicine and Society in the Renaissance, II, pp. 121–130.
WEAR, Andrew W. "Contingency and logic in Renaissance anatomy and physiology." Ph.D. London, 1975.
— "The spleen in Renaissance anatomy". Medical History 1977, vol. 21(1): 43–60.
WICKERSHEIMER, C. A. Er. La Médecine et Les Médecins en France à l'époque de la Renaissance. Paris, 1906.
WIGHTMAN, W. P. D. Science and the Renaissance, Edinburgh and London, 1962.
— "Quid sit methodus? 'Method' in Sixteenth Century Medical Teaching and 'Discovery'" J.H.M. 1964, 19: 360–376.
— "Myth and Method in Seventeenth Century Biological Thought." J.H.B. 1969, 2(2): 321–36.
— The Emergence of Scientific Medicine. Edinburgh, 1971.
— Science in a Renaissance Society, London, 1972.
WILSON, Leonard G. "Erasistratus, Galen and the Pneuma" Bulletin of the History of Medicine vol. 33(1), 1959, 293–314.
WOLFSON, Harry Austryn. Studies in the History of Philosophy and Religion. 2 volumes. Cambridge, Massachusetts and London, 1973.
ZAMBELLI, Paolo. "I problemi metodologici nel necromante Agostino Nifo", Medioevo, 1975, 1: 129–171.

Bibliographical

ARISTOTLE: Index rerum omnium, quae in Aristotelis operibus continentur, absolutissimus. Venice, 1573.
BIBLIOTHÈQUE NATIONALE Catalogue des Ouvrages de Claude Galien conservés au Département des Imprimés. Paris, 1913.
BRASAVOLUS, Antonius Musa. Index refertissimus in omnes Galeni libros. Hildesheim and New York, 1975. Facsimile of the Venice edition of 1625.
CRANZ, F. E. A Bibliography of Aristotle Editions, 1501–1600. Baden-Baden, 1971.
— A check-list of Aristotle publications to 1520 (priv. circ. only).

COSENZA, Mario Emilio. Biographical and Bibliographical Dictionary of the Italian Humanists and of the World of Classical Scholarship in Italy, 1300–1800. Boston, 1967.
DURLING, Richard. A catalogue of Sixteenth-century printed Books in the National Library of Medicine. Compiled by Richard Durling, Bethesda, Maryland, 1967.
FERRARI, Luigi. Onomasticon. Repertorio Biobibliografico degli scrittori Italiani dal 1501 al 1850. Milan, 1947.
GORDON, B. L. Medieval and Renaissance Medicine, London, 1960.
KLEBS, Arnold C. Incunabula Scientifica et Medica. Short title list. Bruges, 1938.
KRISTELLER, P. O. (ed.) Catalogus Translationum et Commentariorum: Medieval and Renaissance Latin Translations and Commentaries. Annotated Lists and Guides. Volume I. Editor in chief, Paul Oskar Kristeller, 1960; volume II do. Associate editor F. Edward Cranz. 1971. Washington D.C.
LEITNER, Helmut. Bibliography to the Ancient Medical Authors. Bern, 1973.
LIPENIUS, M. Martinus. M. Martini Lipenii Bibliotheca Realis Medica. Omnium Materiarum, Rerum, et Titulorum, in Universa Medicina Occurentium. Frankfurt, 1679.
POYNTER, F. N. L. A catalogue of printed books in the Wellcome Historical Medical Library. I. Books printed before 1641. London, 1962.
RENOUARD, Philippe. Bibliographie des editions de Simon de Colines, 1526–1546. Paris, 1894.
SARTON, George. Appreciation of ancient and medieval science during the Renaissance. Philadelphia, 1955.
SCHMITT, Charles. A Critical Survey and Bibliography of Studies on Renaissance Aristotelianism 1958–1969. Padua, 1971.
WINGATE, S. D. The Medieval Latin Versions of the Aristotelian Scientific Corpus with special reference to the Biological Works. London, 1931.

Notes

1. Victor Cardelinus, De origine foetus.. Vincenza, 1628; ch. 1, p. 5.
2. Candidus Decembrius, De genitura hominis [Rome] c. 1495; Charles de Bouelles, Liber de generatione (with other works) [Paris, 1510]; Fernel, op. cit.; Jacobus Sylvius, de..hominis generatione, Paris, 1555; Peramato, Opera medicinalia, op. cit.; Dominicus Terelius, De generatione..Lyons, 1578; Andreas Helepyrus, Concordia..de Humano Conceptu.., Ingolstadt, 1582; Hieronymus Mercurialis, in his Praelectiones Pisanae, Venice, 1597; Joannes Costaeus, De Humani Conceptus..Pavia, 1604; Joannes Sisinius, De Natura Foetus..Rome, 1615; Cardelinus, ibid.; Joannes Gallego de la Serna, in Opera Physica, Medica, Ethica..,Lyons, 1634.
3. See, for example, Schmitt, Critical Survey and "Paduan Averroism"; Cranz, "Editions"; Kristeller, esp. "Renaissance Aristotelianism"; Lohr; Siraisi, Arts, for the natural philosophers, and Lewis, op. cit.; Durling, "Census"; Cunningham and Bylebyl, op. cit. for the medical writers; Temkin, Galenism, considers both.
4. Recent work by Frances Yates, D. P. Walker, P. Rattansi, Walter Pagel, Allen Debus, Keith Thomas, among others, might be considered to come into this category.
5. R. G. Lewis, op. cit., Conclusion, p. 352.
6. Op. cit. as in note 3.

7. Fernel, De naturali parte medicinae, 1542; Medicina, 1554, Book I, esp. ch. 1.
8. Riolan, Opera Omnia, 1610; Anatome Joannis Riolani Filii Anatomici Regii, (sep. pag.), Praefatio, vo, on the parentally-inspired vocation to complete his father's work with an anatomy, pharmacy and work on diet.
9. See the references in note 44, Introduction, above for anatomical works, and note 43 for 'women's books'. Anatomical treatises on the organs serving generation were published by Aranzio (1564) and Plazzoni (1664), among others.
10. See references in note 2, Introduction and, for Vesalius, Charles O'Malley, Andreas Vesalius of Brussels, 1514–1564 Berkeley, 1965, Wightman, op. cit. and others, including the articles cited below. For Harvey, see especially the writings of Pagel, Webster and Bylebyl.
11. Lewis, op. cit. p. 289.
12. Berengario da Carpi, A Short Introduction to Anatomy, tr. Lind, Chicago, 1959. p. 68; Lind, Anatomists, op. cit., for example the discussions of Achillini (pp. 49–51) on the right v. left testicle, cells of the uterus, place of elaboration of the seed; Benedetti on the uses of the testes (in which he follows Aristotle) and female seed, pp. 96–101; Massa on the contents of the seminal vessels, the determination of sex according to the right or left side of the uterus (a suggestion which he rejects, in favour of a single cavity), pp. 198–209.
13. Leonardo da Vinci: Anatomical Drawings from the Royal Collection Royal Academy of Arts [Exhibition Catalogue] London, 1977, pp. 58–61, Figs 16–18, esp. 16A.
14. See Charles Mayo Goss, "On the Anatomy of the Uterus" (a translation of Galen's De uteri dissectione, Anatomical Record, October 1962, vol. 144, no. 2.
15. Figures 1a and b are taken from the 1603 edition of Guibelet in the Wellcome Library; Fig. 1a, b are based on the same source. They are similar to the arrangement shown in the plates of Laurentius' Historia Anatomica (1599), pp. 194–5 and 196–7 and are clearly based on earlier, pre-Fallopian models. Needham, op. cit. p. 110 fig. 9, shows an illustration of the female genital organs and the passages connecting veins, arteries, female testes and uterus which is very similar; it comes from Walther Ryff, Anatomia, 1544 and was attributed to Hans Baldung Grien. The illustrations in De Fabrica, reproduced in Robert Herrlinger and Edith Feiner, "Why did Vesalius not discover the Fallopian tubes?" Med. Hist. 1964, 8: 335–41, are similar, as their sketch makes clear (Fig. 2).
16. See Herrlinger and Feiner, op. cit. and Charles Singer, "Some Vesalian problems". B.H.M. 1945, 17: 426–428.
17. Benedetti, in Lind, op. cit., pp. 96–101.
18. Selected bibliography of references used and cited in this thesis, not including standard reference works or general reading.

Printed by Printforce, the Netherlands